QUANTUM ELECTRONICS: A TREATISE

VOLUME I
Nonlinear Optics, Part B

CONTRIBUTORS

S. A. AKHMANOV
ROBERT L. BYER
A. I. KOVRYGIN
A. P. SUKHORUKOV
JOHN WARNER

QUANTUM ELECTRONICS: A TREATISE

Edited by

HERBERT RABIN / C. L. TANG

Naval Research Laboratory
Washington, D. C.

Department of Electrical Engineering
and Materials Science Center
Cornell University
Ithaca, New York

VOLUME I
Nonlinear Optics, Part B

ACADEMIC PRESS   New York   San Francisco   London   1975
A Subsidiary of Harcourt Brace Jovanovich, Publishers

ACADEMIC PRESS, INC.
111 Fifth Avenue, New York, New York 10003

United Kingdom Edition published by
ACADEMIC PRESS, INC. (LONDON) LTD.
24/28 Oval Road, London NW1

Library of Congress Cataloging in Publication Data
Main entry under title:

Quantum electronics.

       CONTENTS: v. 1.   Nonlinear optics.   2 v.
       1.   Quantum electronics.   I.   Rabin, Herbert.
II.   Tang, Chung Liang, (date)
QC680.Q8   537   74-29323
ISBN 0–12–574041–7 (v. 1, pt. B)

PRINTED IN THE UNITED STATES OF AMERICA

Contents

Part III.   APPLICATIONS

8.   Optical Harmonic Generation and Optical Frequency Multipliers

S. A. Akhmanov, A. I. Kovrygin, and A. P. Sukhorukov

9.   Optical Parametric Oscillators

Robert L. Byer

10.   Difference Frequency Generation and Up-Conversion

John Warner

List of Contributors

Numbers in parentheses indicate the pages on which the authors' contributions begin.

S. A. AKHMANOV (475), Department of Physics, Moscow State University, Moscow, USSR

ROBERT L. BYER (587), Department of Applied Physics, Stanford University, Stanford, California

A. I. KOVRYGIN (475), Department of Physics, Moscow State University, Moscow, USSR

A. P. SUKHORUKOV (475), Department of Physics, Moscow State University, Moscow, USSR

JOHN WARNER (703), Royal Radar Establishment, Malvern, Worcestershire, United Kingdom

Preface

Following the advent of the laser, quantum electronics has emerged as a multidisciplinary subject of great breadth and richness, attracting the interests of basic as well as applied research workers. This subject, accordingly, is characterized by numerous specialized research reported in a wide range of original literature sources. In this treatise, "Quantum Electronics," through review articles written by principal workers in their respective fields, it is planned to present unified discussions of major quantum electronics topics. Through this approach it is hoped to stimulate understanding and progress in quantum electronics by making it relatively easy for an advanced student or investigators with limited prior background to survey topics of interest, as well as by providing reviews containing original material, both in content and organizational style, of benefit to advanced workers in the field.

Volumes IA and IB deal with the topic of nonlinear optics, with review articles on the subjects of nonlinear optical susceptibilities, nonlinear optical processes, and applications. The topics selected in these first volumes represent a compromise between the large number of subjects now embodied by nonlinear optics, an attempt to represent a balance between theoretical and experimental interests, and, of course, the general availability of authors. We are indebted to our colleagues, the authors, for their contributions.

Contents of Volume I, Part A

Part III

APPLICATIONS

8

Optical Harmonic Generation and Optical Frequency Multipliers

S. A. AKHMANOV, A. I. KOVRYGIN, A. P. SUKHORUKOV

Department of Physics
Moscow State University
Moscow, U.S.S.R.

I. INTRODUCTION

A. Optical Harmonic Generation and Optical Frequency Multipliers. Review of the Current Literature

Optical harmonic generation occupies a special place in nonlinear optics. Optical second harmonic generation (SHG) was the first nonlinear effect discovered with the help of lasers.

In connection with research on harmonic generation important methods of phase matching of light waves were developed. These methods soon found numerous applications in the creation of such important devices as optical parametric oscillators, optical image and signal converters, difference frequency generators, etc. Experiments with optical harmonics have stimulated the development of the theory of nonlinear wave interactions. It should be noted also that optical multipliers produced the first practical implementation of the principles of nonlinear optics. Figure 1 shows a schematic diagram of the experimental setup which Franken and co-workers (1961) used in their pioneer work. The ruby laser beam (with a peak power of several kilowatts) was passed through crystalline quartz. At the output of the crystal, beside the main beam with frequency ω_1 there appeared also a weak beam of second harmonic radiation $\omega_2 = 2\omega_1$. In these experiments the power conversion efficiency $\eta_{p_2} = P_2/P_1$ was of the order 10^{-11}. Franken *et al.* also offered a physical interpretation of the second harmonic generation effect.

The appearance of the second harmonic is due to the nonlinear polarization of the crystal. The linear dependence between the polarization and the light electric field, which was assumed as the basis of the whole of prelaser optics is, in actual fact, no more than an approximation.

A true picture is presented by the following formulas:

$$\mathscr{P} = \mathscr{P}^L + \mathscr{P}^{NL} \tag{1}$$

$$\mathscr{P}^L = \hat{\kappa}E, \qquad \mathscr{P}^{NL} = \hat{\chi}^{(2)}EE + \hat{\chi}^{(3)}EEE + \cdots \tag{2}$$

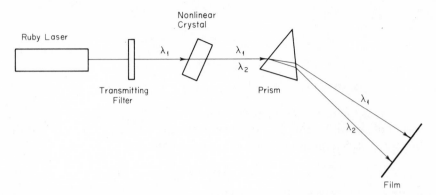

Fig. 1. Experimental arrangement for the detection of second harmonic generation of light (from Franken *et al.*, 1961).

where the nonlinear susceptibilities $\hat{\chi}^{(2)}, \hat{\chi}^{(3)}, \ldots$ are tensors of the third, fourth, and higher ranks. The order of magnitude of these tensors is determined by the characteristic atomic field E_{at}, $\chi^{(N)} \propto E_{\text{at}}^{-(N-1)}$ (the linear susceptibility being $\kappa \simeq E_{\text{at}}^0 \simeq 1$). The typical values of $\chi^{(2)}$ and $\chi^{(3)}$ for the usual kind of crystals are $\chi^{(2)} \simeq 10^{-9}$ esu and $\chi^{(3)} \simeq 10^{-14}$ esu. The second harmonic light is described by the first term in the nonlinear polarization $\mathscr{P}^{\text{NL}}$ in Eq. (2). If the laser wave is plane and monochromatic,

$$\mathbf{E}_1 = \mathbf{U}_1 A_1 \sin(\omega_1 t - k_1 z), \qquad k_1 = n_1 \omega_1/c = 2\pi n_1/\lambda_1 \tag{3}$$

where $\mathbf{U}_1$ is the unit vector of the wave polarization. From (2), the nonlinear polarization of the medium then has the following form:

$$\mathscr{P}^{\text{NL}}(2\omega_1) = -(\tfrac{1}{2})\hat{\chi}^{(2)}\mathbf{U}_1\mathbf{U}_1 A_1^2 \cos(2\omega_1 t - k_{\text{p}_2} z), \qquad k_{\text{p}_2} = 2k_1 \tag{4}$$

The SHG process in this classical picture may be treated as the radiation of the dipoles, oscillating with the frequency $2\omega_1$ and moving in space with the velocity

$$v_{\text{p}} = 2\omega_1/k_{\text{p}_2} = v_1 \tag{5}$$

The intensity of the second harmonic as excited at some point in the nonlinear medium (the "local" nonlinear effect) is proportional to the square of the nonlinear susceptibility $\chi^{(2)}$ and the square of the intensity of the fundamental wave.

The intensity of the second harmonic which is radiated by a volume of nonlinear medium (linear size z) depends essentially on its dispersion properties. This is one of the most fundamental problems and we will discuss it in detail in this chapter.

The second harmonic field is the result of interference of two waves: the forced wave which propagates synchronously with the nonlinear polarization wave

$$E_2^{(forced)} = U_2 A_2^{(forced)} \cos(2\omega_1 t - k_p z) \tag{6}$$

and the free wave* whose wave number is k_2:

$$E_2^{(free)} = U_2 A_2^{(free)} \cos(2\omega_1 t - k_2 z), \qquad k_2 = 2\omega_1 n_2/c \tag{7}$$

where n_2 is the refractive index at frequency $\omega_2 = 2\omega_1$. It is evident that the period of spatial beating L_{coh} (coherence length) between the forced and the free waves is determined by the following relation:

$$L_{coh} = \lambda_1/4 |n_2 - n_1| \tag{8}$$

where λ_1 is the wavelength in a vacuum.

The classical model of radiation of moving dipoles can be used in calculations of second harmonic (SH) intensity. For the conversion efficiency η_{I_2} (i.e., the ratio of the second harmonic power density at the output I_2 to the first harmonic power density at the input I_1, if $I_2 \ll I_1$) we obtain (see also Section II)

$$\eta_{I_2} = |\chi_2 A_1|^2 \left(\frac{8\pi L_{coh}}{n_1^{1/2} n_2^{1/2} \lambda_1} \right)^2 \sin^2 \left(\frac{\pi z}{2L_{coh}} \right) \tag{9}$$

Thus, in the dispersive medium ($n_2 \neq n_1$), the intensity of the second harmonic increases only when $z < L_{coh}$, and oscillates with distance for $z > L_{coh}$.

It should be pointed out that an exhaustive theory of the frequency doubling of plane monochromatic waves in a dispersive medium was developed by Khokhlov in 1961 before the Franken experiment. In particular, Khokhlov predicted the possibility of achieving a 100% efficiency of conversion into the second harmonic when $n_1 = n_2$ and $L_{coh} \to \infty$. This condition is usually termed the "phase-matching" condition. In this connection the next important step in the study of optical harmonic generation was the work of Giordmaine (1962) and Maker et al. (1962) who undertook experimental investigation of the interference of free and forced waves and proposed an effective method of phase matching in optics.

Figure 2 illustrates the results obtained by Maker et al. (1962) who observed for the first time the oscillations of the SH intensity described by Eq. (9). The distance between the two consecutive maxima corresponds to $2L_{coh}$. This value for quartz is very small (14 μm). Subsequent work of other authors has shown that this value is typical for many crystals in the visible region. Giordmaine and Maker have shown that color dispersion may be

---

* In formal terms the general solution for the inhomogeneous wave equation is composed of the respective solutions of homogeneous and inhomogeneous ones.

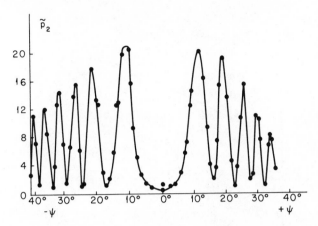

Fig. 2. The second harmonic intensity is a periodic function of the optical thickness of the piezoelectric crystal when there is a mismatch in the phase velocities of the fundamental and harmonic waves (from Maker *et al.*, 1962).

compensated by the anisotropy of the phase velocities in a birefringent crystal (see also Section I, B). The experimental realization of this idea has made it possible to obtain a conversion efficiency $\eta_{p_2} \simeq 10^{-6}$ with a free running ruby laser (compared with Franken's first experiment in which $\eta_{p_2} \simeq 10^{-11}$). It thus became evident that the harmonic generation effect is of interest not only as a method of measuring nonlinear susceptibilities, but also as the basis for techniques of laser frequency conversion in the short-wave range. Research in this direction developed very quickly. The main principles of the latter are the simple consequences of relation (9).

To construct efficient harmonic generators, it is necessary:

(1) to possess media in which phase-matching conditions can exist and nonlinear susceptibilities are assured sufficiently high, and

(2) to obtain sufficiently high electric light fields in nonlinear media.

As far back as 1963 Terhune *et al.* obtained a 20% conversion of the output of the Q-switched ruby laser to the ultraviolet second harmonic ($\lambda_2 = 0.347$ μm) (see Fig. 3).

In 1962 and 1963 the first reports on frequency multiplication of Nd-glass laser radiation ($\lambda_1 = 1.06$ μm) were published. The high power of the Q-switched Nd-glass second harmonic ($\lambda_2 = 0.53$ μm) enabled Johnson (1964) and Akhmanov *et al.* (1965a) to construct an efficient cascade generator (using two successive doublings in KDP crystals) of the fourth harmonic ($\lambda_4 = 0.265$ μm). As far back as 1965 Akhmanov *et al.* (1965d, e, f) used optical harmonic generators to investigate a number of nonlinear effects, such as optical breakdown, stimulated Raman and Brillouin scattering, etc.

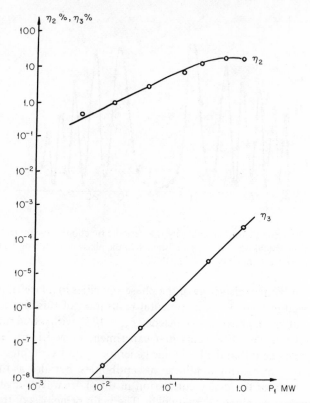

Fig. 3. The experimental curves of conversion efficiency to second and third harmonic (from Terhune *et al.*, 1963).

Giordmaine and Miller (1965) and Akhmanov *et al.* (1965d) have demonstrated the usefulness of the second harmonic of the Nd laser as a pump for optical parametric amplifiers and oscillators. In 1969 Akmanov *et al.* obtained the fifth harmonic ($\lambda_5 = 0.21$ μm) of a Nd glass laser by means of cascade frequency multiplication in a train of KDP crystals (see Fig. 4). It should be

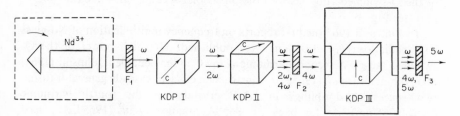

Fig. 4. Experimental arrangement for the cascade fifth harmonic generation of Nd laser radiation (from Akmanov *et al.*, 1969).

noted that the successes in the development of optical harmonic generators were connected to a considerable extent with the rapid progress in the theory of wave interactions in dispersive nonlinear media. The theory of harmonic generation by plane monochromatic waves, which was worked out from 1961 to 1964 (Khokhlov, 1961; Armstrong *et al.*, 1962; Akhmanov and Khokhlov, 1964) was generalized for monochromatic finite aperture beams by Kleinman (1962), Akhmanov *et al.* (1965d), Boyd *et al.* (1965), Francois and Siegman (1965), Akhmanov *et al.* (1966a), Bjorkholm (1966), Sukhorukov (1966), Kleinman *et al.* (1966a), Boyd and Kleinman (1968), Ward and New (1969), Sukhorukov and Tomov (1970a), and on nonmonochromatic waves (powerful optical noise, ultrashort pulses) by Akhmanov and Chirkin (1966), Miller (1968), Comly and Garmire (1968), and Akhmanov *et al.* (1968a, b). In these papers it was shown that the spatial and temporal modulation of waves decreases the efficiency of harmonic generators. Ways were therefore suggested of doing away with this detrimental effect. At the same time, the interesting interaction of spatial and temporal modulation in processes of harmonic generation was pointed out (Akhmanov *et al.*, 1968b; Orlov *et al.*, 1969; Sukhorukov and Tomov, 1970a).

It goes without saying that one of the most important results of these studies was the discovery of the principles of optimum focusing. The expediency of the fundamental beam focusing by harmonic generation is evident because of the increasing of the intensity I_1. The problem, however, is not as simple as it seems at first sight, for with increasing intensity I_1 through focusing, the extension of the focal region decreases (the focal length of the beam is $L_f \propto R^2$, where R is the focal length of the lens). Sukhorukov (1966), Bjorkholm (1966), Boyd and Kleinman (1968), and Bjorkholm and Siegman (1967) have shown that for optimum focusing, $L_f \simeq L$ in the case SHG. Extensive analysis of harmonic generation was made under various conditions of phase matching and double refraction. By using the principle of optimum focusing effective harmonic generators has made it possible to set up low-power cw lasers (with power less than 1 W). The advantages of optimum focusing of the fundamental radiation were made strikingly clear by Geusic *et al.* (1968) who used it inside a laser cavity for doubling radiation of a YAG:Nd^{3+} laser (see Fig. 5). In their experiment, the conversion efficiency was 100%, in the sense that SH output with $\lambda_2 = 0.53$ μm was the same as the laser power without the nonlinear crystal. An efficient second harmonic doubling of an argon-ion laser was demonstrated by Labuda and Johnson (1967) and Dowley (1968).

Although the quartz used in the pioneering work of Franken and co-workers proved to be unsuitable for constructing frequency multipliers (because of nonfulfilment of phase-matching conditions), efficient materials for multipliers in the visible and near ultraviolet ranges were found as far

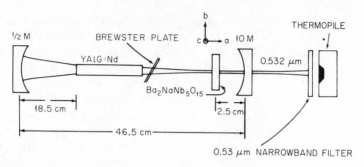

Fig. 5. Arrangement of the intracavity laser harmonic generation (from Geusic *et al.*, 1968).

back as 1962, in particular, KH_2PO_4 (KDP) and $NH_4H_2PO_4$ (ADP) crystals used by Giordmaine (1962) and Maker *et al.* (1962) in their first experiments. These crystals, because of sufficiently large nonlinear susceptibilities $[\chi^{(2)}_{312} = 1.36 \cdot 10^{-9}$ esu (ADP)] and the possibility of phasematching in the 1.70–0.517-μm fundamental wave region, have played an important role in nonlinear devices up to the present day. Many other nonlinear crystals have been found in the course of the last decade, starting with $LiNbO_3$ $(\chi^{(2)} = 10\chi^{(2)}_{KDP})$, $Ba_2NaNb_5O_{15}$ $(\chi^{(2)} = 31\chi^{(2)}_{KDP})$, $LiIO_3$ $(\chi^{(2)} = 11.9\chi^{(2)}_{KDP})$, and HIO_3 $(\chi^{(2)} = 10.5\chi^{(2)}_{KDP})$. Interesting results were obtained by Patel (1965) who conducted an experiment on SHG with a CO_2 laser. He found that the nonlinear susceptibility of Te is approximately 10^3 times larger than that for KDP. [McFee *et al.* (1970) later published data showing $\chi^{(2)}_{Te} = 1460\chi^{(2)}_{KDP}$.]

It is interesting to observe that all these achievements were made before a well-grounded microscopic theory of nonlinear polarization was established. Elaboration of this theory has been completed quite recently (see Chapter 2 by Flytzanis). It must be pointed out, however, that previous research was not totally devoid of theoretical thinking. Thus Miller (1964a) had already formulated a semiphenomenological rule according to which

$$\chi^{(2)}_{ijk}(2\omega_1) = \kappa_{ii}(\omega_1)\kappa_{jj}(\omega_1)\kappa_{kk}(2\omega_1)\delta_{ijk} \simeq (n^2-1)^3\delta_{ijk} \tag{10}$$

where κ is the linear susceptibility and δ_{ijk} the universal tensor. In later publications this rule was derived in detail.

The years 1968–1972 saw the publication of a series of papers on the frequency multiplication of picosecond pulses. It was found that harmonic generation can be used not only for the obvious task of conversion of the medium frequency of a picosecond pulse, but also for modulation of its duration (Shapiro, 1968; Orlov *et al.*, 1969) and the improvement of the picosecond train structure (DeMaria *et al.*, 1969; Akhmanov *et al.*, 1972). A schematic diagram to illustrate the improvement of the picosecond pulses

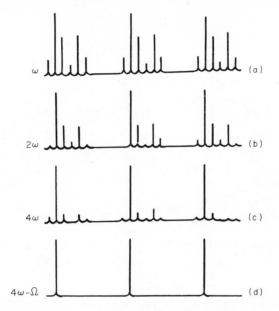

Fig. 6. Schematic diagram of enhancement of contrast in a train of picosecond pulses by cascade harmonic generation. (d) is SRS Stokes radiation in liquid N_2.

structure by successive nonlinear transformations (cascade multiplication) is given in Fig. 6.

Armstrong (1967) and Maier et al. (1966) proposed methods of measuring the intensity correlation functions of laser beams with the help of frequency doublers (see Fig. 7). These studies have had considerable influence on the development of methods for measuring picosecond pulses duration.

Studies in the field of frequency multiplication of picosecond pulses have renewed an interest in statistical effects in the harmonic generation process. Work in the field began as far back as 1963 and 1964 with that of Ducuing and Bloembergen (1964) and Akhmanov et al. (1963) with solid state lasers operating in the free running regime. The difficulties encountered in full mode synchronization were the cause of a return to this subject in 1968 and 1969 (Grütter et al., 1969; Akhmanov et al., 1970; Akhmanov and Chirkin, 1971).

The introduction of reliably powerful picosecond lasers has made possible an altogether new approach to the potentialities of higher-order nonlinearities in the technique of optical frequency multipliers. It should be noted although the phase-matched third harmonic generation had already been demonstrated by Terhune et al. in 1962, this effect was not practically used in harmonic generators. The high electric field of picosecond pulses compensates for the comparatively low value of the cubic susceptibility. Tunkin et al. (1972) obtained the fifth harmonic of the picosecond Nd-glass laser radiation in a

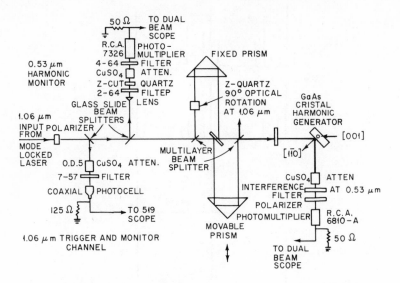

Fig. 7. The apparatus for measuring picosecond pulse widths with the help of second harmonic generation in a GaAs crystal. The light from a mode-locked laser is divided into two roughly equal beams with ortogonal polarizations. When the two beams from the picosecond laser overlap during their arrival at the nonlinear crystal, a second harmonic is produced. The delay between the beams can be varied in steps as small as 10^{-13} sec, so the pulse duration can be measured with the same accuracy (from Armstrong, 1968).

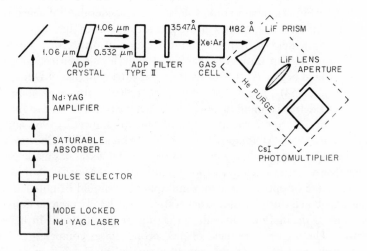

Fig. 8. Schematic diagram of the experimental apparatus for generation of 1182 Å (from Kung et al., 1973).

cascade scheme by tripling in $CaCO_3$ and then mixing the third harmonic and the fundamental wave in the other $CaCO_3$ crystal. The output power at $\lambda_5 = 2120$ Å was comparable with that produced by crystals with second-order susceptibility [on KDP crystals in the Akmanov 1969 experiment (Akmanov et al., 1969)].

The new opportunities, however, which were offered by techniques of harmonic generation with the employment of cubic susceptibility, were brilliantly demonstrated by Kung et al. (1972, 1973) who used alkali metal vapors and inert gases for harmonic generation in the vacuum ultraviolet range (see Fig. 8). The intense ninth harmonic of picosecond pulses ($\lambda_9 = 1182$ Å) from a YAG:Nd^{3+} laser were then obtained, which showed conclusively that the physics and the technology of optical harmonic generation still had a long way to go.

B. Nonlinear Optical Susceptibilities. Nonlinear Dispersion

A detailed description of the properties of nonlinear susceptibilities is found in Chapter 2 and 3 by Flytzanis and Kurtz, respectively. In the present chapter we shall therefore confine ourselves to a classification and a brief discussion of the interactions of different types of nonlinear effects. In a medium with polarization of the following kind:

$$\mathscr{P}^L = \hat{\kappa}E, \qquad \mathscr{P}^{NL} = \hat{\chi}^{(2)}EE + \hat{\chi}^{(3)}EEE \tag{11}$$

in a light field

$$E = \tfrac{1}{2}E_c + \tfrac{1}{2}E_c{}^*, \qquad \text{where} \quad E_c = UA \exp\{i(\omega t - kz)\} \tag{12}$$

with $\mathscr{P} = \tfrac{1}{2}\mathscr{P}_c + \tfrac{1}{2}\mathscr{P}_c{}^*$. There are several cases:

(i) Linear polarization of frequency:

$$\mathscr{P}_c{}^L(\omega) = \hat{\kappa}(\omega)E_c \tag{13}$$

(ii) Nonlinear polarization of zero frequency (dc effect or optical rectification):

$$\mathscr{P}_c^{NL}(0) = \tfrac{1}{2}\hat{\chi}^{(2)}(0 = \omega - \omega)E_c E_c{}^* \tag{14}$$

(iii) Nonlinear polarization at the second harmonic 2ω:

$$\mathscr{P}_c^{NL}(2\omega) = \tfrac{1}{2}\hat{\chi}^{(2)}(2\omega = \omega + \omega)E_c E_c \tag{15}$$

(iv) Nonlinear polarization at the third harmonic 3ω

$$\mathscr{P}_c^{NL}(3\omega) = \tfrac{1}{4}\hat{\chi}^{(3)}(3\omega = \omega + \omega + \omega)E_c E_c E_c \tag{16}$$

(v) Nonlinear polarization at the fundamental frequency

$$\mathscr{P}_c^{NL}(\omega) = \tfrac{3}{4}\hat{\chi}^{(3)}(\omega = \omega + \omega - \omega)E_c E_c E_c{}^* \tag{17}$$

In general, the tensor $\hat{\chi}^{(2)} \neq 0$ in media without a center of inversion. The tensor $\hat{\chi}^{(3)} \neq 0$ in all media. The effects of second harmonic generation (mixing of frequency) and optical rectification can be considered as mutually independent.* A more complex interaction is found between the nonlinear effects stimulated by tensors $\hat{\chi}^{(2)}$ and $\hat{\chi}^{(3)}$. An influence of the third harmonic on phase-matched second harmonic generation is not essential because usually there is no simultaneous phase matching for both harmonics and vice versa. Self-actions, however, can be very significant. According to Eq. (17), the high-intensity light field changes the dispersion properties of a medium (Fig. 9) and not only the nonlinear polarization, but also, the coherence length depends on the intensity of light.

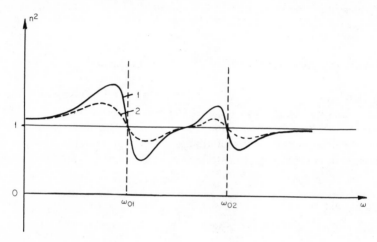

Fig. 9. Dispersion of the optical media in the neighborhood of the two resonant transitions. Curve 1 is linear dispersion ($E \ll E_{sat}$), and curve 2 is nonlinear dispersion (saturation only is taken into account). If optical Stark shift is sufficient, dispersion curves near resonant frequencies become asymmetric in shape.

If the dispersion of a medium in linear optics near some frequency may be represented by the following expression for the wave number:

$$k(\omega+\Omega) = k(\omega) + \frac{\partial k}{\partial \omega} \Omega + \frac{\partial^2 k}{\partial \omega^2} \frac{\Omega^2}{2} + \cdots \qquad (18)$$

where $u = \partial \omega / \partial k$ is the group velocity dependences on frequency ω, then in nonlinear optics the cofficients in (18) depend on the intensity of the light

---

* The static polarization arising due to optical rectification slightly changes the indices of refraction of crystals and consequently can influence the second harmonic generation. In actual fact, however, this effect is very weak (see Moldavskaya, 1967).

wave. Correspondingly for phase and group velocities we have

$$v = v(\omega) + v^{NL}(\omega, I) \tag{19a}$$

$$u = u(\omega) + u^{NL}(\omega, I) \tag{19b}$$

When investigating harmonic generation in crystals it is usually sufficient to take into account only the nonlinear addition to phase velocity; in the case of vapors and gases not only the nonlinear correction to group velocity, but also dependence on the intensity in $\partial^2 k/\partial\omega^2$ is found to be essential.

C. Phase-Matching Methods in Optics

First let us generalize the phase-matching conditions. The phase-matching conditions may be written for the second harmonic process as

$$\mathbf{k_1}' + \mathbf{k_1''} = \mathbf{k_2} \tag{20}$$

and for the third harmonic process as

$$\mathbf{k_1}' + \mathbf{k_1''} + \mathbf{k_1'''} = \mathbf{k_3} \tag{21}$$

where the wave vectors $\mathbf{k_1}'$, $\mathbf{k_1''}$, $\mathbf{k_1'''}$ may be different not only in direction but also in length (in an anisotropic media). From (20) and (21) one may formulate the requirements for dispersion of media to realize phase matching. Indeed, if for simplicity we assume $|\mathbf{k_1}'| = |\mathbf{k_1''}|$ from (20) it follows that $2|\mathbf{k_1}'| \geqslant |\mathbf{k_2}|$ and the phase-matching conditions for second harmonic generation can be satisfied only if the indices in the medium are:

$$n_1 \geqslant n_2 \tag{22}$$

From (21) it follows that for phase-matched third harmonic generation it should be $n_1 \geqslant n_3$. This corresponds to the anomalous dispersion. Real anomalous dispersion was used effectively for vapors and gases; in solids, however, different methods of simulating anomalous dispersion are much more widely used. The most fruitful idea is that the simulation of anomalous dispersion can best be achieved by setting up an interaction between different normal waves of the medium. A short review of matching methods in optics is given below; only the linear dispersive properties are taken into account. A discussion of nonlinear dispersion is found in Section II.

1. SIMULATION OF THE ANOMALOUS DISPERSION IN BIREFRINGENT CRYSTALS

This method was the first in the field and even to this day continues to play an important role in optical harmonic generators. Its general idea can be understood from Fig. 10 where the indices of refraction of the ordinary wave

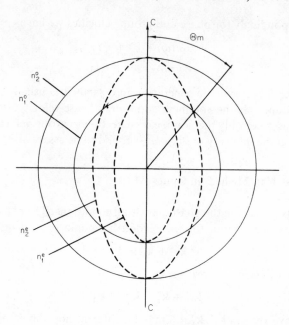

Fig. 10. Index surfaces n^o and n^e for the uniaxial crystal KDP.

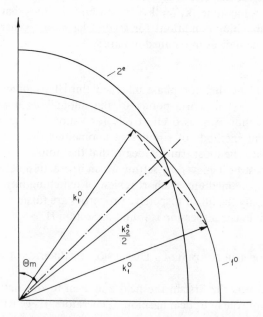

Fig. 11. Noncollinear phase-matched interaction in a KDP crystal.

$n_1°$ of frequency ω_1 and the extraordinary wave n_2^e of frequency $2\omega_1$ are shown in a birefringent KDP crystal (uniaxial negative crystal of class $4m$). At an angle of θ_m to the optic axis $n_1° = n_2^e(\theta_m)$. This direction is usually termed "the direction of collinear phase matching" (in reality there is of course a cone of such directions).[†] With $\theta > \theta_m$ and $n_1° > n_2^e(\theta)$, the crystal acts as a medium with anomalous dispersion. Here two noncollinear fundamental waves can generate phase-matched second harmonic waves (see Fig. 11). In uniaxial crystals, the above mentioned interaction can be denoted as

$$\gamma°(\omega_1) + \gamma°(\omega_1) \rightarrow \gamma^e(2\omega_1) \qquad (23)$$

or in short $O_1 O_1 \rightarrow E_2$. It is also possible to observe phase-matched interaction:

$$\gamma°(\omega_1) + \gamma^e(\omega_1) \rightarrow \gamma^e(2\omega_1) \qquad (24)$$

(in short $O_1 E_1 \rightarrow E_2$). Both the collinear and noncollinear second harmonic generation processes were studied in Giordmaine's pioneering research in 1962; his experimental set up is shown in Fig. 12 and the angular distribution of second harmonic radiation in Fig. 13. The interactions of types (23) and (24) are widely used in frequency doublers in the visible, ultraviolet, and infrared ranges.[*]

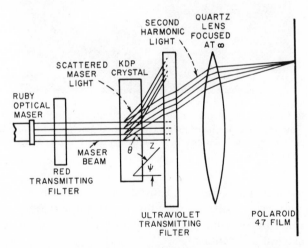

Fig. 12. Experimental arrangement for collinear and noncollinear second harmonic generation. The focal length of the lens is 15 cm. The angle ψ could be varied by rotating the crystal about an axis perpendicular to the page (from Giordmaine, 1962). (Original Giordmaine notations for the angles are used; ψ on the figure corresponds to θ in this chapter.

* The additional selection of directions from this cone is determined by symmetry properties of the nonlinear susceptibility $\chi_{ijk}^{(2)}$.

†Phase matching is possible also in biaxial crystals (see Hobden, 1967).

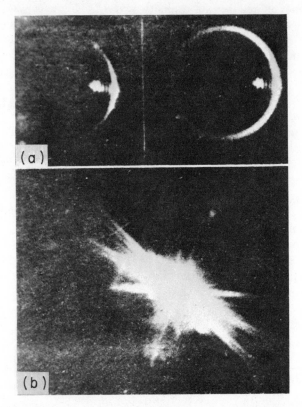

Fig. 13. Photographs of angular distribution of harmonic emission. (a) $\psi = 50.9°$; (b) $\psi = 49.6°$; the angular diameter of the ring in the photographs is $3.9°$ (from Giordmaine, 1962). (Original Giordmaine notations for the angles are used; ψ on the figure corresponds to θ in this chapter.)

The phase-matching condition can be met by large birefringence of a number of uniaxial crystals, not only for the second harmonic, but also for the higher ones. For example, in the $CaCO_3$ crystal it is possible to realize the phase-matched interactions

$$O_1 O_1 O_1 \rightarrow E_3, \qquad O_1 O_1 E_1 \rightarrow E_3, \qquad O_1 E_1 E_1 \rightarrow E_3 \qquad (25)$$

Akhmanov et al. (1974) have shown that birefringence in LFM crystal is so large that phase-matching conditions may be fulfilled for fourth harmonic generation of Nd^{3+} laser radiation. This process is of great interest for measurements of the fourth-order nonlinearity; unfortunately there is strong competition with the cascade processes, which are excited due to lower-order nonlinearity. It should be emphasized that high efficiency of nonlinear interaction can be assured only where there is phase-matched interaction of plane monochromatic waves. For real, finite-aperture, nonmonochromatic laser beams, however, group velocity matching is also very important. It should be noted that if phase matching is achieved by using linear birefringence, the group velocities are typically mismatched in magnitude and direction. In the next subsection we will discuss this important problem in detail.

(a) *Mismatching and Synchronism of Group Velocities.* Figure 14 shows the geometry of optical doubling using the $O_1 O_1 \to E_2$ interaction. The angle between the Poynting vectors s_1 and s_2 differ from zero,

$$\angle \, s_1 s_2 = \beta = \arctan \frac{1}{n_2} \frac{\partial n_2}{\partial \theta} \bigg|_{\theta_m} \tag{26}$$

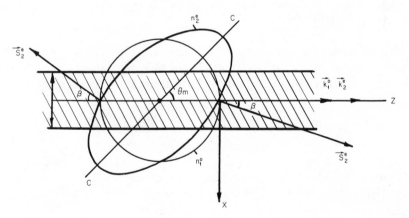

Fig. 14. Geometry of the phase-matched SHG in an uniaxial birefringent crystal for the $O_1 O_1 \to E_2$ interaction, where β is the walk-off angle.

As a result the energy flow of the extraordinary second harmonic wave walks off from the fundamental beam. This decreases the interaction path. The walk-off angle is seen to be a maximum for propagation at 45° to the optic axis and is zero for propagation either along or at 90° to the optic axis (group velocities are matched in direction). The 90° phase matching can be obtained by the temperature tuning of refractive indices. The absolute values of group velocities are usually mismatched in birefringent crystals. The group velocity mismatch parameter

$$v = (1/u_2) - (1/u_1) \tag{27}$$

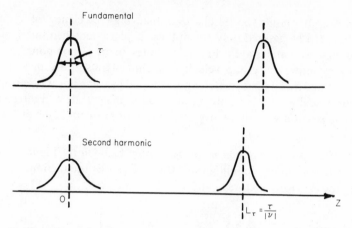

Fig. 15. Illustration of the group velocity mismatch effect in the optical harmonic generation process. In media with normal dispersion the SH pulse travels slower than the fundamental pulse. Accordingly, at the distance $z = L_\tau = \tau/v$ ($v = u_2^{-1} - u_1^{-1}$), the retardation time $T_r = vz$ equals the pulse duration τ.

is very important for the theory of doubling of the picosecond pulses (see Fig. 15). The values of β and v also determine the sensitivity of phase matching to fluctuation of direction and frequency of the fundamental wave.

2. PHASE-MATCHING DUE TO ROTARY DOUBLE REFRACTION

In optically active media, the normal waves propagating along the optic axis are two waves with circular polarizations of opposite sense of rotation. The phase velocities of these waves are different, which is the reason for optical activity of the medium. Rotatory double refraction can also be used for phase matching as in the case of the double refraction of the linearly polarized waves considered in the previous section.

Attention was drawn to this possibility for the first time by Rabin and Bey (1967). They showed that in the second harmonic generation process the phase mismatching $\Delta_2 = k_2 - 2k_1$ caused by color dispersion can be compensated in an optically active medium, in which

$$\Delta_2 = \rho_2 + 2\rho_1 \tag{28}$$

where ρ_1 and ρ_2 are specific rotations of frequencies ω_1 and $2\omega_1$. For the third harmonic the phase-matching condition in an optically active medium is

$$\Delta_3 = \mathbf{k}_3 - 3\mathbf{k}_1 = \rho_3 - \rho_1 \tag{29}$$

Unfortunately in the visible range specific rotations of most optical active

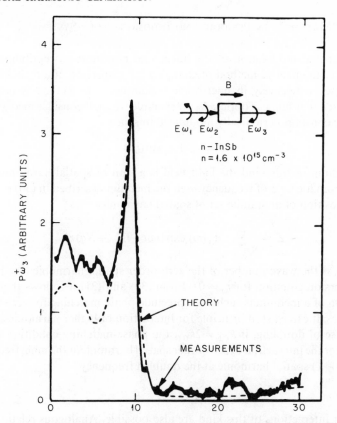

Fig. 16. Power of the right-hand circular polarized wave $P_{\omega_3}^+$ versus B for n-InSb having $n_c = 1.6 \times 10^{15}$ cm^{-3}, when P_{ω_1} is left-hand circularly polarized and P_{ω_2} is right-hand circularly polarized. Phase-matching occurs at $B = 10$ kg. Dashed line shows theoretical results and solid line shows experimental results (from Patel and Van Tran, 1969).

media are not enough to compensate the normal color dispersion* (typical values are $\Delta = 10^3$ cm^{-1} and $\rho = 10^2$ cm^{-1}. This is the reason that, experimentally, only the modification of the interference fringes for the different types of waves can be observed (Simon and Bloembergen, 1968). At the same time, Patel and Van Tran (1969) have demonstrated experimentally that the rotatory double refraction in semiconductors, connected with magnetic plasma effects, is sufficient for phase matching by difference frequency generation. In their experiments the difference frequency radiation at $\lambda_3 = 100$ μm was obtained by mixing the two lines $\lambda_1 = 9.6$ μm and $\lambda_2 = 10.6$ μm of a CO$_2$ laser in the InSb crystal in a magnetic field (see Fig. 16).

---

* An exception is made for cholesteric liquid crystals, where specific rotations are sufficiently high. For phase matching in this case, however, it is preferable (see Shelton and Shen, 1970) to use the periodic structure of these crystals and not the rotatory double refraction.

3. Phase-Matching in Periodic and Inhomogeneous Systems

The spatial modulation of the linear and nonlinear susceptibilities of a medium is an effective method of changing the dispersion of a medium. This fact is used extensively in microwave techniques; the method is now also widely used in nonlinear optics. As an example we shall consider media with a dielectric constant periodically modulated in space:

$$\varepsilon = \varepsilon_0 + \varepsilon_q \sin(qz) \tag{30}$$

In a medium of this kind the light field is a sum of spatial harmonics. The monochromatic field of frequency ω in the medium described in (30) becomes a superposition of an infinite set of spatial harmonics

$$E = \sum_{N=-\infty}^{\infty} A_N(\omega) \exp\{i\omega t - i(k_0 + Nq)z\} \tag{31}$$

where k_0 is the wave number of the zero-order spatial harmonic (the root of the dispersion equation for $\varepsilon_q = 0$). From (30) and (31) it follows that if the dispersion of a medium is such as to preclude phase matching for zero spatial harmonics, it can exist in principle for interaction of higher harmonics. Thus, in the case of doubling, if $k_{02} \neq 2k_{01}$, the phase-matching condition can be fulfilled for the interaction of zero-order spatial harmonic at the main frequency and the (-1) spatial harmonic at the doubled frequency

$$2k_{01} = k_{02} - q \tag{32}$$

Other interactions of this kind are also possible. Analogous relations can be written for the media with nonlinear susceptibilities modulated in space,

$$\chi^{(2)} = \chi_0^{(2)} + \chi_q^{(2)} \sin(qz) \tag{33}$$

It follows from (30) and (31) that the phase-matching conditions can be fulfilled not only for waves propagating in the same direction, but also for opposite traveling waves, if the modulation of a medium is fast enough. The period of modulation must be $\Lambda_F = 2\pi/(k_2 - 2k_1)$ for waves propagated in one direction and $\Lambda_B = 2\pi/(k_2 + 2k_1)$ for the waves propagating in the opposite direction.

In periodic media the directions of the group velocities of interacting waves are also matched. The magnitudes of group velocities can be controlled to a large extent. The main shortcoming of such systems is that the relative intensities of a high number of spatial harmonics are small. Thus the efficiency of phase-matched interactions of this kind are also small.

Armstrong et al. (1962) were the first to propose the employment of periodic and inhomogeneous media for phase matching. Their arguments were very clear. In so far as energy exchange between two waves in SHG processes is defined by the generalized phase $\phi = 2\phi_1 - \phi_2$ [see formulas (6)–(9)] in a medium in which $k_2 \neq 2k_1$, phase correctors should be placed along the medium at a distance of L_{coh} from one another. Three types of phase correctors are shown in Fig. 17. The correctors pictured in Fig. 17a are practically

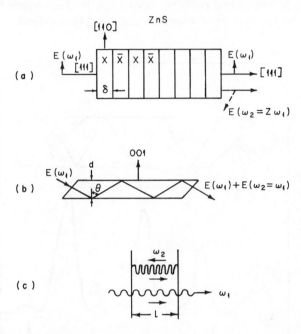

Fig. 17. Three experimental arrangements to provide phase correction if the phase velocities of the fundamental and second harmonic are not perfectly matched. (a) After the distance $\delta = \pi(k_2 - 2k_1)^{-1}$, the crystal is replaced by its inversion image, the nonlinear susceptibility χ_{xyz} changes sign, while the linear optical properties remain the same. (b) Both fundamental and second harmonic undergo multiple total reflection in a crystal of thickness $d = (k_2 - 2k_1)^{-1}\pi\cos\theta$. On each reflection E_1 and E_2 undergo a 180° phase shift; the product $E_2 E_1^2$ changes sign. (c) The traveling wave at ω_1 pumps the interferometer cavity, which contains a nonlinear dielectric and is resonant at ω_2, $l < \pi|k_2 - 2k_1|^{-1} = L_{\mathrm{coh}}$. The backward harmonic wave does not interact with the pump, and on each forward pass it has the correct phase for amplification (from Armstrong *et al.*, 1962).

identical to those described by formula (33); here, however, $\chi_0^{(2)} = 0$. Thus the idea of phase correctors may be regarded as one of the physical interpretations of phase-matched interactions in periodic media. Miller (1964b) showed that a system of the type in Fig. 17a is realized in a polydomain crystal. The system in which the phase correction is effected with complete internal reflection on the facets of a plane-parallel slab was realized by Boyd and Patel

in 1966. The experimental setup and curves are shown in Fig. 18. The system
with an optical resonator was studied by Akhmanov *et al.* (1965c) and Ashkin
et al. (1966). An elaborate investigation of harmonic generation in media
described by relations (30) and (33) was carried out by Shelton and Shen
(1970, 1972). They investigated the third harmonic generation in cholesteric
liquid crystals. The spatial period of crystals of this kind equals fractions of
a micron, and interactions of waves traveling in opposite directions were also
possible.

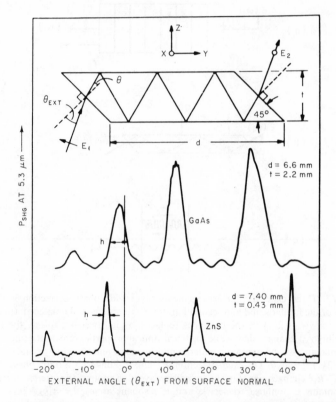

Fig. 18. SHG power as a function of external angle. Inset shows experimental geometry;
$N = d/t = 16$ and 3 for ZnS and GaAs, respectively (from Boyd and Patel, 1966).

An interesting variant of an optical periodic medium is presented by one
whose parameters are modulated by acoustic waves. The period of modulation
Λ_a equals several microns if the acoustic frequency is $\Omega_a/2\pi = 10^6$ Hz. The
photon–phonon interaction

$$\omega_2 = 2\omega_1 + \Omega_a \tag{34}$$

is phase matched if

$$k_2 = 2k_1 + k_a \tag{35}$$

In so far as $\Omega_a/\omega \simeq 10^{-9} \ll 1$, $\omega_2 \simeq 2\omega_1$, and the interaction under consideration is practically a doubling of the light frequency, relation (35) may be regarded as an analog of (32). An interaction of such kind was first observed by Boyd *et al.* (1970) (see Fig. 19). It should be emphasized that this effect is no longer described by a third-rank tensor, but by an elastooptic tensor of the

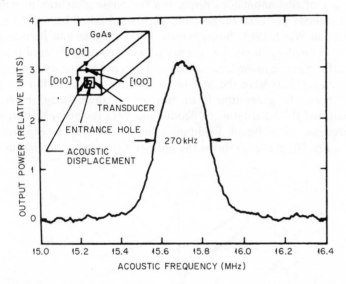

Fig. 19. Output power of the SH near 5.3 μm versus acoustic frequency. Inset shows crystal geometry (from Boyd, 1970).

fifth rank. The component of nonlinear polarization which gives rise to radiation of frequency ω_2 equals

$$\mathscr{P}_i = f_{ijklm} E_j E_k (\partial u_l / \partial x_m) \tag{36}$$

where E_j, E_k are the components of the electric field of fundamental frequency and **u** the acoustic displacement. In actual fact the tensor of the fifth rank $\hat{f}^{(4)}$ is determined by the dependence of the quadratic nonlinear susceptibility $\hat{\chi}^{(2)}$ on strain in the medium. According to Boyd *et al.* the nonlinear susceptibilities $f^{(4)}$ and $\chi^{(2)}$ have the same order of magnitude. In connection with the development of integral optical schemes, in recent years the harmonic generation in optical waveguides and thin films has been exciting an even greater interest. A strong dependence of dispersion on the geometry of such systems gives rise to many possibilities of phase matching. Anderson and Boyd (1970) investigated the harmonic generation of a CO_2 laser in a GaAs dielectric waveguide. Tien *et al.* (1970) excited the second harmonic of a Nd-glass laser in ZnS thin film optical waveguide. They obtained second harmonic generation in the form of coherent Čerenkov radiation.

4. Phase-Matching in Homogeneous Isotropic Media with Anomalous Dispersion

The use of real anomalous dispersion for phase matching in nonlinear optical processes has been proposed in papers published from 1963 to 1965 (Franken and Ward, 1963; Bloembergen, 1965). Zernike and Berman (1965) employed anomalous dispersion associated with a strong infrared resonance in quartz for phase-matched, far-infrared, difference frequency generation.

Bey *et al.* (1967) were the first to use anomalous dispersion for phase-matched harmonic generation in an isotropic medium (they observed the phase-matched third harmonic of neodimium laser radiation by introducing dye molecules in a liquid medium). The idea of their experiment is shown in Fig. 20; curve (a) shows the normal dispersion of a medium (index

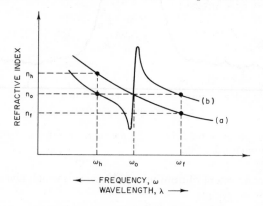

Fig. 20. Principle of phase matching by anomalous dispersion. Curve (a) shows a normally dispersive medium with $n_h > n_f$ at the frequencies ω_h and ω_f. Curve (b), the same medium after the introduction of anomalous dispersion centered at ω_0, results in identical indices n_0 at the two frequencies (from Bey *et al.*, 1967).

mismatched for the frequencies ω_1 and ω_3 associated with the color dispersion $\Delta n = n_3 - n_1$). After introduction of an appropriate resonant absorption at an intermediate frequency $\omega_3 > \omega_0 > \omega_1$, anomalous dispersion modified the dispersion curve; as a result, dispersion curve (b) was obtained, and phase-matching conditions may be fulfilled. Bey *et al.* obtained collinear phase matching by using the optimal concentration of the dye fuchsin red dissolved in hexafluoroacetone. The conversion efficiency was sufficiently reduced by a strong absorption of third harmonic. In later papers, Bey *et al.* (1968, 1971) tried to enhance the third harmonic generation by coupled nonlinear absorption. The idea was that by increasing the laser power and taking advantage of the third power law enhancement of THG, the absorption could be suppressed by exceeding the threshold for saturable absorption.

The technique of phase matching in mixtures of media with anomalous and normal dispersion was of great importance in frequency multipliers and mixers in the vacuum ultraviolet range. Harris and Miles (1971) have predicted that in mixtures of alkali vapors and inert gases very effective phase-matched third harmonic generation and four-photon frequency mixing may be obtained. Phase matching is possible because alkali metal vapors are anomalously dispersive if the frequency of the main radiation is below and its third harmonic above the primary resonance lines. A normally dispersive buffer gas may be added to vary the indicies of refraction until the phase-matched conditions are fulfilled (so the alkali vapor acts as a dye, and a buffer gas as a solvent in the Bey *et al.* experiments). It is very significant that the small absorption cross sections of the alkali vapors above their ionization potentials assure that the third harmonic is not reabsorbed by the vapor itself. This ultraviolet transparency allows harmonic generation in spectral regions which are not accessible to nonlinear optical crystals.

Young *et al.* (1971) have demonstrated the usefulness of mixtures of alkali vapors and inert gases in phase-matched third harmonic generation in obtaining the third-harmonic of the Nd:YAG laser ($\lambda_3 = 3547$ Å) in a Xe–Rb mixture. (The phase matching was obtained when the Xe to Rb concentration ratio was 412:1.) In the paper by Kung *et al.* (1972) the generation of 1773, 1520, and 1182 Å radiation by frequency tripling and summing in a phase-matched mixture of Cd and Ar was reported. For the third harmonic process ($\lambda_1 = 5320$ Å, $\lambda_3 = 1773$ Å) phase matching occurs for Cd and Ar atoms in the ratio 1:25. The energy conversion efficiency to 1773 Å was about 10^{-4}. Inert gases were very useful for harmonic generation at the wavelengths shorter than 1500 Å. The lowest excited state for inert gases corresponds to $\lambda_0 = 1495$ Å; so far, for $\lambda_1 > \lambda_0$ and $\lambda_3 < \lambda_0$, phase-matched third harmonic generation may be realized. Kung *et al.* (1973) have demonstrated the usefulness of this possibility, obtaining the powerful third harmonic with the wavelength $\lambda_3 = 1182$ Å in a phase-matched mixture of Xe and Ar. (Here Xe acts as the medium with the anomalous dispersion.) Apparently this method of phase matching can be efficient down to a wavelength on the order of $\lambda = 500$ Å. Group velocity mismatch in such a system is relatively small, so effective multipliers of picosecond pulses may be constructed. The important thing here is the saturation effects. In some cases breaking of phase-matching conditions as a result of atomic saturation may be significant (Javan and Kelly, 1968; Miles and Harris, 1973).

5. Nonlinear and Diffraction Corrections to Phase-Matching Conditions

In concluding this subsection we emphasize that considerations in Subsections 1–4 are true for plane monochromatic waves, which do not change the dispersion of a medium. Spatial and temporal modulation of real light beams and the nonlinear correction to phase and group velocities will be taken into account in Sections II and III. It should be mentioned here that the influence of these effects on harmonic generation can be very significant. For example, the wave number decreases in the focal region of the beam; this causes an additional phase shift between the interacting waves. This phenomena can be regarded as the "diffraction phase mismatch." As a result, in focused laser beams maximal harmonic power is emitted in the direction that differs from the direction of collinear phase matching; this effect is especially noticeable for the generation of high harmonics. As a consequence of relation (19) phase-matching conditions cannot be fulfilled for a whole beam with inhomogeneous intensity distribution. This circumstance is essential for doublers of cw radiation [here the dependence in (9) is connected with thermal effects] and for frequency multipliers in which resonance interactions are used.

D. Scope of the Chapter

This chapter is devoted to only one topic of the general problem of optical harmonic generation, namely to the theory and design of the optical frequency multipliers. As a result, many interesting phenomena such as harmonic generation in reflection (Bloembergen, 1966), nonlinear light scattering (Terhune *et al.*, 1965), nonlinear diffraction (Freund, 1968), etc., are only mentioned. Their role in nonlinear optics is connected first of all with nonlinear susceptibility measurements. At the same time an attempt is made to present the present-day state of the art both in the theory and design of optical frequency multipliers. In Section II the steady-state and quasi-steady-state theory of optical multipliers is considered. The theoretical approach used here is based to a large extent on the solutions of nonlinear parabolic equations (this corresponds to a quasi-optical approximation). Optical multipliers based on isotropic media and birefringent crystals are at the center of attention because they find at present the only actual practical applications. It goes without saying that the methods and also some of the results described in Section II can be used for other types of optical multipliers (for example, in optically active media, in periodic systems, etc.). Along with traditional problems such as aperture effects, optimal focusing, etc., the harmonic generation inside the laser resonator and the influence of self-focusing and self-defocusing on harmonic generation are also considered.

In Section III the transient harmonic generation problem is discussed, in particular, the effects connected with group velocity mismatching and the dispersion broadening of wave packets. In crystals these effects are very important if frequency multiplication of picosecond pulses and powerful optical noise (multimode radiation with independent phases of superluminescence modes) is considered. Strong linear and nonlinear dispersion in vapors and gases in the vicinity of resonance transitions causes sufficient transients in harmonic generation of relatively narrowband radiation in these media. Experimental devices and properties of harmonic generators are described in Section IV, where a review of experimental works published before the middle of 1973 is given. Section IV, D is devoted to some physical applications of optical noise (multimode radiation with independent phases of modes superluminescence) is considered. Strong linear and nonlinear dispersion in vapors coherent radiation in different physical experiments. In this section devices are described in which dispersion of the phase-matching directions is utilized for purposes of spectroscopy (nonlinear spectrographs). Another topic treated here is the higher-order correlation of optical field measurements with the help of harmonic generation. The cascade harmonic generators have enabled us to obtain radiation of several frequencies with identical characteristics. Such devices are very useful in measurements of nonlinear susceptibility dispersion; this aspect is also discussed briefly in Section IV.

II. THEORY OF STEADY-STATE AND QUASI-STEADY-STATE OPTICAL HARMONIC GENERATION. DESIGN CONSIDERATIONS

The propagation of light in a nonlinear dielectric is described by the wave equation

$$\text{rot rot } \mathbf{E} + \frac{1}{c^2} \frac{\partial^2}{\partial t^2} (\mathbf{E} + 4\pi \mathscr{P}) = 0 \tag{37}$$

where $\mathscr{P} = \mathscr{P}^{L} + \mathscr{P}^{NL}$ [see Eq. (2)]. Let us consider direct* Nth harmonic generation; in this case the electric field is a superposition of the fundamental (with frequency ω_1) and harmonic ($\omega_N = N\omega_1$; $N = 2, 3 \cdots$) fields

$$\mathbf{E} = \mathbf{E}_1 + \mathbf{E}_N, \qquad \mathbf{E}_j = \tfrac{1}{2}\mathbf{A}_j(\mathbf{r}, t) \exp\{i(\omega_j t - \mathbf{k}_j \mathbf{r})\} + \text{c.c.} \tag{38}$$

For the direct process, only nonlinear polarizations with the same frequencies

---

* In principle the Nth harmonic generation process may be connected with the Nth order nonlinear term in Eq. (2) (direct process) or with cascade processes of the lower-order nonlinearities. In this Section special attention is given to an analysis of direct second and third harmonic generation, which are of greatest practical importance now.

(ω_1, ω_N) should be taken into account

$$\mathscr{P}^{\mathrm{NL}} = \mathscr{P}_1^{\mathrm{NL}} + \mathscr{P}_N^{\mathrm{NL}}, \quad \mathscr{P}_j^{\mathrm{NL}} = \tfrac{1}{2}\mathbf{A}_{pj}^{\mathrm{NL}}(\mathbf{r}, t) \exp\{i(\omega_j t - \mathbf{k}_{pj}\mathbf{r})\} + \text{c.c.} \quad (39)$$

Thus, from the wave equation (37) two coupled nonlinear equations may be obtained:

$$\operatorname{rot} \operatorname{rot} \mathbf{E}_1 + \frac{1}{c^2}\frac{\partial^2}{\partial t^2}(\mathbf{E}_1 + 4\pi\mathscr{P}_1^{\mathrm{L}}) = -\frac{4\pi}{c^2}\frac{\partial^2 \mathscr{P}_1^{\mathrm{NL}}}{\partial t^2} \qquad (40a)$$

$$\operatorname{rot} \operatorname{rot} \mathbf{E}_N + \frac{1}{c^2}\frac{\partial^2}{\partial t^2}(\mathbf{E}_N + 4\pi\mathscr{P}_N^{\mathrm{L}}) = -\frac{4\pi}{c^2}\frac{\partial^2 \mathscr{P}_N^{\mathrm{NL}}}{\partial t^2} \qquad (40b)$$

For monochromatic fields (40) reduces to:

$$\operatorname{rot} \operatorname{rot} \mathbf{E}_1 + \hat{\varepsilon}(\omega_1)(\omega_1/c)^2 \mathbf{E}_1 = 4\pi(\omega_1/c)^2 \mathscr{P}_1^{\mathrm{NL}} \qquad (41a)$$

$$\operatorname{rot} \operatorname{rot} \mathbf{E}_N + \hat{\varepsilon}(\omega_N)(\omega_N/c)^2 \mathbf{E}_N = 4\pi(\omega_N/c)^2 \mathscr{P}_N^{\mathrm{NL}} \qquad (41b)$$

A. Nonlinear Effects at the Boundary and Inside of Nonlinear Media. Harmonic Generation in Reflection

When a monochromatic wave is incident on a plane boundary of a non-linear medium (the plane $z = 0$), the reflected and refracted waves contain not only radiation at the fundamental frequency but at the harmonic frequencies too (Bloembergen and Pershan, 1962) (see Fig. 21). Let us consider optical harmonic generation at the boundary of isotropic nonlinear media.

1. NORMAL INCIDENCE

Close to the boundary of the nonlinear medium the harmonic intensity is relatively small, $|A_N| \ll |A_1|$; so the nonlinear polarization at the frequency of the fundamental wave can be neglected ("fixed field" or parametric approximation). In this "fixed field" approximation the amplitudes of reflected and refracted fundamental waves is determined by Fresnel laws for the linear medium. For the reflection and transmission coefficients we obtain:

$$F_{\mathrm{T},1} = 2n_{\mathrm{L},1}/(n_{\mathrm{L},1} + n_1), \qquad F_{\mathrm{R},1} = 1 - F_{\mathrm{T},1} \qquad (42)$$

The transmitted fundamental wave excites the harmonic polarization wave (39). As one can see from Eq. (41b) polarization $\mathscr{P}_N^{\mathrm{NL}}$ creates the harmonic light field which is a superposition of free and forced waves with harmonic frequency:

$$\mathbf{E}_{\mathrm{c},N}^{\mathrm{T}} = 4\pi\mathbf{A}_{\mathrm{p},N}^{\mathrm{NL}} F_{\mathrm{T},N} \exp\{i(\omega_N t - k_N z)\}$$
$$+ 4\pi\mathbf{A}_{\mathrm{p},N}^{\mathrm{NL}} F_{\mathrm{p},N} \exp\{i(\omega_N t - k_{\mathrm{p},N} z)\} \qquad \mathbf{A}_{\mathrm{p},N}^{\mathrm{NL}} = 2^{1-N}\chi^{(N)}\mathbf{A}_1^{\,N}$$

$$(43a)$$

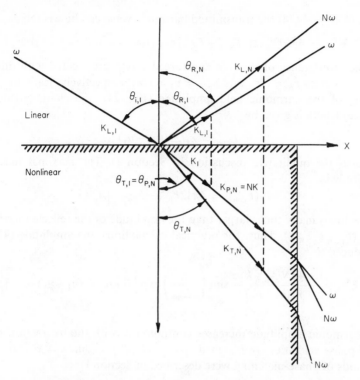

Fig. 21. The creation of the harmonic waves ($\omega_N = N\omega_1$) ($N = 2, 3, ...$) at the boundary of a nonlinear medium with $\mathbf{P}^{NL} = \chi^{(N)}\mathbf{E}^N$. The fundamental wave excites one reflected wave at the angle $\theta_{R,N}$ and two refracted waves (free one at $\theta_{T,N}$ and forced one at $\theta_{P,N} = \theta_{T,1}$) with harmonic frequencies.

The free wave [first term in (43a)] is a general solution of the homogeneous equation [Eq. (41b) without the term on the right-hand side], and the forced wave is a particular solution of the inhomogeneous equation (41b). The tangential components of the electric and magnetic fields should be continuous everywhere on the boundary at all times; thus gives rise to the harmonic reflected wave:

$$\mathbf{E}_{c,N}^{R} = 4\pi A_{p,N}^{NL} F_{R,N} \exp\{i(\omega_N t + k_{L,N} z)\} \tag{43b}$$

The nonlinear Fresnel formulas can be obtained from the boundary conditions; we have:

$$F_{R,N} = 1/[(n_N + n_1)(n_N + n_{L,1})] \tag{44a}$$

$$F_{p,N} = 1/(n_N^2 - n_1^2) \tag{44b}$$

$$F_{T,N} = F_{R,N} - F_{p,N} \tag{44c}$$

From (43a) and (44) the transmitted harmonic wave can be written as

$$\mathbf{E}_{c,N}^{T} = 4\pi \mathscr{P}_{c,N}^{NL}(z)\{F_{R,N} + F_{p,N}[\exp[i(n_N - n_1)\omega_N z/c] - 1]\} \tag{45}$$

The interference of the free and forced waves propagated with different phase velocities $v_N = c/n_N$ and $v_{p,N} = v_1 = c/n_1$, respectively, leads to spatial beating of the harmonic field amplitude (Fig. 24); the characteristic (coherence) length is given by

$$L_{coh} = \lambda_1/2N(n_N - n_1) \tag{46}$$

(see also the qualitative discussion in Section I). The maximal harmonic amplitude is

$$|A_N(L_{coh})| = 4\pi |2F_{p,N} - F_{R,N}| \cdot |A_{p,N}^{NL}| \tag{47}$$

This value is many times larger than the amplitude of the reflected harmonic wave $(F_{p,N} \gg F_{R,N})$. Thus, for practical calculations, the amplitude (45) can be reduced to

$$\mathbf{E}_{c,N}^{T} = \frac{-4\pi N i A_{p,N}^{NL} k_1 z}{n_1(n_1 + n_N)} \, \text{sinc}\left(\frac{\pi z}{2L_{coh}}\right) \exp\left\{i\left[\omega_N t - (n_1 + n_N)\frac{\omega_N z}{2c}\right]\right\} \tag{48a}$$

The harmonic amplitude increases continuously with the interaction length if the phase velocities are matched $v_1 = v_2$ $(L_{coh} \to \infty, \, \text{sinc} \, x = x^{-1}\sin x \to 1)$ (methods of phase-matching were described in Section I):

$$\mathbf{E}_{c,N}^{T} = \frac{-2\pi N i A_{p,N}^{NL} k_1 z}{n_1^2} \exp\{i(\omega_N t - k_N z)\} \tag{48b}$$

Formula (48b) has the obvious restriction $|A_N| \ll |A_1|$ (the fixed field approximation or $z \ll L_{NL} \simeq \lambda_1/\chi^{(N)}|A_1|^{N-1}$ (see Section II, B). It may be noted that the harmonic wave has a phase shift $\pi/2$ with respect to the polarization wave.

2. ARBITRARY INCIDENCE

In this case, besides the Fresnel coefficients the angles of reflection and refraction need to be computed too (Fig. 21). From the boundary conditions we have:

$$n_{L,1}\sin\theta_{i,1} = n_{L,1}\sin\theta_{R,1} = n_{L,N}\sin\theta_{R,N} = n_1\sin\theta_{T,1} = n_N\sin\theta_{T,N} \tag{49}$$

Some physical consequences of (49) should be mentioned:

(1) The angles of reflection at the fundamental and harmonic frequencies are different from one another if the linear medium is dispersive $n_{L,1} \neq n_{L,N}$.

(2) Inside a nonlinear medium the free and forced harmonic waves propagate in different directions, $\theta_{T,N} \neq \theta_{p,N} = \theta_{T,1}$, and consequently, there are two harmonic beams at the exit of a nonlinear medium (a nonlinear prism).

(3) The angles of total reflection are different for all waves:

$$\sin\theta_{T,1}^{cr} = n_1/n_{L,1}, \qquad \sin\theta_{T,N}^{cr} = n_N/n_{L,N}, \qquad \sin\theta_{p,N}^{cr} = n_1/n_{L,N}$$

If it was possible that $\theta_{T,N}^{cr} < \theta_{p,N}^{cr} < \theta_{T,1}^{cr}$, then only the free harmonic wave would propagate in a nonlinear medium (Bloembergen et al., 1969).

3. LOSSY MEDIUM

All formulas just given also remain valid for an absorbing medium when the dielectric constant is a complex quantity $\varepsilon = \varepsilon' + i\varepsilon''$. Taking into account the absorbtion coefficient of the waves, $\delta = k\varepsilon''/2\varepsilon'$, the wave numbers k in (48a) have to be replaced by $k - i\delta$. Now the harmonic amplitude is

$$|A_N| = \frac{4\sqrt{2}\,\pi N |A_{p,N}^{NL}|}{n_1(n_1 + n_N)} \left| \frac{\cosh[(\delta_N - N\delta_1)z] - \cos(\Delta_N z)}{(\delta_N - N\delta_1)^2 + \Delta_N{}^2} \right|^{1/2} \exp\frac{-(\delta_N + \delta_1)z}{2} \tag{50}$$

For the fixed parameters $(\delta_1, \delta_N, \Delta_N)$, the length of a nonlinear medium can be optimized to maximize harmonic power. If $\Delta_N = 0$, the optimum length is

$$L_0 = \frac{2}{\delta_N - N\delta_1}\, \text{arctan}\, h\, \frac{\delta_N - N\delta_1}{\delta_N + \delta_1}$$

In the case of large absorbtion ($\delta z \gg 1$), for example in metals, the harmonic generation in reflection can be more efficient than that inside the nonlinear medium. (In this case the "surface" nonlinear effect is much more pronounced than the "bulk" effect.)

B. Truncated Equations for a Nonlinear Medium

1. QUASI-OPTICAL EQUATIONS

In the foregoing subsection the idealized case of harmonic generation by the plane monochromatic waves was considered; the problem is simplified when the fixed field (parametric) approximation is valid. The exact analytical solution of system (40) for waves possessing spatial and temporal modulation apparently can not be obtained. The problem, however, can be sufficiently simplified because real laser beams differ only slightly from ideal mono-chromatic waves. Such a quasi-monochromatic wave can be presented as

$$\mathbf{E}_c = \mathbf{U} A (\mu_1 kx, \mu_2 ky, \mu_3 kz, \mu_4 \omega t) \exp \{i(\omega t - kz)\} \tag{51}$$

where the factors $\mu_j \ll 1$ indicate the "slow variables"; for the quasi-mono-chromatic waves, the complex amplitudes vary much more slowly than $\exp\{i(\omega t - kz)\}$. Expression (51) is valid for relatively wide light beams for which characteristic dimension of the cross section $a_M \gg \lambda$, or $\alpha \ll 1$ (where α is the angular divergence in the far field), and broad wave packets ($\tau_M c \gg \lambda$, where τ_M is the characteristic modulation time, or $\Omega \ll \omega$, where Ω is the bandwidth). These parameters of spatial and temporal modulation generally have different orders of magnitudes, so we will use different kinds of truncated equations for different situations.

(a) *Waves with Spatial Modulation* (*Finite Aperture Beams*). Let us consider a wave which propagates at a small angle β with respect to the Z-axis in a linear medium. From the fact that its amplitude varies more rapidly in the transverse direction than along the Z-axis (transition from light to shadow in XY-plane) we may put $\mu_1 = \mu_2 = \mu$ and $\mu_3 = \mu^2$, $\beta = \mu$ (because β is small and the beam axis almost coincides with Z-axis). Substituting (51) in the wave equation (41) leads to the truncated equation for the amplitude

$$\frac{\partial A}{\partial z} + \beta \frac{\partial A}{\partial x} = \frac{1}{2ik} \left(\frac{\partial^2 A}{\partial x^2} + \frac{\partial^2 A}{\partial y^2} \right) \tag{52}$$

This equation corresponds to the quasi-optical approximation in the theory of diffraction of optical beams; accordingly diffraction spreading is treated as the transverse diffusion of ray amplitudes. Because the diffusion coefficient is imaginary, this parabolic equation describes behavior of both an amplitude profile and a phase front. Equation (52) is very useful in the theory of optical resonators.

Formally, Eq. (52) is the same in the case of a weak anisotropic medium with the angle β equal to the angle $\angle \mathbf{sz}$, where $\mathbf{s} = \partial\omega/\partial\mathbf{k}/|\partial\omega/\partial\mathbf{k}|$ (Akhmanov *et al.*, 1966a). If $\mathbf{k}$ is parallel to $\mathbf{z}$ (normal incidence), the angle β is equal to the double refraction angle $\angle \mathbf{sk} = \beta$, where

$$\beta = \arctan \left[\frac{1}{n(\theta)} \frac{\partial n(\theta)}{\partial \theta} \right]$$

The nature of the approximations in deriving Eq. (52) can be better under-stood when the spectral domain is used. Let

$$A_q = \exp(-i\mathbf{qr}) \tag{53}$$

The surface of the wave vectors is a sphere (isotropic media or ordinary waves

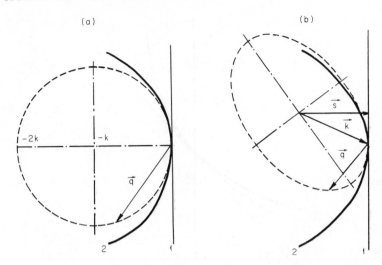

Fig. 22. The cross sections of the wave vector surfaces given by Eq. (54) of an uniaxial crystal for ordinary (a) and extraordinary (b) waves. The dashed curve corresponds to the wave equation (37) with polarization (56); line 1 (the tangents $\mathbf{qs} = 0$, $\mathbf{s}$ is the ray vector) is the geometrical optics; line 2 [parabolics, Eq. (55)] is the quasi-optics approximation Eq. (52).

in uniaxial crystals) or an ellipsoid (an extraordinary wave in crystals) (see Fig. 22)

$$|\mathbf{k+q}| = n(\theta)\omega/c \tag{54}$$

In the quasi-optical approximation these surfaces are described by paraboloid

$$2kq_z + 2k\beta q_x + q_x^2 + q_y^2 = 0 \tag{55}$$

In Fig. 23 the sections of the wave vector surfaces [Eqs. (54) and (55)] and the plane $q_y = 0$ are shown. This approximation is valid because the angular spectrum of wide beams is sufficiently narrow, $\alpha = |\mathbf{q}|/k \simeq \mu$. It should be mentioned that in geometrical optics these surfaces are approximated by the tangent planes $q_z + \beta q_x = 0$.

(b) *Plane Quasi-monochromatic Waves.* The truncated equation for broad wave packets in the medium with temporal dispersion

$$\mathscr{P}^{\mathrm{L}} = \int_0^\infty \hat{\kappa}(t_1)\mathbf{E}(t-t_1)\,dt_1 \tag{56}$$

can be derived by employing similar considerations [see (51)]. Taking into account dispersion only in the first approximation ($\mu_3 = \mu_4$), we derive the

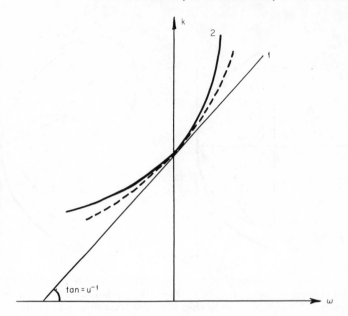

Fig. 23. The wave number as a function of the frequency in a dispersive medium. The dashed curve $k = n(\omega)\omega/c$ corresponds to wave equation (37) with (56); line 1 [tangent (60), $u = (\partial\omega/\partial k)^{-1}$ is the group velocity] is the geometrical optics (57), line 2 [parabolic (61)] is the quasi-optics approximation (58).

following truncated equation in a linear medium:

$$\frac{\partial A}{\partial z} + \frac{1}{u}\frac{\partial A}{\partial t} = 0 \tag{57}$$

which describes the propagation of wave packets with the group velocity

$$u = \partial\omega/\partial k, \qquad A = A(\eta) \qquad \text{where} \quad \eta = t - (z/u)$$

In the second approximation the form of the wave packet changed during the propagation. The truncated equation has the form:

$$\frac{\partial A}{\partial z} = \frac{i}{2}\left(\frac{\partial^2 k}{\partial\omega^2}\right)\frac{\partial^2 A}{\partial\eta^2} \tag{58}$$

which is also a parabolic equation [compare Eq. (52)]. Diffusion of the amplitude of a wave packet takes place relative to its center propagating with the group velocity. The diffusion coefficient depends on the dispersion of the group velocity $\partial^2 k/\partial\omega^2 = \partial u^{-1}/\partial\omega$. In the spectral domain,

$$A_q = \exp\{i(\Omega t - qz)\} \tag{59}$$

We may find the dispersion characteristics corresponding to different approximations (Fig. 23). In the geometrical optics approximation, the dispersion is linear

$$q = \Omega/u \tag{60}$$

and, in the quasi-optical approximation, it is parabolic

$$q = \frac{\Omega}{u} + \frac{\partial^2 k}{\partial \omega^2} \frac{\Omega^2}{2} \tag{61}$$

(c) *Interactions of Waves and Wave Packets in Nonlinear Media.* The truncated parabolic equations can be used also in studying wave interactions in weakly nonlinear media. The condition of a weak nonlinearity is $|A_p^{NL}/A_p^{L}| = \mu_0 \ll 1$, so the terms on the right-hand side in (40) are of the order of μ. Substituting (51) in (40) we derive the system of quasi-linear parabolic equations:

$$\frac{\partial A_j}{\partial z} + \mathscr{L}_j(x, y) A_j + \mathscr{L}_j(\eta) A_j = -\delta_j A_j + \mathscr{F}_j(A_1, ..., A_N; \mathbf{r}; t) \tag{62}$$

where δ_j is the linear absorbtion coefficients and the operators $\mathscr{L}_j$ describe diffraction

$$\mathscr{L}_j(x, y) = \beta_j \frac{\partial}{\partial x} + \frac{i}{2k_j} \left(\frac{\partial^2}{\partial x^2} + \frac{\partial^2}{\partial y^2} \right) \tag{63}$$

and dispersion

$$\mathscr{L}_j(\eta) = v_j \frac{\partial}{\partial \eta} - \frac{i}{2} \left(\frac{\partial^2 k_j}{\partial \omega_j^2} \right) \frac{\partial^2}{\partial \eta^2} \tag{64}$$

where $\eta = t - (z/u)_1$ is the time corresponding to the first wave, and $v_j = u_j^{-1} - u_1^{-1}$ the group velocity mismatch. The nonlinear terms are expressed through the Fourier components of a nonlinear polarization

$$\mathscr{F}_j = \frac{-i4\pi\omega_j \mathbf{U}_j \mathbf{A}_{pj}^{NL}}{n_j c} \exp\{i[k_j - k_{p,j}]z\} \tag{65}$$

The relative role of diffraction and dispersion effects depends critically on the values of a_M and τ_M, respectively. The characteristic diffraction length is $R_d = Ka_M^2/2$ and the dispersion length is $L_{dis} = \tau_M^2/2(\partial^2 k/\partial \omega^2)$. The second derivative with respect to the transverse coordinates may be omitted in (63) if $L \ll R_d$ (diffraction may be neglected): $\mathscr{L}_j(x, y) = \beta_j \partial/\partial x$; the second derivative with respect to the η coordinate may be omitted in (64) if $L \ll L_{dis}: \mathscr{L}_j(\eta) = v_j \partial/\partial \eta$. If $L \ll R_d, L_{dis}$ simultaneously, the wave propagation is described by the first-order equation (the geometric optics approximation) $\mathscr{L}_j = \beta_j \partial/\partial x + v_j \partial/\partial \eta$.

(d) *Space–Time Analogy in Nonlinear Optics.* The purpose of this subsection is to draw attention to a very useful mathematical analogy that exists between the equations describing nonlinear interactions of spatially modulated waves (beams) and time-modulated waves (wave packets). This analogy was established by Akhmanov *et al.* (1968b). Indeed, on the one hand, nonlinear interactions of spatially modulated beams in the quasi-optical approximation are described by the parabolic equations

$$\frac{\partial A_j}{\partial z} + \mathscr{L}_j(x, y) A_j = -\delta_j A_j + \mathscr{F}_j \tag{66}$$

On the other hand, the equations describing the interaction of quasi-monochromatic plane waves are

$$\frac{\partial A_j}{\partial z} + \mathscr{L}_j(\eta) A_j = -\delta_j A_j + \mathscr{F}_j \tag{67}$$

Equation (67) is equivalent to (66), if in the latter, the two-dimensional case is under consideration. Accordingly, we can directly establish one-to-one correspondence between the characteristic parameters in the theories of nonlinear interactions of spatially modulated waves and wave packets. For example, the beam diameter a is a direct analog of the pulse duration τ. The angular divergence of the beam in a "spatial problem" corresponds to the frequency spread Ω_0 in the "temporal problem." The angle β in the "spatial problem" is a direct analog of the group velocity mismatch parameter v in the "temporal problem." Hence, the spatial splitting of beams propagating in a medium in different directions is compared to the temporal splitting wave packets propagating with different group velocities. The characteristic lengths of these effects are $L_a = a/\beta$ and $L_\tau = \tau/v$, respectively.

Further the dispersion spreading of a wave packet can be compared with the diffraction spreading of a beam. Here, the longitudinal diffusion coefficient of a wave packet $(i/2)(\partial^2 k/\partial\omega^2)$ is an analog to the transversal diffusion coefficient of a beam $1/(2ik)$, and the length of dispersion spreading $L_{\rm dis} = \tau^2/(2\,\partial^2 k/\partial\omega^2)$ is an analog to the diffraction length $R_{\rm d} = ka^2/2$. This analog also takes place for nonlinear effects, so the optimal focusing in harmonic generation can be compared to the optimal compression of a pulse (see Section III), and so on.

2. Coupling Coefficients (Effective Nonlinearities)

Let us consider the Fourier amplitudes $A_{\rm p}^{\rm NL}$ of the nonlinear polarization, which determine in details the effectiveness of nonlinear wave interactions. Here we will compute $A_{\rm p}^{\rm NL}(\omega_j)$ for harmonic generation in media with quadratic and cubic nonlinearities and for self-actions.

(a) *Three Coupled Waves (Media with Quadratic Nonlinearity)*. The nonlinear polarization quadratic in $\mathbf{E}$ can be expressed as

$$\mathscr{P}^{(2)} = \int\limits_0^\infty\!\!\int dt_1\, dt_2\; \hat{\chi}^{(2)}(t_1, t_2)\, \mathbf{E}(t-t_1)\, \mathbf{E}(t-t_1-t_2) \qquad (68)$$

In such a medium a single fundamental wave generates the second harmonic wave (degenerate three wave interaction); so the total electric field is

$$\mathbf{E}_c = \mathbf{U}_1 A_1(\mathbf{r}, t)\exp\{i(\omega_1 t - k_1 z)\} + \mathbf{U}_2 A_2(\mathbf{r}, t)\exp\{i(\omega_2 t - k_2 z)\} \qquad (69)$$

Substituting (69) into (68) ,we obtain expressions for the polarization waves at the frequencies ω_1 and $2\omega_1$. Since the nonlinear optical susceptibilities which describe SHG are connected with very fast electronic processes, dispersion of nonlinear susceptibility within the spectra of fundamental and second harmonic radiation may be neglected (in a weakly nonlinear medium this dispersion would have had the order of magnitude of μ^4). Therefore even for temporal-modulated waves (down to very short picosecond pulses) nonlinearity can be described quasi-statically.

Slowly varying amplitudes A_j may be extracted from the integral (67) for $t_1 = t_2 = 0$. Taking this into account for $\mathscr{F}_1(\omega_1)$ and $\mathscr{F}_2(\omega_2)$ we obtain:

$$\mathscr{F}_1 = \frac{-2i\pi\omega_1}{cn_1}\left[\mathbf{U}_1\hat{\chi}^{(2)}(\omega_1 = 2\omega_1 - \omega_1)\mathbf{U}_2\mathbf{U}_1\right]A_2 A_1{}^*e^{-i\Delta_2 z}$$

$$\mathscr{F}_2 = \frac{-2i\pi\omega_1}{cn_2}\left[\mathbf{U}_2\hat{\chi}^{(2)}(2\omega_1 = \omega_1 + \omega_1)\mathbf{U}_1\mathbf{U}_1\right]A_1{}^2 e^{i\Delta_2 z} \qquad (70)$$

where $\Delta_2 = k_2 - 2k_1$. The spectral components of the tensor $\hat{\chi}^{(2)}$ are

$$\chi_{ijk}^{(2)}(\omega_3 = \omega_1 + \omega_2) = \int\limits_0^\infty\!\!\int dt_1\, dt_2\; \chi_{ijk}^{(2)}(t_1, t_2)\exp[-i(\omega_3 t_1 + \omega_2 t_2)] \qquad (71)$$

Tables of the nonzero components of the tensor $\hat{\chi}^{(2)}$ can be found in Chapter 2 by Flytzanis. If we take into account the permutation symmetry relation (see also Chapter 2 by Flytzanis)

$$\chi_{ijk}^{(2)}(\omega_3 = \omega_1 + \omega_2) = \chi_{jik}^{(2)}(\omega_1 = \omega_3 - \omega_2) = \chi_{kij}^{(2)}(\omega_2 = \omega_3 - \omega_1) \qquad (72)$$

The quasi-optical equations describing second harmonic generation become

$$(\partial A_1/\partial z) + \mathscr{L}_1 A_1 = -\delta_1 A_1 - i(\gamma_2/n_1)A_2 A_1{}^* e^{-i\Delta_2 z} \qquad (73a)$$

$$(\partial A_2/\partial z) + \mathscr{L}_2 A_2 = -\delta_2 A_2 - i(\gamma_2/n_2)A_1{}^2 e^{i\Delta_2 z} \qquad (73b)$$

where $\gamma_2 = 2\pi\omega_1\chi_2/c$, $\chi_2 = \mathbf{U}_1\hat{\chi}^{(2)}\mathbf{U}_2\mathbf{U}_1 = \mathbf{U}_2\hat{\chi}^{(2)}\mathbf{U}_1\mathbf{U}_1$ is the effective nonlinearity, which takes into account the polarizations of phase-matched waves. The tensor elements $\chi_{ijk} = 2d_{ijk}$ for uniaxial crystals are listed in Table III in Chapter 3 by S. Kurtz. If the O_1E_1–E_2 phase-matched interaction is under consideration, the total electric field in the light wave may be presented as

$$\mathbf{E}_c = \mathbf{U}_1^{\circ}A_1^{\circ}\exp\{i(\omega_1 t - k_1^{\circ}z)\} + \mathbf{U}_1^{e}A_1^{e}\exp\{i(\omega_1 t - k_1^{e}z)\}$$
$$+ \mathbf{U}_2^{e}A_2^{e}\exp\{i(\omega_2 t - k_2^{e}z)\}$$

and the truncated equations for the complex amplitudes are

$$(\partial A_1^{\circ}/\partial z) + \mathscr{L}_1^{\circ}A_1^{\circ} = -\delta_1^{\circ}A_1^{\circ} - i(\gamma_2/n_1^{\circ})A_2^{e}A_1^{e*}e^{-i\Delta_2 z} \tag{74a}$$

$$(\partial A_1^{e}/\partial z) + \mathscr{L}_1^{e}A_1^{e} = -\delta_1^{e}A_1^{e} - i(\gamma_2/n_1^{e})A_2^{e}A_1^{0*}e^{-i\Delta_2 z} \tag{74b}$$

$$(\partial A_2^{e}/\partial z) + \mathscr{L}_2^{e}A_2^{e} = -\delta_2^{e}A_2^{e} - i(2\gamma_2/n_2^{e})A_1^{\circ}A_1^{e}e^{i\Delta_2 z} \tag{74c}$$

where $\Delta_2 = K_2^{e} - K_1^{\circ} - K_1^{e}$, $\gamma_2 = 2\pi\omega_1\chi_2/c$, $\chi_2 = \mathbf{U}_1^{\circ}\hat{\chi}^{(2)}\mathbf{U}_2^{e}\mathbf{U}_1^{e}$. In ideal optically transparent media $(\delta_j = 0)$, Eq. (74) has the conservation integral $W(z) = \text{const}$:

$$W = (c/8\pi)\int\int\int_{-\infty}^{+\infty} d\eta\, dx\, dy\{n_1^{\circ}|A_1^{\circ}|^2 + n_1^{e}|A_1^{e}|^2 + n_2^{e}|A_2^{e}|^2\} \tag{75a}$$

For plane monochromatic waves $(\mathscr{L}_j A_j = 0)$, the integral conservation law (75a) reduces to the algebraic relation between intensities $I(z) = \text{const}$:

$$I = (c/8\pi)\{n_1^{\circ}|A_1^{\circ}|^2 + n_1^{e}|A_1^{e}|^2 + n_2^{e}|A_2^{e}|^2\} \tag{75b}$$

(b) *Four Coupled Waves.* The nonlinear polarization cubic in $\mathbf{E}$ can be expressed as

$$\mathscr{P}^{(3)} = \int\int\int_0^{\infty} dt_1\, dt_2\, dt_3\, \hat{\chi}^{(3)}(t_1, t_2, t_3)\mathbf{E}(t-t_1)\mathbf{E}(t-t_1-t_2)\mathbf{E}(t-t_1-t_2-t_3) \tag{76}$$

In the degenerate case (only two waves with frequencies ω_1 and $3\omega_1$ interact in the nonlinear medium) the nonlinear terms are

$$\mathscr{F}_1 = -i(\gamma_3/n_1)\{A_3 A_1^{*2}e^{-i\Delta_3 z} + |A_1|^2 A_1 + 2|A_3|^2 A_1\} \tag{77}$$

$$\mathscr{F}_2 = -i(\gamma_3/n_3)\{A_1^3 e^{i\Delta_3 z} + 3|A_3|^2 A_3 + 6|A_1|^2 A_3\}$$

where $\Delta_3 = k_3 - 3k_1$; $\gamma_3 = 3\pi\omega_1\chi_3/2c$; $\chi_3 = \chi^{(3)}$. Truncated equations with coupling coefficients in form (77) describe the third harmonic generation process in isotropic media (for example, in gases and vapors). In the same way $O_1O_1O_1$–E_3, $O_1O_1E_1$–E_3, or $O_1E_1E_1$–E_3 interactions may be treated in uniaxial negative crystals.

(c) *Nonlinear Terms Describing Self-actions.* The influence of self-actions (self-focusing, self-defocusing, and self-modulation) on the harmonic generation process are described in the truncated equations by the Fourier components of the nonlinear polarization

$$\mathscr{F}_j = -ik_j n_j^{\mathrm{NL}}/n_j^{\mathrm{L}}$$

where n_j^{NL} is the nonlinear part of the refractive index (see Akhmanov *et al.*, 1967a). In practice self-actions are very important in considering frequency multiplication of cw or quasi-cw laser radiation; here the self-actions are connected with the thermal effects (heating of nonlinear crystal in laser beam). In this case

$$n_j^{\mathrm{NL}} = (dn_j/dT)\, T' \tag{78a}$$

where T' is a variation of the temperature of the sample; in the laser beam

$$\frac{\partial T'}{\partial t} = \chi_T\left(\frac{\partial^2 T'}{\partial x^2} + \frac{\partial^2 T'}{\partial y^2} + \frac{\partial^2 T'}{\partial z^2}\right) + \frac{cn\delta|A|^2}{4\pi\rho c_{\mathrm{P}}} \tag{78b}$$

The influence of thermal self-actions on second harmonic generation will be discussed in Section II, F.

3. Exact Solutions of Truncated Equations for Second and Third Harmonics. Reaction of Harmonic Waves on the Amplitude and Phase of the Fundamental Wave

In this subsection exact solutions of the coupled wave equations are given. Special attention is paid to the SHG process. It should be emphasized that exact solutions of Eq. (73) for the general case of modulated waves do not exist. It is very instructive, however, to discuss the exact solutions for plane monochromatic waves; these were first obtained by Khokhlov (1961) and Armstrong *et al.* (1962).

(a) *Second Harmonic Generation. Two Coupled Waves.* It is convenient to introduce real variables (amplitude and phase):

$$A_j = A_{0j}\exp(i\psi_j)$$

so that Eq. (73) may be rewritten as

$$(\partial A_{01}/\partial z) + \delta_1 A_{01} + (\gamma_2/n_1) A_{01} A_{02} \sin\phi = 0 \tag{79a}$$

$$(\partial A_{02}/\partial z) + \delta_2 A_{02} - (\gamma_2/n_2) A_{01}^2 \sin\phi = 0 \tag{79b}$$

$$\frac{\partial\phi}{\partial z} = \Delta_2 + \gamma_2\left[\frac{2A_{02}}{n_1} - \frac{A_{01}^2}{n_2 A_{02}}\right]\cos\phi \tag{79c}$$

where $\phi = 2\phi_1 - \phi_2 + \Delta_2 z$ and $\Delta_2 = k_2 - 2k_1$.

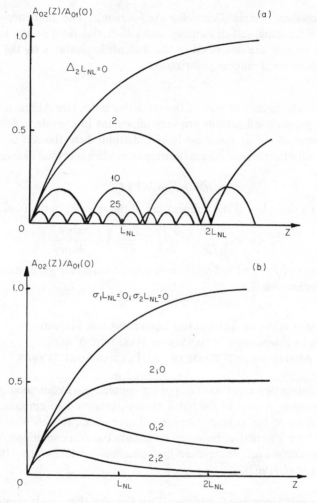

Fig. 24. The SH amplitude normalized to the initial amplitude of the fundamental wave as a function of the distance inside the nonlinear medium [L_{NL} given by Eq. (80c)] (a) for different phase mismatching $\Delta_2 = k_2 - 2k_1$, and (b) for various absorptions of light waves. [Part (a) is from Bloembergen, 1965].

(i) *Lossless medium* ($\delta_1 = \delta_2 = 0$). If at the boundary of nonlinear media, $A_{02}(0) = 0$, the second harmonic intensity in this case (see Fig. 24a) is

$$I_2(z) = KI_1(0)\, sn^2\left[(z/L_{NL}\,K^{1/2}); K\right] \tag{80a}$$

where

$$K = \{(\Delta_2 L_{NL}/4) + [1 + (\Delta_2 L_{NL}/4)^2]^{1/2}\}^{-2} \tag{80b}$$

and the characteristic nonlinear length L_{NL} is

$$L_{NL} = (n_1 n_2)^{1/2}/\gamma_2 A_{01}(0) = n_1 (n_2 c)^{1/2}/\gamma_2 [8\pi I_1(0)]^{1/2} \qquad (80c)$$

For the exact phase-matching ($\Delta_2 = 0$), formula (80a) reduces to

$$I_2(z) = I_1(0) \tanh^2 (z/L_{NL}) \qquad (81)$$

In the last case, the harmonic intensity grows monotonically and in the limit ($Z \to \infty$) the conversion efficiency is equal to 100%. In the characteristic length L_{NL} about 60% of the fundamental intensity will be converted to the second harmonic.

If $\Delta_2 \neq 0$, spatial modulations of the harmonic and fundamental waves take place. The period of this modulation depends on the parameter $\Delta_2 L_{NL}$. If $\Delta_2 L_{NL} \ll 1$, almost 100% conversion to the second harmonic occurs at the distance $z = L_{NL} \ln(1/\Delta_2 L_{NL})$. If $\Delta_2 L_{NL} \gg 1$, conversion efficiency reduces significantly and the second harmonic intensity can be written as

$$I_2(z) = I_1(0) (z^2/L_{NL}^2) \, \text{sinc}^2 (\Delta_2 z/2) \qquad (82)$$

(the result is the same as that in the parametric approximation; compare Section II, A). Now the period of the interference pattern depends only on the dispersion of the media and equals $2L_{coh} = 2\pi/\Delta_2$. From the energy conservation law, intensity of the fundamental wave can be written as

$$I_1(z) = \frac{I_1(0)}{K} \left[K - 1 - dn^2 \left(\frac{z}{L_{NL} K^{1/2}} ; K \right) \right] \qquad (83)$$

If $K = 1$, the elliptic function dn reduces to $1/\cosh(z/L_{NL})$ and equals a constant when $K = 0$.

The relation between the phases ϕ satisfies the equation

$$\sin \phi = (cn - dn)/1 - K \, dn^2 \qquad (84)$$

when $K = 1$, $\phi = \pi/2$ and when $K \ll 1$, $\phi = (\pi/2) + (\Delta_2 z/2)$. The derivatives of the phases of the fundamental and the second harmonic

$$\frac{\partial \phi_1}{\partial z} = \frac{1 - K}{L_{NL} K^{1/2}} \frac{I_2(z)}{I_1(z)}, \qquad \frac{\partial \phi_2}{\partial z} = \frac{1 - K}{L_{NL} K^{1/2}} \qquad (85)$$

show that the phase velocities of fundamental wave and the second harmonic change as a result of the nonlinear interaction. For the harmonic wave this effect is the result of the interference between free and forced waves: $\partial \phi_2/\partial z \simeq \Delta_2/2$. The phase of a fundamental wave is changing very rapidly near $z = L_{NL}$: $\partial \phi_1/\partial z \simeq K^{1/2}/L_{NL}$. Thus in the case of large conversion efficiency there is the specific self-action effect, which can lead to self-focusing of the fundamental wave (see also Ostrovsky, 1967; Karamzin and Sukhorukov, 1975).

(ii) *Lossy medium.* If $\delta_1 = \delta_2 = \delta$, exact solution of Eqs. (79) may be obtained, if we introduce the new variables:

$$A_{0j}(z) = \tilde{A}_{0j}(\tilde{z})e^{-\delta z}, \qquad \tilde{z} = (1-e^{-\delta z})\delta^{-1}$$

In these new variables, the solution is of form (80). If $\Delta_2 = 0$, for the second harmonic intensity we obtain [compare with (81), see Fig. 24b].

$$I_2(z) = I_1(0)e^{-2\delta z}\tanh^2\left(\frac{1-e^{-\delta z}}{\delta L_{NL}}\right) \tag{86}$$

Maximum conversion occurs at $z = \delta^{-1}$; conversion efficiency in a lossy medium is high, if $\delta L_{NL} \ll 1$. If attenuation is large only at the fundamental frequency ($\delta_1 \neq 0$, $\delta_2 = 0$), the solution of Eqs. (79) can be obtained by substituting $A_2(z) = \tilde{A}_2(z) - (\delta_1 n_1/\gamma_2)$. The limiting value of the second harmonic intensity in this case ($Z \to \infty$) is

$$I_{2,\lim} = I_1(0)\{[1+(\delta_1 L_{NL})^2]^{1/2} + \delta_1 L_{NL}\}^{-1} \tag{87}$$

Finally, if $\delta_1 = 0$ and $\delta_2 \neq 0$, absorption of the SH radiation is much more pronounced, a situation which is very typical in the ultraviolet doublers. In the case of large absorption $\delta_2 Z \gg 1$, we have

$$I_2(z) \simeq I_1(0)[\delta_2 L_{NL} + (2z/L_{NL})]^{-2} \tag{88}$$

where the optimal length of the doubler satisfies the relation $\delta_2^{-1} \lesssim L_0 \lesssim \delta_2 L_{NL}^2$.

(c) *Three Coupled Waves.* This situation occurs in the second harmonic generation process when the $O_1E_1-E_2$ interaction is used. Conversion efficiency in the nonlinear regime is maximal, if the intensities of the ordinary and extraordinary fundamental waves are equal, i.e., $I_1^\circ = I_1^e$. If the phase-matching conditions are fulfilled, harmonic intensity for the $O_1E_1-E_2$ interaction grows in a similar manner as for the $O_1O_1-E_2$ interaction [see Eq. (81)]. If, however, $I_1^\circ \neq I_1^e$ spatial beats occur, even for $\Delta_2 = 0$. General solution of this problem was obtained by Armstrong *et al.* (1962); they also obtained solutions for third harmonic generation.

4. CONVERSION EFFICIENCY OF OPTICAL MULTIPLIERS EXCITED WITH BEAMS OF FINITE CROSS SECTION AND WAVE PACKETS; QUASI-STEADY-STATE APPROXIMATION

As was mentioned above, in the general case of modulated waves, exact solutions of the equations describing harmonic generation do not exist. Accordingly, the theory of optical multipliers excited with real laser beams which possess spatial and temporal modulations in most cases is based on the parametric approximation. There are several exceptions, however. One of them is under consideration in this section: we present here some results of the geometric-optical theory of the optical doubler, a theory in which diffraction and dispersion effects are neglected (quasi-steady-state approximation).

(a) *Limiting Efficiency of the Optical Doubler Excited with a Monochromatic Beam of Finite Aperture.* If the length of a nonlinear medium $L < R_d = ka^2/2$ and $L < L_a = a/\beta$, it may be considered that all light rays are parallel to the Z-axis in the near field. Therefore, in (52), $\mathscr{L}_j(x, y)A_j = 0$ and the transverse coordinates are included in the solutions of the harmonic generation equations only as a parameters. It means that in this case the exact solutions obtained in Section II,B,3 may be used, but the constant value $I_1(0)$ should be replaced by the initial intensity profile of the fundamental wave $I_1(x, y, 0)$. Now the harmonic generation efficiency depends on the transverse coordinate of the ray; the nonlinear length (80c)

$$L_{NL} = (n_1 n_2)^{1/2}/[\gamma_2 A_{01}(x, y, 0)]$$

is a function of x, y. As a result, the power conversion efficiency for the whole beam is less than the intensity conversion efficiency in the center of a beam. For example, the second harmonic power

$$P_2 = \int\int\limits_{-\infty}^{\infty} I_2(x, y, z) \, dx \, dy$$

of the Gaussian beam $I_1(x, y, 0) = (16P_1/cn_1 a^2) \exp(-2r^2/a^2)$ is

$$P_2 = P_1 \left[1 + \frac{2 \ln \cosh \tilde{z}}{\tilde{z}^2} - \frac{2 \tanh \tilde{z}}{\tilde{z}} \right], \qquad \tilde{z} = \frac{z}{L_{NL}(0)} \qquad (89)$$

Thus, for example, when $\eta_{P_2} = P_2/P_1 = 0.8$, the intensity conversion efficiency at the center of the beam, $I_2(0, z) = 0.9 I_1(0, 0)$ (see Fig. 25). Taking into account diffraction, $\mathscr{L}_j(x, y)A_j \neq 0$, the theory of limiting efficiency was developed by Karamzin and Sukhorukov (1974).

(b) *Wave Packets.* If $L < L_{dis} = \tau^2/[2 \, \partial^2 k/\partial \omega^2]$ and $L < L_\tau = \tau/v$, the dispersion spreading of a pulse may be neglected and the group velocities of the fundamental and second harmonic may be put equal, $u_1 = u_2 = u$. In this case the time is included in the solutions only as a parameter $A_j = A_j(z, t-(z/u))$. We illustrate the influence of temporal modulation on the efficiency of the optical doubler including simultaneous spatial modulation. Let us consider the fundamental radiation which has the form $I_1(r, 0, t) = (8/cn_1 a^2) P_1(t)(1 + r^2/a^2)^{-2}$ where $P_1(t) = P_0$ at $|t| \leqslant \tau$ and $P_1(t) = P_0 \exp\{-(|t| - \tau)/\tau_w\}$ at $|t| \geqslant \tau$. The calculation yields for energy conversion

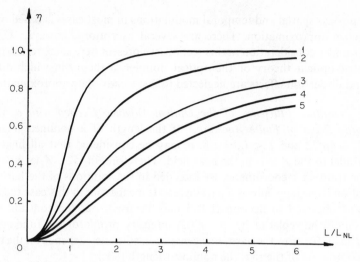

Fig. 25. The SH energy conversion as a function of the ratio of the crystal length to nonlinear length: curve 1 is a plane monochromatic fundamental wave (2.45); 2 is Gaussian beam (2.52); 3–5 are Lorentian beams (2.54) with temporal modulation (exponential "wings" energy equals $W_{w1} = 0(3)$, $W_{w1} = 0.5W_1(4)$, $W_{w1} = W_1(5)$, where W_1 is total pulse energy).

$$w_2 = w_1\left[1 - \frac{w_{01}}{w_1}\frac{\tanh \tilde{z}}{\tilde{z}} - \frac{2w_{w1}}{w_1}\frac{\ln\cosh \tilde{z}}{\tilde{z}}\right], \qquad \tilde{z} = \frac{z}{L_{NL}(0,0)} \qquad (90)$$

where $w_1 = w_{01} + w_{w1}$ is the total energy, w_{01}, w_{w1} are the energies of the center part of pulse and the "wings" of pulse, respectively. In Fig. 25 the energy conversion is represented for different ratios of w_{w1}/w_1. One can see that temporal modulation decrease the efficiency of a frequency doubler in comparison with the ideal monochromatic wave.

C. Harmonic Generation by Gaussian Beams in Isotropic and Quasi-isotropic Media

First of all let us consider harmonic generation in a nonlinear medium in which walk-off of the Poynting vector is absent. This condition is fulfilled in isotropic media or in anisotropic media when the waves are phase matched along a principal axis of the crystal. Harmonic generation in such isotropic and quasi-isotropic media can be described by the parabolic equations for ray amplitudes [see (63)–(77)]:

$$\frac{\partial A_1}{\partial z} - \frac{1}{2ik_1}\left(\frac{\partial^2 A_1}{\partial x^2} + \frac{\partial^2 A_1}{\partial y^2}\right) = 0 \qquad (91)$$

$$\frac{\partial A_N}{\partial z} - \frac{1}{2iNk_1}\left(\frac{\partial^2 A_N}{\partial x^2} + \frac{\partial^2 A_N}{\partial y^2}\right) = -i\left(\frac{\gamma_N}{n_N}\right)A_1^{N}e^{i\Delta_N z} \qquad (91b)$$

where N is a harmonic number and $\Delta_N = k_N - Nk_1$; $\gamma_N = 4\pi N\chi_N \omega_1/2^N c$. The boundary conditions are

$$A_1 = A_1(x, y, 0), \qquad A_N(x, y, 0) = 0 \tag{92}$$

Generally speaking the harmonic amplitude at the boundary differs from zero because of the additional boundary conditions connected with nonlinear polarization (see Section II, A); optical harmonics in reflection also should be taken into account. For "bulk" harmonic generation, however, which is under consideration here, the reflected harmonics may be treated as negligible; thus, we may put $A_N(x, y, 0) = 0$.

1. Harmonic Generation by Gaussian Beams with Spherical Phase Fronts

The fundamental laser radiation in the TEM_{oo} mode propagating in a nonlinear medium can be written as a Gaussian beam:

$$A_1 = \left(\frac{16P_1}{cn_1 a^2}\right)^{1/2} \psi^{-1}(z) \exp\left\{-\frac{r^2}{a^2\psi(z)}\left(1 - \frac{i\alpha_0}{\alpha_d}\right)\right\} \tag{93}$$

where $\psi(z) = 1 - i(z/R_d)[1 - (i\alpha_0/\alpha_d)]$; $R_d = k_1 a^2/2$; $f = |\psi|$ is the dimensionless beam radius. The minimum beam radius (waist) $a_w = a(z_w)$, the confocal parameter b, and the angular divergence in the far field α satisfy the relations: $\alpha = (\alpha_0^2 + \alpha_d^2)^{1/2}$, $a_w = 2/(k_1\alpha)$; $b = 4/(k_1\alpha^2) = k_1 a_w^2$; $z_w = a\alpha_0/\alpha^2$, where $\alpha_0 = a/R$ and $\alpha_d = 2/ka$ are the initial and the diffraction angular divergence, respectively. Our primary interest is in focused laser beams (Fig. 26). It should be noted that the phase of a wave propagated through the focus changes by π. This corresponds to a decrease of the wave number by

$$k_{1,d}(z) = k_1 - (2/b)[1 + 4(z - z_w)^2/b^2]^{-1} \tag{94}$$

In the parametric approximation, with the boundary conditions mentioned above, solutions of Eq. (91) may be obtained with the help of the Green's functions. It may be shown that for a fundamental wave of form (93) the harmonic is also a Gaussian beam, $N^{1/2}$ times narrower than the fundamental. The harmonic power that may be obtained depends significantly on the focusing parameter

$$m = L/b = Lk_1\alpha^2/4 \tag{95}$$

(where L is the length of nonlinear medium) and the position of the focus (position of the waist) in the nonlinear media is given by

$$\mu = (2z_w/L) - 1 \tag{96}$$

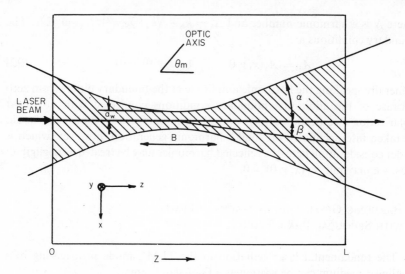

Fig. 26. A schematic diagram of the Gaussian laser beam focused in a crystal at $X = Y = 0$, $Z = Z_w$ (shaded area), when α_w is the beam radius at waist, $b = k_1 \alpha_w^2$ the confocal parameter, $\alpha = 2/(k_1 \alpha_w)$ the divergence half-angle in the far field, β the double refraction angle, θ_m the matching angle, and L the crystal length.

Thus the harmonic power P_N can be written as

$$P_N = \mathscr{K}_N P_1{}^N (Lk_1)^{3-N} h_N(m, \mu, \sigma_N) \qquad (97a)$$

$$\mathscr{K}_N = 4^N N\pi^2 \omega_1{}^2 \chi_N{}^2 k_1^{2N-4}/c^{N+1} n_N n_1{}^N \qquad (97b)$$

where the aperture function h_N describes diffraction, focusing, and the degree of phase matching $\sigma_N = \Delta_N b/2$. These aperture functions were calculated for second harmonic generation ($N = 2$) by Sukhorukov (1966), Bjorkholm (1966), Kleinman et al. (1966), and Boyd and Kleinman (1968). For the third harmonic ($N = 3$) calculations were done by Ward and New (1969) and Sukhorukov and Tomov (1970a)

$$h_N(m, \mu, \sigma_N) = (m^{N-3}/4) \left| \int_{-m(1-\mu)}^{m(1+\mu)} (1 - iz')^{1-N} e^{i\sigma_N z'} \, dz' \right|^2 \qquad (98)$$

In the case of symmetric focusing, μ equals zero; in the semisymmetric case $\mu = \pm 1$. The function h_N can be expressed through the well-known special functions (the integral sin and cos of complex arguments). In the limit $m \ll 1$ (weak focusing) and $m \gg 1$ (strong focusing; $L \gg b$) there are no problems with the estimation of (98).

In Fig. 27 the aperture functions h_N are shown as functions of the mismatch parameter $\sigma_N m$. For a weakly divergent beam ($m \ll 1$) the harmonic power decreases symmetricly relative to the point of exact phase matching

$\Delta_N = 0$ (see Fig. 27a). We have

$$h_N(m, 0, \sigma_N) = m^{N-1} \mathrm{sinc}^2(m\sigma_N), \qquad m\sigma_N = \Delta_N L/2 \qquad (99)$$

(Maker *et al.*, 1962; Ashkin *et al.*, 1963). For strong focusing ($m \gg 1$), mis-

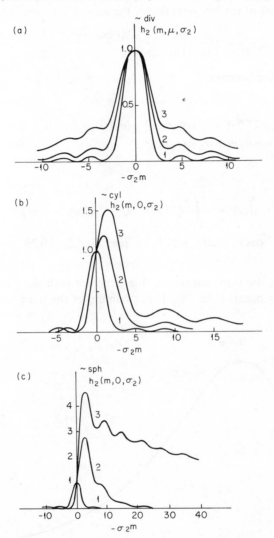

Fig. 27. The SHG aperture functions versus the mismatching parameter $\sigma_2 m = \Delta_2 L/2$. The crystal is placed (a) in the far field of a Gaussian beam ($\mu \gg m^{-1}$); (b) in the focal spot of a cylindrical lens ($\mu = 0$); (c) in the focal spot of a spherical lens ($\mu = 0$). Curve 1 is $m \ll 1$, curve 2 is $m = 10$, curve 3 is $m = 100$, where $m = L/b$ is the focusing parameter. The curves are normalized to ones at $\sigma_2 = 0$ (from Sukhorukov, 1966).

match due to diffraction $\sim 2(N-1)/b$, Eq. (94) appears which is compensated for in the region $\Delta_N < 0$ and becomes more phase mismatched in the region $\Delta_N > 0$. As a result there is an asymmetry in the dependence of the harmonic power on the mismatch parameter $\sigma_N m$ (see Figs. 27b and c). Asymptotic approximations of the integral (98) for the second harmonic had to

$$h_2(m,0,\sigma_2) = \begin{cases} m^{-1} \left| \pi e^{-|\sigma_2|} + \mathrm{si}(|\sigma_2|m) \right|^2 & \text{for} \quad \sigma_2 < 0 \\ m^{-1} \left| \mathrm{si}(\sigma_2 m) \right|^2 & \text{for} \quad \sigma_2 > 0 \end{cases} \tag{100}$$

and for the third harmonic

$$h_3(m,0,\sigma_3) =$$

$$\begin{cases} \left| \pi |\sigma_3| e^{-|\sigma_3|} + m^{-1} \cos(\sigma_3 m) + |\sigma_3| \, \mathrm{si}(|\sigma_3|m) \right|^2 & \text{for} \quad \sigma_3 < 0 \\ \left| m^{-1} \cos \sigma_3 m + \sigma_3 \, \mathrm{si}(\sigma_3 m) \right|^2 & \text{for} \quad \sigma_3 > 0 \end{cases}$$

$$\tag{101}$$

where

$$\mathrm{si}(x) = -\int_x^\infty \xi^{-1} \sin \xi \, d\xi, \qquad \mathrm{si}(0) = -\pi/2$$

$$\mathrm{si}(x) \simeq -x^{-1} \cos x \quad \text{at} \quad x \gg 1, \qquad \max |\mathrm{si}(x_m)| = 0.28 \quad \text{at} \quad x_m = \pi \tag{102}$$

In the case of the third harmonic, fine structure with the period $\sigma_3 m = \pi$ completely disappears (Fig. 28). The maximum of the third harmonic takes

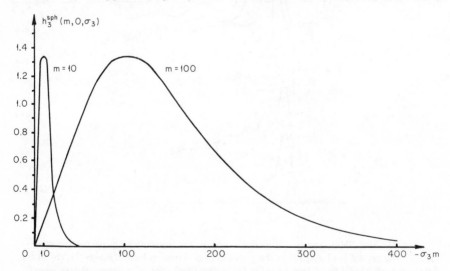

Fig. 28. The THG aperture functions versus the mismatching parameter $\sigma_3 m = \Delta_3 L/2$ for two values of focusing parameter. The Gaussian laser beam is strongly focused by a spherical lens in the crystal center ($\mu = 0$) (from Sukhorukov and Tomov, 1970a).

place which corresponds to a finite phase mismatch $\sigma_{3,m} = -1$ and $h_{3,m} = 1.33$. For strong focusing, the third harmonic power under conditions of exact phase matching is practically zero, $h_3(m, 0, 0) = m^{-2}$. Note that for the second harmonic these effects are not so pronounced.

Using (98) we can solve the very important problem of optimum focusing. The question is how to choose the focusing parameter to obtain the maximum

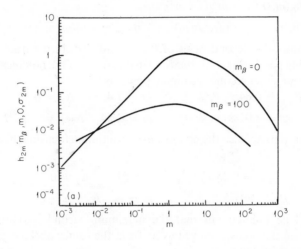

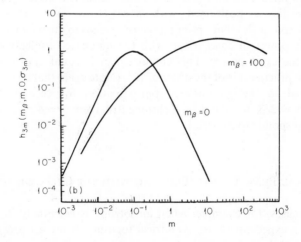

Fig. 29. (a) SHG and (b) THG power represented by the aperture functions $h_{Nm}(m_\beta, m, 0, \sigma_m)$ as a function of the focusing parameter $m = L/b$ for two values of double refraction $m_\beta = Lk_1\beta^2/4$ (β is the birefringent angle). The cases $m_\beta = 0$ correspond to an isotropic medium and noncritical matching $\theta_m = 90°$ in an uniaxial crystal, $m_\beta \neq 0$ correspond to critical matching $\theta_m \neq 90°$.

harmonic power for a fixed amount of power at the fundamental wavelength. We can write, for the second harmonic,

$$h_2(m, 0, 0) = (\arctan m)^2/m$$

$$h_2(m, 0, \sigma_{2,m}) \simeq (\text{arcsinh}\, m)^2/m, \qquad \sigma_{2,m} \simeq -(2/m)\arctan(m/2) \quad (103)$$

and for the third harmonic:

$$h_3(m, 0, 0) = m^2/(1+m^2)^2 \tag{104}$$

$$h_3(m, 0, \sigma_{3,m}) \simeq (\arctan m)^2, \qquad \sigma_{3,m} \simeq -(4/m)\arctan(m/2)$$

Figure 29 shows the dependencies of the second and third harmonic power on the focusing parameter m for focusing in the center of the nonlinear medium ($\mu = 0$). The maximal second harmonic power is

$$P_{2,m} = 0.64\mathcal{K}_2 P_1^{\,2} k_1 L, \qquad m = 1.4, \qquad \Delta_2 = 0 \tag{105a}$$

$$P_{2,mm} = 1.07\mathcal{K}_2 P_1^{\,2} k_1 L, \qquad m = 2.84, \qquad \Delta_{2,m} = -3.2/L \tag{105b}$$

The last conditions are also the conditions for optimum focusing. For the third harmonic:

$$P_{3,m} = 0.25\mathcal{K}_3 P_1^{\,3}, \qquad m = 1, \qquad \Delta_3 = 0 \tag{106a}$$

$$P_{3,mm} = 1.5\mathcal{K}_3 P_1^{\,3}, \qquad m \simeq 4, \qquad \Delta_{3,m} \simeq -8/L \tag{106b}$$

It should be pointed out that the maximum second harmonic power decreases when the focus is moved away from the center of the medium, but the maximum third harmonic power remains the same even for $m \gg 1$, if the nonlinear media is positioned with $\mu = \pm(1-m^{-2})^{1/2}$. It should be emphasized (see Fig. 28) that optimum focusing leads to great enhancement in the harmonic power in comparison with the unfocused fundamental beam ($m_d \ll 1$) of the same power. This enhancement is equal approximately to $(R_d/L)^{N-1}$; in practice it is of the order of 10^2. (Note again that all these results were obtained in the parametric approximation for second and third harmonics, $N = 2, 3$. It should be mentioned that for $N = 4, 5$ the optimum focusing corresponds to $m \to \infty$.)

2. CYLINDRICAL PHASE FRONTS (FOCUSING WITH THE CYLINDRICAL LENS)

In some cases, for example, when anisotropy is essentially in only one direction, it is expedient to use cylindrical focusing. If $m_d \ll 1$, the Gaussian beam focused in the XZ-plane can be presented as

$$A_1 = \left(\frac{16P_1}{cn_1 a^2}\right)^{1/2} \psi^{-1/2}(z) \exp\left\{-\frac{X^2}{a^2\psi(z)}\left(1 - \frac{i\alpha_{0x}}{\alpha_d}\right) - \frac{Y^2}{a^2}\left(1 - \frac{i\alpha_{0y}}{\alpha_d}\right)\right\} \tag{107}$$

The harmonic power is described by the formula (97) with the aperture function ($m_d = L/R_d$)

$$h_N(m_d, m, \mu, \sigma_N) = \tfrac{1}{4}m_d^{(N-1)/2}m^{(N-5)/2}\left|\int_{-m(1-\mu)}^{m(1+\mu)} dz'(1-iz')^{-(N+1)/2}e^{i\sigma_N z'}\right|^2 \tag{108}$$

(Sukhorukov, 1966; Sukhorukov and Tomov, 1970a). From comparing formulas (98) and (108), it may be concluded that $m_d h_2^{sph} = h_3^{cyl}$. For second harmonic generation, formula (99) is valid if $m \ll 1$; in the other limiting case $m \gg 1$ ($\mu = 0$), the aperture function equals

$$h_2(m_d, m, 0, \sigma_2) = \frac{\pi m_d^{1/2}}{4|\sigma_2|m^{3/2}}\begin{cases} [2e^{-|\sigma_2|}-2+C((m|\sigma_2|)^{1/2})+S((m|\sigma_2|)^{1/2})]^2 \\ \qquad\qquad\qquad\qquad\qquad\qquad\text{for } \sigma_2 < 0 \\ [C((m\sigma_2)^{1/2})-S((m\sigma_2)^{1/2})]^2 \quad \text{for } \sigma_2 > 0 \end{cases} \tag{109}$$

where $C(x) = (2/\pi)^{1/2}\int_0^x \cos\xi^2\, d\xi$, and $S(x) = (2/\pi)^{1/2}\int_0^x \sin\xi^2\, d\xi$ are the Fresnel integrals. The last function with $m \to \infty$ has a narrow peak with the width $\sigma_2 m = \pi$ (see Fig. 27b). If $\Delta_N = 0$, $\mu = 0$ the second and third harmonic power equals, respectively,

$$P_2 = 2\mathcal{K}_2 P_1^2 k_1 L m_d^{1/2}m^{1/2}[1+(1+m^2)^{1/2}]^{-1} \tag{110}$$

$$P_3 = \mathcal{K}_3 P_1^3 m_d m^{-1}(\tan^{-1} m)^2 \tag{111}$$

The optimum focusing condition remains practically the same as that for spherical focusing [compare (110) and (111) with (105) and (106)]. Enhancement in harmonic power by optimum cylindrical focusing equals approximately $(R_d/L)^{(N-1)/2}$. This value is smaller than the enhancement which can be obtained by spherical focusing, since with a cylindrical lens the light beam contracts only in one plane.

The excellent agreement with the theory of SHG in a focused Gaussian beam was demonstrated in the experiments both on critically phase-matched ($\theta_m < 90°$) SHG (Bjorkholm, 1966; Francois and Siegman, 1965) and on noncritically phase-matched ($\theta_m = 90°$) SHG (Kleinman and Miller, 1966b; McGeoch and Smith, 1970).

D. Harmonic Generation in Anisotropic Media

1. Equations Describing the Excitation of Harmonics with Ordinary Waves

Let us consider harmonic generation in an uniaxial crystal when the phase-matching angle θ_m (see Fig. 10) differs from 90° with respect to the axis of the crystal. In this case, phase-mismatch parameter Δ_N grows with angular detuning θ, $\Delta_N \propto \theta$,* and directions of energy flow for the fundamental radiation and harmonic are generally different. For a divergent (or convergent) fundamental optical beam with a finite cross section these effects lead to a decreasing of the efficiency of the harmonic generation process. In general case, both these effects (in this chapter they are termed as "angular aperture effect" and "diaphragm aperture effect") act simultaneously. We begin with the very useful situation where the fundamental radiation propagates as an ordinary wave and the harmonic as an extraordinary wave (O_1O_1–E_2 interaction for second harmonic generation and $O_1O_1O_1$–E_3 interaction for third harmonic generation). In this case, the relevant parabolic equations may be written as

$$\frac{\partial A_1}{\partial z} - \frac{1}{2ik_1}\left(\frac{\partial^2 A_1}{\partial x^2} + \frac{\partial^2 A_1}{\partial y^2}\right) = 0, \tag{112a}$$

$$\frac{\partial A_N}{\partial z} + \beta\frac{\partial A_N}{\partial x} - \frac{1}{2ik_N}\left(\frac{\partial^2 A_N}{\partial x^2} + \frac{\partial^2 A_N}{\partial y^2}\right) = -i\frac{\gamma_N}{n_N}A_1{}^N e^{i\Delta_N z} \tag{112b}$$

Here β is the walk-off angle (see Fig. 14) and Δ_N is phase mismatch for the center of the beam.

2. ANGULAR SPECTRA

Equations (112) can be solved by the spectral method. Let us introduce the spectra of complex amplitudes

$$A_N = \int\!\!\!\int_{-\infty}^{\infty} S_N(\theta, \phi)\exp(-ik_N\theta x - ik_N\phi y)\,d\phi\,d\theta \tag{113}$$

Here θ is the angle in the plane of the optical axis, and ϕ the angle in the perpendicular plane. Angular intensity distribution $I_N{}^S(\theta, \phi) = (cn_N/8\pi)|S_N|^2$ of the harmonic excited with the Gaussian fundamental beam [shown by Sukhorukov (1966) and Kleinman et al. (1966a)] follows from the Gaussian distribution and the aperture functions (98) and (108):

$$I_N{}^S(\theta_1\phi) \simeq \exp\left\{\frac{-2\theta^2}{N\alpha_\theta{}^2} - \frac{2\phi^2}{N\alpha_\phi{}^2}\right\} h_N(m, \mu, \sigma_N(\theta)) \tag{114}$$

One of the arguments of the aperture function is now

$$\sigma_N(\theta) = (\Delta_N b/2) - (\pi\theta/2m\theta_{\text{coh}}), \qquad \theta_{\text{coh}} = \pi/Nk_1\beta L \tag{115}$$

where θ_{coh}, the phase-matching angular width, is a very important parameter.

---

* For 90° phase matching $\Delta_N \propto \theta^2$, and may be neglected for beams with small angular divergence.

Obviously when the angular divergence of the fundamental beam is less than θ_{coh}, the harmonic maintains the Gaussian distribution (see Section II, D, 3). In the opposite case the angular spectrum of the harmonic in the plane of the optical axis (angle θ) is determined by the aperture function h_N and is essentially asymmetric when $m \gg 1$. This assymmetry is due to a decrease in the wave number in the focal region, as mentioned above. This asymmetry can be treated also as a result of the vector interaction of the crossing rays (Maker et al., 1964; Akhmanov et al., 1965f, 1966a; Francois and Siegman, 1965).

It should be pointed out that in anisotropic media there is the possibility of amplitude modulation and scanning the harmonic radiation direction by changing the phase-matching condition. Different ways of varying the dispersion are suitable; for example, heating or cooling, rotation of the crystal, and application of an electric field (Akhmanov et al., 1965b; Van der Ziel, 1964).

The angular spectrum of the second harmonic was carefully studied in a KDP crystal using a helium–neon laser (Kleinman et al., 1966b) and a multimode Nd-glass laser (Kovrygin et al., 1967, 1969). For weakly divergent and convergent beams ($m \ll 1$) the harmonic far field had a stripe structure (see Fig. 30) described by the aperture function (99). For strong focusing ($m \gg 1$), the form of the harmonic angular spectrum is asymmetric (Fig. 30b) which agrees well with formula (100). Finally, the angular spectra of the second harmonic is shown in the Fig. 30c, which was obtained with cylindrical focusing of the fundamental beam in the optical axis plane (angle θ) and in the perpendicular plane (angle ϕ). The experimental data are compared with the aperture function h_2^{cyl} [see formula (99)]. It should be noted that in spite of the multimode nature of the fundamental beam the harmonics angular spectra are described well with the aperture functions computed for the Gaussian beam (Fig. 27). This is due to the weak dependence of the aperture function on the angular spectrum of the fundamental radiation. In particular, the theoretical analysis of the harmonic generation of a cylindrical wave with a constant amplitude leads to the same expressions as those for the Gaussian beam (Akhmanov et al., 1966a; Zel'dovitch, 1966). Excellent experimental data concerning the angular spectra of the second harmonic were obtained with the single-mode gas laser by Kleinman et al. (1966a).

The characteristic asymmetry of the angular spectra of the third harmonic generated in focused beams have been observed in the first experiments with calcite by Maker et al. (1964). Here, the fine angular strip structure is practically absent. [Compare h_2^{sph} in (100) and h_3^{sph} in (101)]. The reason for this difference is the strong dependence of the nonlinear effect on the fundamental field $A_N \simeq A_1^N$; as a result the effective wave mismatch in the focal region $\Delta_{N,d} \simeq 2(N-1)b^{-1}$ increases.

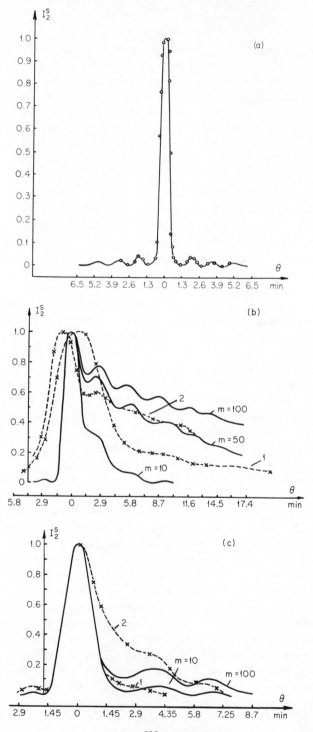

3. Harmonic Power

As was mentioned above, the walk-off of the Poynting vector decreases the efficiency of the optical multiplier. The total harmonic power P_N has been given as an integral of the angular intensity distribution (114) by Sukhorukov (1966) and Sukhorukov and Tomov (1970a); the same results were obtained by Bjorkholm (1966), and Boyd and Kleinman (1968). For the anisotropic medium it is convenient to introduce a new parameter

$$m_\beta = Lk_1\beta^2/4 \tag{116}$$

("anisotropy parameter" in the harmonic generation theory). Aperture effects, connected with the anisotropy are significant if $\alpha_\theta > \theta_{\text{coh}}$ or $mm_\beta > 1$; the quasi-isotropic media case treated above corresponds to $m_\beta \ll 1$ (if $m \lesssim 1$).

Let us consider now some results concerning the total harmonic power. Of great practical significance is a case of a weakly divergent Gaussian beam. Here the expression for the harmonic power may be obtained in the geometric optics approximation (Boyd et al., 1965; Bokuth and Khatkhevich, 1964; Sukhorukov and Tomov, 1970a):

$$P_N = \mathcal{K}_N P_1{}^N (L/a)^2 (k_1 a)^{4-2N} [\sqrt{\pi}\,\xi^{-1}\,\text{erf}\,\xi + \xi^{-2}(e^{-\xi^2}-1)] \tag{117a}$$

Here $\xi = L/L_\beta$ is a normalized length parameter and

$$L_\beta = 2(2/N)^{1/2}/(k_1\beta\alpha) \tag{118a}$$

is the characteristic aperture length. If $L \ll L_\beta$, anisotropy is unimportant and

$$P_N = \mathcal{K}_N P_1{}^N (k_1 a)^{4-2N} L^2/a^2 \tag{117b}$$

so $P_N \simeq L^2$ as for plane waves. If $L \gg L_\beta$

$$P_N = \mathcal{K}_N P_1{}^N (k_1 a)^{4-2N} \sqrt{\pi}\, L_\beta L/a^2 \tag{117c}$$

As an example, for second harmonic generation

$$P_2 = \frac{32\pi^2\omega_1{}^2\chi_2{}^2 P_1{}^2}{n_1{}^3 c^3} \begin{cases} L^2/a^2 & \text{for} \quad L \ll L_\beta \\ \sqrt{\pi}\,L_\beta L/a^2 & \text{for} \quad L \gg L_\beta \end{cases} \tag{117d}$$

Formulas (117) take into account simultaneously both aperture effects: the diaphragm aperture effect and the angular aperture effect. There are, however, practical situations when only one of them is significant.

Fig. 30. The far field of the second harmonic wave excited by a multimode Nd-laser radiation ($\alpha_0 = 10'$) in a KDP crystal at critical matching. (a) Unfocused beam; (b) focused by spherical lens; (c) focused by cylindrical lens. The experimental curves (dashed) are: (1) for $m = 50$; (2) for $m = 300$. The aperture functions of the Gaussian coherent beam (solid lines) are plotted also (see Fig. 27); Arbitrary units are used for the vertical scale (from Kovrygin et al. 1967, 1969).

Let us first consider harmonic generation in the near field of the diffraction-limited Gaussian beam. Here $\alpha = \alpha_d$ ($\alpha_0 = 0$) and

$$L_\beta = L_a = (2/N)^{1/2} a/\beta \tag{118b}$$

In this situation only the difference in the directions of the harmonic and fundamental wave energy flows is significant (Kleinman, 1962) and diaphragm aperture effect takes place in a pure form. In the opposite case $\alpha \gg \alpha_d$, the angular aperture effect dominates. Now

$$L_\beta = L_a = 2(2/N)^{1/2}/(k_1 \beta \alpha_0) \tag{118c}$$

where L_α is the coherent length for a divergent beam. The angular effect take place in a pure form if

$$\alpha_0 \gg \theta_{coh}^2 \alpha_d^{-1} \quad \text{or} \quad L \ll L_a(\alpha_d/\alpha_0)^{1/2} \tag{119}$$

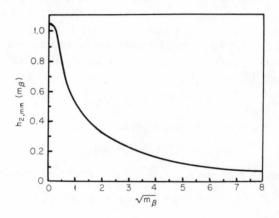

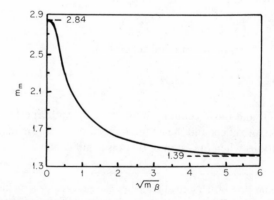

Fig. 31. Optimized second harmonic power [represented by $h_{mm} = h(m_\beta, m_m, 0, \sigma_m)$] and optimum focusing parameter $m_m(m_\beta)$ as functions of the parameter m_β (from Boyd and Kleinman, 1968).

If $m_\beta > 1$, the efficiency of optical multipliers excited with focused beams also decreases. Second harmonic power in a birefringent crystal, when the funamental beam is focused in the center of the crystal ($\mu = 0$) is now represented by the aperture function (see Fig. 29a)

$$h_2(m_\beta, m, 0, \sigma_{2m}) = (\sqrt{\pi} \arctan m)/2 \, (mm_\beta)^{1/2} \tag{120}$$

Optimum focusing in isotropic media corresponds to $m_m = 1.4$; however, the maximum harmonic power decreases $h_m \simeq 0.7(m_\beta)^{-1/2}$. In anisotropic media ($m_\beta \gg 1$), the maximum harmonic power is $0.7m_\beta^{-1/2}$ times less than in isotropic media (Fig. 31).

Cylindrical focusing in the ϕ-plane leads to approximately the same harmonic power: $h_{2m} \simeq (m_m m_\beta)^{-1/2}$. Optimum focusing in anisotropic media by third harmonic generation was treated by Sukhorukov and Tomov (1970a). They showed that in this case the conditions of optimum focusing corresponds to

$$m \simeq m_\beta, \qquad \alpha_0 \simeq \beta$$

It is interesting that by such strong focusing the maximum harmonic power is the same as in isotropic media (Fig. 29b).

4. HARMONIC GENERATION DUE TO MIXING OF ORDINARY AND EXTRAORDINARY FUNDAMENTAL WAVES

For the phase-matched interactions $O_1E_1-E_2$ and $O_1O_1E_1-E_3$, Poynting vector walk-off limited not only the interaction length between the harmonic and fundamental waves but also the interaction length between the different (O and E) fundamental waves too. This fact leads to some new features of aperture effects in harmonic generation.

The parabolic equations describing the process under consideration may be written as

$$\frac{\partial A_1{}^o}{\partial z} - \frac{1}{2ik_1}\left(\frac{\partial^2 A_1{}^o}{\partial x^2} + \frac{\partial^2 A_1{}^o}{\partial y^2}\right) = 0 \tag{121}$$

$$\frac{\partial A_1{}^e}{\partial z} + \beta_1 \frac{\partial A_1{}^e}{\partial x} - \frac{1}{2ik_1}\left(\frac{\partial^2 A_1{}^e}{\partial x^2} + \frac{\partial^2 A_1{}^e}{\partial y^2}\right) = 0 \tag{121b}$$

$$\frac{\partial A_N{}^e}{\partial z} + \beta_N \frac{\partial A_N{}^e}{\partial x} - \frac{1}{2iNk_1}\left(\frac{\partial^2 A_N{}^e}{\partial x^2} + \frac{\partial^2 A_N{}^e}{\partial y^2}\right) = \frac{-i\,\gamma_N}{n_N} A_1{}^e (A_1{}^o)^{N-1} e^{i\Delta_N z} \tag{121c}$$

Here β_1 is the walk-off angle for the fundamental wave and β_N is for the harmonic. Now the angular spectrum $I_N{}^S(\theta, \phi)$ of the harmonic excited by the

Gaussian fundamental beam (93) is described by formula (114); the aperture function is

$$h_N(m_1\mu_1\sigma_N(\theta)) = \left| \int_{-m(1+\mu)}^{m(1-\mu)} dz'(1-iz')^{1-N} \exp\left\{ -\frac{(N-1)L^2 z'^2}{NL_{a,1}^2 m^2} + i\sigma_N(\theta)z' \right\} \right|$$

(122)

with

$$\sigma_N(\theta) = \{\Delta_N + (N\beta_N - \beta_1)k_1\theta\}b/2$$

Here the integrand has an additional exponential term describing the difference in the energy flows of the fundamental beams. It should be pointed out that now the harmonic power saturates in a length $L > L_{a,1} = a/\beta_1$ even if the reaction of the harmonic on the fundamental wave is negligible and the parametric approximation is still valid ($m \ll 1$, $\beta_1 = \beta_2$, $\Delta_N = 0$)

$$P_N = \mathcal{K}_N P_1^e (P_1^\circ)^{N-1} (k_1 a)^{4-2N} \frac{N^{1/2}\alpha_d}{2(N-1)\alpha\beta^2} \left\{ \pi - \arctan\frac{2N^{1/2}\alpha_d}{(N-1)\alpha} \right\}$$

(123)

It follows from (123) that the harmonic power is proportional to α_0^{-1} if $\alpha_0 > \alpha_d$. The angular width of phase matching is defined by the usual formula $\theta_{coh} = [\pi/(N-1)k_1\beta L]$ if $L \ll L_a$, but if $L \gg L_a$, the interaction length contracts and L in this formula must be replaced by L_a and $\theta_{coh} = \pi(N/2)^{1/2}\alpha_d/(N-1)$.

When $L > L_a$ and $\alpha_0 > \alpha_d$, the harmonic beam is of oval shape; the ratio of the diameters is equal to $2\sqrt{N}$ and the angular spectrum can be written as

$$I_N^S(\theta,\phi) \simeq \begin{cases} \exp\left\{ -\frac{2N^2\theta^2}{\alpha_0^2} - \frac{2N\phi^2}{\alpha_0^2} \right\} & \text{for } \alpha_0\theta > 0 \\ 0 & \text{for } \alpha_0\theta < 0 \end{cases}$$

(124)

From (2.88) it follows that the harmonic radiation is concentrated in the region $\theta > 0$, if $\alpha_0 > 0$ (convergent beam), and in the region $\theta < 0$ if $\alpha_0 < 0$ (divergent beam). Sukhorukov and Tomov (1970a) have studied this effect in the third harmonic generation of the Nd laser in a calcite crystal. In a focused beam the asymmetry mentioned above is much more pronounced. It should be noted that by focusing inside the crystal, even for $m \simeq 1$, the focal regions of ordinary and extraordinary fundamental beams do not overlap, and for strong focusing the harmonic power decreases sharply (Sukhorukov and Tomov, 1970a).

E. Limiting Efficiency of Optical Multipliers with Birefringent Crystals

In this section the limiting efficiency connected with the reaction of an harmonic wave on the fundamental is considered. As was mentioned above, for ideal plane monochromatic waves, full conversion of the fundamental power into harmonic is possible (100% conversion). For a real laser beam, however, the limiting efficiency may be significantly less. In isotropic and quasi-isotropic media it can be connected with the relative weak nonlinear interactions in the wings of the optical beam and optical pulse (see Section II,B,4). In birefringent crystals the limiting efficiency may be connected with aperture effects. In this section we will present the nonlinear theory of second harmonic generation taking into account the finite aperture and divergence of the fundamental beam and the reaction of the harmonic radiation on the fundamental wave. Let us consider in the geometric optics approximation the SHG for the $O_1O_1-E_2$ interaction. The truncated equations may be written as

$$\partial A_1/\partial z = -i(\gamma_2/n_1) A_2 A_1^* e^{-i\Delta_2 z} \tag{125a}$$

$$(\partial A_2/\partial z) + \beta(\partial A_2/\partial x) = -i(\gamma_2/n_2) A_1^2 e^{i\Delta_2 z} \tag{125b}$$

1. LIMITING DUE TO THE DIAPHRAGM APERTURE EFFECT

Let us consider SHG in the near field of the Gaussian beam ($\alpha_0 = 0$) under ideal phase-matching ($\Delta_2 = 0$). In this case the optimum phase relations between the fundamental and harmonic wave remains the same for all points of the nonlinear media $\phi = 2\psi_1 - \psi_2 = \pi/2$. In this case the two equations in (125) reduce to the single equation (Akhmanov et al., 1967b)

$$\frac{\partial^2 A_{01}^{-1}}{\partial z^2} - \frac{\gamma_2^2}{n_1 n_2} G(x - \beta z) A_{01}^{-1} = 0 \tag{126a}$$

where $G(x)$ is defined from the boundary conditions

$$G(x) = A_{01}^2(x, 0) \quad \text{at} \quad A_{02}(x, 0) = 0 \tag{126b}$$

Transformation from (125) to (126) greatly simplifies the problem. As an example, the exact solution of Eq. (126) may be obtained if the laser beam is Lorentzian, $A_{01}(x, 0) = A_0(1 + (x^2/a^2))^{-1}$. This solution may be written as

$$A_{01}(x, z) = A_{01}^{1/2}(x, 0) A_{01}^{1/2}(x - \beta z, 0)\{\cosh \zeta + (x/a)(g_\beta^2 - 1)^{1/2} \sinh \zeta\}^{-1}$$

$$\zeta = (g_\beta^2 - 1)^{1/2}\{\arctan(x/a) - \arctan[(x - \beta z)/a]\} \tag{127}$$

Now the SHG process depends critically on the characteristic parameter

$$g_\beta = L_a/L_{NL} = \gamma_2 n_1^{-1/2} n_2^{-1/2} A_0(a/\beta) \tag{128}$$

When $g_\beta \ll 1$, aperture effects are very much pronounced in the fixed field approximation. In the opposite limit $g_\beta \gg 1$, the influence of the aperture effect decreases. When power transfers from the fundamental to the harmonic beam, the amplitude modulation of this wave changes: the centers of the beams move in the opposite direction and the width of the beams change too (see

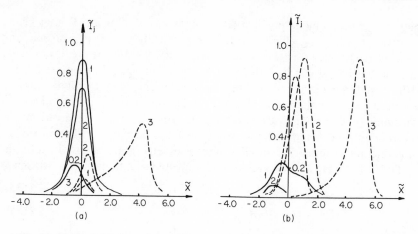

Fig. 32. Plots of the intensity profiles of fundamental (solid) and second harmonic (dashed) beams for several distances in a uniaxial crystal. Curve 1: $z = 0.5L_a$; curve 2: L_a; curve 3: $5L_a$, where $L_a = a/\beta$ is the aperture length and β is the double refraction angle. The fundamental beam was Lorentzian at the entrance. (a) $g_\beta = (L_a/L_{NL})^2 = 2$; and (b) $g_\beta^2 = 10$ (from Akhmanov et al., 1967).

Fig. 32). The harmonic efficiency grows with the parameter g_β. It should be noted that the parameters g_β do not depend on focusing: $g_\beta \propto P_1^{1/2}$.

The influence of the aperture effect can be reduced in a stratified medium having opposite walk-off directions (Akhmanov et al., 1967b). In such a structure the aperture effect is defined by the length of one nonlinear anisotropic slice. For the proper phase shift it was possible to use the dispersion

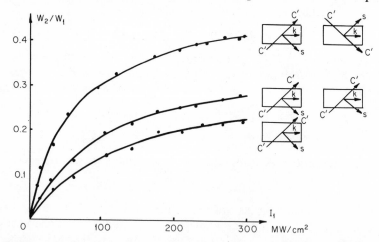

Fig. 33. Measured dependence of the SHG efficiency on Nd laser (Gaussian) beam intensity. The second harmonic was excited in a single KDP crystal ($L \simeq L_a$) and in two KDP crystals with parallel or rotated optic axes. In the last scheme the walk-off effect is significantly decreased (from Akmanov et al., 1967).

of air and placing the slices a distance $L_{\text{coh}}^{\text{air}} = (\lambda_1/4)\,|n_2^{\text{air}} - n_1^{\text{air}}|^{-1}$. Figure 33 illustrates how the beam shape of the fundamental radiation and its second harmonic transforms during the nonlinear interaction. The diaphragm aperture effect may be sufficiently reduced in stratified media that consist of slabs of inverted birefringent crystals separated by air gaps. In Fig. 33 experimental results are presented which illustrate the enhancement of the second harmonic power in a system of two inverted birefringent crystals (Akmanov *et al.*, 1967).

2. LIMITING DUE TO THE ANGULAR APERTURE EFFECT

For strongly divergent beams the angular aperture effect dominates [see Eq. (119)]. In this case the limiting efficiency of the optical doubler can be estimated from the equations:

$$\partial A_1/\partial z = -i(\gamma_2/n_1)\,A_2 A_1^{*} e^{-i\Delta_2(x)z} \tag{129a}$$

$$\partial A_2/\partial z = -i(\gamma_2/n_2)\,A_1^{2} e^{i\Delta_2(x)z}, \qquad \Delta_2(x) = \Delta_2 - (\pi\alpha_0 x/2\theta_{\text{coh}} a) \tag{129b}$$

Equations (129) are similar to the equations describing the interactions of plane waves [compare (179)], but in (129) variation of the phase-mismatch parameter $\Delta_2(X)$ has been taken into account. Solutions of (129) have been given by Akhmanov *et al.* (1965c) and Volosov (1969).

F. Thermal Self-Actions in Optical Multipliers; Influences of Self-Induced Thermal Phase Mismatching and Self-Defocusing on Second Harmonic Generation

Self-actions connected with the cubic nonlinear susceptibility (see Section I, B) are very important limiting factors in optical harmonic generation. In optical multipliers, excited with cw and quasi-cw lasers self-induced thermal effects are much more pronounced. The nonuniform heating of the nonlinear crystal with the laser beam causes nonuniform phase matching across the beam; in several cases, self-defocusing (or self-focusing) of the fundamental and harmonic waves are also important. Let us consider the influence of self-induced thermal effects on second harmonic generation. Neglecting aperture and dispersion effects, we can write for the complex amplitudes A_1, A_2 and the temperature T' of the nonlinear media [compare with (91)]:

$$\frac{\partial A_1}{\partial z} - \frac{1}{2ik_1}\left(\frac{\partial^2 A_1}{\partial x^2} + \frac{\partial^2 A_1}{\partial y^2}\right) = -i\frac{k_1}{n_1}\frac{dn_1}{dT}T'A_1 - \delta_1 A_1 \tag{130a}$$

$$\frac{\partial A_2}{\partial z} - \frac{1}{2ik_2}\left(\frac{\partial^2 A_2}{\partial x^2} + \frac{\partial^2 A_2}{\partial y^2}\right) = -i\frac{k_2}{n_2}\frac{dn_2}{dT}T'A_2 - i\frac{\gamma_2}{n_2}A_1^{2}e^{-i\Delta_2 z} - \delta_2 A_2 \tag{130b}$$

$$\frac{\partial T'}{\partial \eta} - \chi_{\text{T}}\left(\frac{\partial^2 T'}{\partial x^2} + \frac{\partial^2 T'}{\partial y^2}\right) = \frac{\delta_1 c n_1}{4\pi\rho c_{\text{p}}}|A_1|^2 \tag{130c}$$

Here $\eta = t - (z/u)$, $u = u_1 = u_2$; ρc_p is the specific thermal capacity, $\chi_T = \kappa_T/\rho c_p$ the thermal conductivity coefficient, and δ_1 the absorption coefficient for the fundamental wave (in practical situations which will be under consideration, heating due to second harmonic absorption is negligible, $\delta_j L \ll 1$).

1. SECOND HARMONIC GENERATION OF CW LASER RADIATION

Nonuniform heating leads to nonuniform phase matching across the beam. Now the coherence length is a function of the transverse coordinate r:

$$L_{coh}(T') = \frac{\lambda_1}{4} \left| n_2 - n_1 + \left(\frac{dn_2}{dT} - \frac{dn_1}{dT} \right) T'(r) \right|^{-1} \tag{131}$$

It is significant that when self-actions are present phase matching in principle cannot be obtained for the whole beam. If the beam is phase mismatched based on linear refractive indices so that $L_{coh}(0) = \infty$ at the center, at the wings of the beam $L_{coh}(a) \neq \infty$. Inside the Gaussian beam $(0 < r < a)$, self-induced thermal mismatch is significant if the fundamental power $P_1 \geqslant P_\Delta$. The power is given by

$$P_\Delta = P_T/[\delta_1 L |(dn_2/dn_1) - 1|] \tag{132}$$

where $P_T = \lambda_1 \kappa_T |dn_1/dT|^{-1}$ is the characteristic power well known in the theory of thermal self-focusing as a "critical dissipated power" and $dn_2/dn_1 = (dn_2/dT)(dn_1/dT)^{-1}$.

Generally, self-induced heating leads not only to nonuniform phase mismatching, but also to self-defocusing or self-focusing. The characteristic nonlinear focal length

$$R_{NL} = R_d (\pi P_T/\delta_1 R_d P_1)^{1/2} \tag{133}$$

The influences of thermal mismatching and self-defocusing on the SHG process are comparable if $L_{coh}(a) \simeq R_{NL}$, or if the fundamental power $P_1 = P_1'$

$$P_1' = P_T(\delta_1 R_d)^{-1} [(dn_2/dn_1) - 1]^{-2} \tag{134}$$

It is convenient (see Fig. 34) to introduce the characteristic "thermal" length L_T, which is defined from the relation $L_T = R_{NL} = L_{coh}$. If $P_1 = P_1'$, we obtain

$$L_T = R_d |(dn_2/dn_1) - 1| \tag{135}$$

It is interesting that L_T is a simple function of beam geometry and material constants only. If $L < L_T$ (see Fig. 34), the nonuniform phase matching dominates. In this case, solution of Eqs. (130) leads to the following formula for second harmonic power (Mikhina et al., 1971; Okada and Ieiri, 1971).

$$P_2 = \mathcal{K}_2 P_1^2 (L/a)^2 h_T(\zeta_p) \tag{136}$$

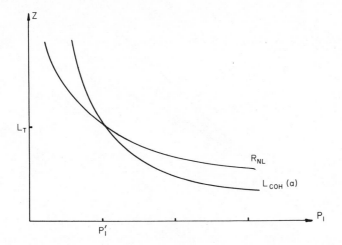

Fig. 34. The coherence length L_{coh} [Eq. (131)] connected with the nonuniform temperature mismatching and the nonlinear defocusing length R_{NL} [Eq. (133)] as the functions of the cw laser beam power P_1.

where $\zeta_{\text{p}} = P_1/P_\Delta$.

The function $h_{\text{T}}(\zeta_{\text{p}})$ is shown in Fig. 35. For low fundamental power $P_1 \ll P_\Delta$, $h_{\text{T}} = 1$ and obviously $P_2 \simeq P_1{}^2$; self-induced nonuniform heating is negligible. If $P_1 \gg P_\Delta$, $\zeta_a \gg 1$, the coherence length for the whole beam decreased sufficiently and $h_{\text{T}} \simeq \zeta_{\text{p}}^{-1}$:

$$P_2 \simeq \mathcal{K}_2 P_1 P_\Delta (L/a)^2$$

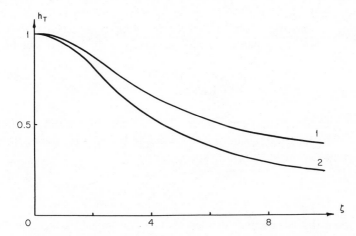

Fig. 35. Thermal mismatching function $h_{\text{T}}(\zeta)$ for cw operation (curve 1) and pulse operation (curve 2) (from Mikhina *et al.*, 1970).

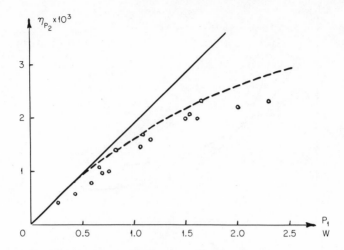

Fig. 36. Thermal mismatching effects in an intracavity doubler (Ar laser, KDP crystal; $\theta_m = 90°$). Dots represent the experimental data of SHG efficiency versus fundamental power. Dashed curve shows the theory, based on (136) and the solid curve shows the theoretical SHG efficiency without thermal effects (Gorokhov *et al.*, 1974).

If $L > L_T$ for the fundamental power $P_1 > P_T/\delta_1 L$, beam shaping due to self-focusing or self-defocusing should be taken into account. When, however, $P_1 > P_\Delta$, nonuniform phase mismatching again dominates. The influence of self-induced thermal effects on second harmonic generation of the Ar laser in KDP and KDP crystals, and the CO_2 laser in proustite has been studied by Gorokhov *et al.* (1974). Their estimates for P_Δ are: If $L = 4$ cm, $\delta_{1,\text{ADP}} = 10^{-3}$ cm^{-1}; $\delta_{1,\text{KDP}} = 3 \cdot 10^{-2}$ cm^{-1}, $P_{\Delta,\text{KDP}} = 0.47$ W, and $P_{\Delta,\text{ADP}} = 1.8$ W. Gorokhov *et al.*, were able to observe the limiting of SHG efficiency by intracavity harmonic generation inside the cavity of an Ar laser (see Fig. 36).

2. Thermal Effects in a Doubler Excited with Laser Pulses

Let us consider short laser pulses whose duration τ is slower than the characteristic thermodiffusion time $\tau_T = a^2/(4\chi_T)$. If $\tau < \tau_T$, the temperature of the nonlinear medium can be expressed as

$$T' = 2\delta_1 \int_{-\infty}^{\eta} I_1(r, z, t_1) \, dt_1$$

It is evident that all thermal self-actions are determined by the value of the energy of the laser pulse. Now the influence of thermal mismatching and self-defocusing on the SHG process are comparable if the energy of the funda-

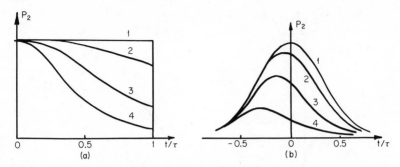

Fig. 37. The SHG pulse distortion due to thermal mismatching. (a) Rectangular fundamental pulse; (b) Gaussian fundamental pulse. Curve 1 is $\zeta_w = 0$; curve 2 is 1.4; curve 3 is 2.8; curve 4 is 5.6 (from Mikhina *et al.*, 1971).

mental wave $w_1 = w_1{}'$ [compare with (134)] is:

$$w_1{}' = w_T(\delta_1 R_d)^{-1} \left| \frac{dn_2}{dn_1} - 1 \right|^{-2}, \qquad w_T = P_T \tau_T \tag{137}$$

If $L < L_T$ (L_T is the same as in steady-state theory, i.e., see Section II, F, 1), the nonuniform phase-mismatching dominates. For a rectangular fundamental pulse, the second harmonic energy is (Mikhina *et al.*, 1971; see also Lokhov *et al.*, 1972)

$$w_2 = \mathscr{K}_2 w_1{}^2 \tau^{-1}(L/a)^2 h_T(\zeta_W) \tag{138}$$

where $\zeta_W = w_1/w_\Delta$, $w_\Delta = P_\Delta \tau_T$, and P_Δ is determined by formula (132). The function $h_T(\zeta_W)$ is shown in Fig. 35. If $w_1 \ll w_\Delta$, $w_2 \propto w_1{}^2$; if $w_1 \gg w_\Delta$ ($\zeta_W \gg 1$), the doubler efficiency is limited by the self-induced thermal phase mismatching and $w_2 \simeq \mathscr{K}_2 w_1 w_\Delta \tau^{-1}(L/a)^2$. There are considerable distortions of the second harmonic pulse when self-induced thermal effects are sufficient (see Fig. 37); thermal phase mismatching leads to pulse shortening. If $L > L_T$, self-defocusing or self-focusing occurs; when $w_1 > w_\Delta$, nonuniform phase mismatching again dominates. Thermal effects by doubling of millisecond pulses of a YAG : Nd^{3+} laser in a CDA crystal has been observed by Golyaev *et al.* (1974). In their experiment, thermal self-defocusing was much more pronounced (Fig. 38). For the laser radiation with $\tau = 6 \cdot 10^{-4}$ sec, $w_1 = 0.1$ J, $\lambda_1 = 1.06~\mu$m, $R_d = 5.7$ cm, and a CDA crystal ($L = 3.6$ cm, $\delta_1 = 0.093$ cm^{-1}, $d_{\text{eff}} = 1.12 \cdot 10^{-9}$ esu, $\theta_m = 83.5°$, and $\theta_{\text{coh}} = 22'$) the characteristic parameters are $L_T = 0.35$ cm and $W' = 3.7 \cdot 10^{-2}$ J. Internal self-defocusing of fundamental pulses is sufficient if $w_1 \geqslant w_{\text{cr}}$, where the critical energy

$$w_{\text{cr}} = \frac{\pi a^2}{\rho c_p |dn_1/dT| L^2} \qquad \text{at} \quad L \ll R_d \qquad (m_d \ll 1)$$

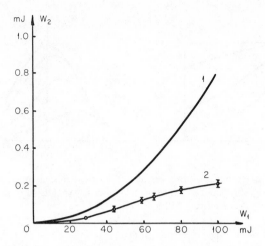

Fig. 38. Energy of the SH excited with a millisecond YAG : Nd^{+3} laser pulse at nearly noncritical matching in a CDA crystal. Curve 1 corresponds to the absence of thermal mismatching $(dn/dT = 0)$. Curve (2) represents the thermal defocusing effect on SHG efficiency (from Golyaev et al., 1974).

It should be mentioned that if thermal self-defocusing is present, the condition of the optimum focusing which was established in Section II,C (focusing parameter $m \simeq 1$) is not valid.

G. Second Harmonic Generation in Optical Resonators

An important problem in the technique of the optical multipliers is the problem of designing effective harmonic generators for cw lasers. Typically, in cw operation the fundamental power available is of the order of 0.1–1 W (Ar–ion laser, cw YAG : Nd^{3+} laser); for these, the traveling wave multipliers treated above are ineffective. In this case, resonant optical harmonic generation is used in order to take advantage of the high fundamental or harmonic power densities available within the optical resonator. There are several schemes of resonant harmonic generators: harmonic generation in resonators external to the laser (here resonance occurs the harmonic field only and simultaneous resonance for harmonic and fundamental is of interest), and intracavity harmonic generators where a nonlinear crystal within a laser cavity acts as an output coupler, in a manner very similar to a transmitting mirror in a normal laser. All these schemes are discussed below; it should be mentioned also that intracavity harmonic generation in Q-switched lasers may be used for the lengthening of the laser pulse.

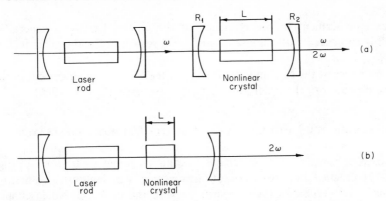

Fig. 39. Two schemes of resonant optical harmonic generation. (a) External resonator; (b) laser with intracavity nonlinear crystal.

1. EXTERNAL RESONATOR

Let us first consider second harmonic generation in an external resonator which is fully transparent for the fundamental radiation (see Fig. 39a; in the case under consideration $R_1(\omega_1) = R_2(\omega_1) = 0$; $R_1(2\omega_1) \neq 0$; $R_2(2\omega_1) \neq 0$). If $\gamma_2 A_{10} L \ll 1$, $|1 - R| \leqslant 1$, for the harmonic amplitude inside the resonator (compare Akhmanov and Khokhlov, 1964; Ashkin *et al.*, 1966) we have

$$T_c(dA_2/dt) + (2\delta_2 L + 1 - R) A_2 = (\gamma_2/n_2) LA_1{}^2 \qquad (139)$$

Here $R = (R_1 R_2)^{1/2}$, L is the crystal length, $T_c \simeq 2L_c/v_1$, L_c is the cavity length, and δ_2 is the absorption coefficient of the nonlinear crystal at the harmonic frequency. From (139) the second harmonic amplitude outside the resonator can be determined:

$$A_2 = (1 - R_2)^{1/2} \gamma_2 L n_2{}^{-1} T_c{}^{-1} \int_0^\infty A_1{}^2 (t - t_1) e^{-t_1/\tau_c} dt_1 \qquad (140)$$

where

$$\tau_c = T_c(2\delta_2 L + 1 - R)^{-1} \qquad (141)$$

For cw fundamental radiation ($A_1 = $ const), the second harmonic intensity outside the resonator is

$$I_2 = 8\pi(1 - R_2)\gamma_2{}^2 L^2 I_1{}^2/n_2 n_1{}^2 c(2\delta_2 L + 1 - R)^2 \qquad (142)$$

From (142), it follows that the optimum coupling of the harmonic resonator corresponds to $R_{1\,\text{opt}} = R_{2\text{opt}} = 1 - 2\delta_2 L$ and the maximal harmonic power density

$$I_2 = \pi\gamma_2{}^2 I_1{}^2/n_2 n_1{}^2 c\delta_2{}^2 \qquad (143)$$

Hence enhancement in the efficiency of the second harmonic generator, which is connected with the use of an optical resonator for the harmonic, is $(1 - R_{2,\text{opt}})^{-1}$.

Resonance at the fundamental frequency leads to the enhancement in the harmonic power of the order $[1 - R_1(\omega_1)]^{-2}$ (see Ashkin et al., 1966).

2. SHG INSIDE THE LASER CAVITY–INTRACAVITY HARMONIC GENERATION

Let us consider now the intracavity second harmonic generation (see Fig. 39b). The practical importance of this arrangement was demonstrated first by Geusic et al. (1968). In their experiment with the cw YAG:Nd laser and $Ba_2NaNb_5O_{15}$ nonlinear crystals the conversion efficiency in the second harmonic was 100%, in the sense that the SH output was equal to the output of the laser at the fundamental frequency without the nonlinear crystal. This section is devoted to an analysis of the intracavity second harmonic generation. Let us consider a three or four level laser with a nonlinear crystal inserted into the cavity. Usually the cavity length is appreciably greater than the length of the laser material; it is assumed also that the cavity is filled homogeneously with a transparent medium of the same index of refraction as that of the laser material. Under these conditions, the following rate equations can be written (Polloni and Svelto, 1968; Smith, 1970):

$$dq/dt = q(Bn - \gamma_L - \gamma_{NL} q) \tag{144}$$

$$dn/dt = -qBn - (n/T) + \gamma_p n_0 \tag{145}$$

where q is the photon density for a four level laser (or doubled photon density for a three level laser) at the laser frequency, n the population inversion between the operating laser levels, n_0 the total number of active ions in the rod, γ_p is proportional to pumping rate, B Einstein coefficient (cubic centimeters per second), T the fluorescent decay time, and γ_L the coefficient, taking into account the linear losses in the laser (reciprocal seconds); $\gamma_L = \gamma_{0L} + (1 - R) T_c^{-1}$, where γ_{0L} is connected with the linear losses in the laser rod; R_1 and R_2 are mirrors reflection coefficients, T_c transit time in the resonator, and γ_{NL} accounts for the nonlinear losses due to second harmonic generation

$$\gamma_{NL} = 8\pi (\gamma_2/n_2)^2 T_c^{-1} L^2 \hbar \omega_1 \tag{146}$$

The intensity of the laser radiation outside the resonator is

$$I_1 = \tfrac{1}{2}(1 - R_2) c \hbar \omega_1 q \tag{147}$$

The second harmonic intensity inside the resonator is

$$I_2 = \tfrac{1}{2}\gamma_{NL} T_c c \hbar \omega_1 q^2 \tag{148}$$

(a) *Steady-State Second Harmonic Generation; 100% Conversion Efficiency.* If the nonlinear crystal is absent ($\gamma_{\mathrm{NL}} = 0$), steady-state laser generation corresponds to $\gamma_{\mathrm{L}} = Bn$, and the intensity of the output radiation is

$$I_1 = \tfrac{1}{2}(1 - R_2)\, c\hbar\omega_1\, q\,[(\gamma_{\mathrm{p}} n_0/\gamma_{\mathrm{L}}) - (BT)^{-1}] \tag{149}$$

For optimum coupling, where $R_1(\omega_1) = 1$ and

$$R_2(\omega_1) = 1 - 2[\gamma_{0\mathrm{L}} + (\gamma_{\mathrm{p}} n_0 T\gamma_{0\mathrm{L}} B)^{1/2}] \tag{150a}$$

the output intensity at the fundamental frequency reaches the maximum

$$I_{1,\,\mathrm{max}} = (\hbar\omega_1 cT_{\mathrm{c}}/2BT\gamma_{0\mathrm{L}})[(\gamma_{\mathrm{p}} n_0 BT\gamma_{0\mathrm{L}})^{1/2} - \gamma_{0\mathrm{L}}]^2 \tag{150b}$$

In the simple traveling wave configuration, the second harmonic intensity at the output of the nonlinear crystal situated outside the resonator is

$$I_2 = (8\pi\gamma_2^{\,2} L^2/n_2 n_1^{\,2} c)\, I_{1,\,\mathrm{max}}^2$$

For typical cw laser, $L < L_{\mathrm{NL}}$, and the conversion efficiency in this external scheme of second harmonic generation is relatively small. Let us consider now the intracavity scheme with mirrors which reflect fully the fundamental radiation $[R_1(\omega_1) = R_2(\omega_1) = 1]$ and are transparent at the harmonic frequency $[R_1(\omega_2) = R_2(\omega_2) = 0]$. Substituting the solutions of (144) into (148), we obtain

$$I_2 = \frac{1}{2}\hbar\omega_1 cT_{\mathrm{c}}\gamma_{\mathrm{NL}}^{-1}\left[-\gamma_{0\mathrm{L}} - \frac{\gamma_{\mathrm{NL}}}{BT} + \sqrt{\left(\gamma_{0\mathrm{L}} - \frac{\gamma_{\mathrm{NL}}}{BT}\right)^2 + 4\gamma_{\mathrm{NL}}\gamma_{\mathrm{p}} n_0} \right] \tag{151}$$

Assuming optimum nonlinear coupling coefficient,

$$\gamma_{\mathrm{NL}} = \gamma_{0\mathrm{L}} BT \tag{152a}$$

then the output second harmonic intensity I_2 is exactly equal to the maximum fundamental power, as defined by (150b). Thus if the optimization condition (152a) is fulfilled, then

$$I_2 = I_{1,\,\mathrm{max}} \tag{152b}$$

In conclusion, it should be mentioned again that, in the optimum intracavity configuration, 100% conversion efficiency can be achieved, even if in the external scheme $L \ll L_{\mathrm{NL}}$.

(b) *Pulse Lengthening via Intracavity Second Harmonic Generation.* Another interesting application of the intracavity second harmonic generation is connected with the possibility of the pulse lengthening of the Q-switched laser. There are several approaches to the problem of pulse lengthening; all of them use nonlinear loss mechanisms It should be mentioned that in applications where the useful output is the second harmonic of the pumping laser, the second harmonic generation mechanism is particularly advantageous because the pulse lengthening is achieved at no expense to the output energy. This problem was studied by Murray and Harris (1970) and by Venkin et al. (1971).

Let us consider briefly the main points of the theory of pulse lengthening. For the Q-switched laser [here $\gamma_L = \gamma_L(t)$], Eq. (145) can be written as

$$dn/dt = -Bnq \tag{153}$$

Initial conditions for $t = 0$ are

$$n = n(0), \qquad q = q(0)$$

For very fast Q-switching (γ_L is practically a stepwise function), Eq. (144) and (153) can be integrated (see Venkin et al., 1971)

$$q = n(B^{-1}\gamma_{NL} - 1)^{-1} - (\gamma_L/\gamma_{NL})$$
$$+ [q(0) - n(0)(B^{-1}\gamma_{NL} - 1)^{-1} + \gamma_L/\gamma_{NL}][n/n(0)]B^{-1}\gamma_{NL} \tag{154}$$

The output energy density of the generated pulse is

$$w_1 = \tfrac{1}{2}(1 - R_2)\hbar\omega_1 c\int_0^\infty q\, dt = \tfrac{1}{2}(1 - R_2)\hbar\omega_1 cB^{-1}\ln[Bn(0)/\gamma_L(\infty)] \tag{155}$$

The peak intensity of the laser pulse is

$$I_{1,t} = \tfrac{1}{2}(1 - R_2)\hbar\omega_1 c\gamma_{NL}^{-1}[Bn(0) - \gamma_L] \tag{156}$$

The rise time of the pulse front is

$$\tau_f = [Bn(0) - \gamma_L]^{-1} \tag{157a}$$

and the decay time of the tail is

$$\tau_t = (\gamma_{NL}/\gamma_L)B^{-1}\ln\{1 + (\gamma_L/\gamma_{NL})n^{-1}(0)[B^{-1}\gamma_{NL} - 1]\} \tag{157b}$$

It is evident from (157) that the laser pulse duration can be varied by varying the nonlinear coupling coefficient γ_{NL}. Venkin et al. (1971) have demonstrated this by rotating the KDR crystal inserted in the cavity of a Nd-glass laser near the phase-matching direction. In this case the nonlinear coupling coefficient (nonlinear losses)

$$\gamma_{NL} \propto \text{sinc}^2[\pi\theta/2\theta_{coh}] \tag{158}$$

Murray and Harris (1970) have treated the case where the nonlinear crystal acts simultaneously as nonlinear loss and an output coupler for the laser. In their paper it was shown that a pulse lengthening factor of about 100 should be obtained.

H. Concluding Remarks. Cascade and Direct Processes in Higher-Order Harmonic Generation

In modern nonlinear optics, not only second and third but also the higher-order harmonics (up to the fifth harmonic) are of practical interest. In general, there are two ways to obtain the Nth optical harmonic. In the first of these, the N order term in the nonlinear polarization is used:

$$\mathscr{P}^{\mathrm{NL}} \simeq \chi^{(N)} E^N$$

An alternative method is based on the use of cascade processes of the lower-order nonlinearities. As an example, the third harmonic may be obtained by the direct process involving the cubic nonlinearity, or by the cascade process of the quadratic nonlinearity. With Q-switched lasers in the visible and infra-red the latter is more effective. Sukhorukov and Tomov (1969, 1970c) have discussed these problems in detail. They have developed the theory of the cascade optical tripler in which two crystals with quadratic nonlinearity (doubler and mixer) are used taking into account aperture effects. It is interesting that in several crystals phase-matched doubling ($\omega_1 + \omega_1 = 2\omega_1$) and mixing ($\omega_1 + 2\omega_1 = 3\omega_1$) may be observed simultaneously. As a result, such a crystal with a quadratic nonlinearity may be used for simultaneous generation of second and third harmonics. The theory of such a cascade process has been given by Akhmanov and Khokhlov (1964) and by Sukhorukov and Tomov (1970b).

Let us consider the simultaneous generation of the second and the third harmonics in the ammonium oxalate crystal. Phase-matching conditions are fulfilled for the interactions

$$\gamma^{\mathrm{o}}(\omega_1) + \gamma^{\mathrm{e}}(\omega_1) \rightarrow \gamma^{\mathrm{e}}(2\omega_1), \qquad \gamma^{\mathrm{o}}(\omega_1) + \gamma^{\mathrm{e}}(2\omega_1) \rightarrow \gamma^{\mathrm{e}}(3\omega_1)$$

These processes are described by the equations

$$\partial A_1{}^{\mathrm{o}}/\partial z = -\gamma_1' A_1{}^{\mathrm{e}} A_2{}^{\mathrm{e}} - \gamma_2' A_2{}^{\mathrm{e}} A_3{}^{\mathrm{e}} \tag{159a}$$

$$\partial A_1{}^{\mathrm{e}}/\partial z + \beta_1 \, \partial A_1{}^{\mathrm{e}}/\partial x = \gamma_3' A_1{}^{\mathrm{o}} A_2{}^{\mathrm{e}} \tag{159b}$$

$$\partial A_2{}^{\mathrm{e}}/\partial z + \beta_2 \, \partial A_2{}^{\mathrm{e}}/\partial x = \gamma_4' A_1{}^{\mathrm{o}} A_1{}^{\mathrm{e}} - \gamma_5' A_1{}^{\mathrm{o}} A_3{}^{\mathrm{e}} \tag{159c}$$

$$\partial A_3{}^{\mathrm{e}}/\partial z + \beta_3 \, \partial A_3{}^{\mathrm{e}}/\partial x = \gamma_6' A_1{}^{\mathrm{o}} A_2{}^{\mathrm{e}} \tag{159d}$$

Sukhorukov and Tomov (1970b) were able to obtain analytic solutions to these equations. These solutions are shown in Fig. 40. Two regimes are of practical interest. If $\eta_{I_1}^{\mathrm{e}} = I_1{}^{\mathrm{e}}/I_1 = 0.6$ ($\eta_I{}^{\mathrm{e}}$ is the fraction of extraordinary wave in the fundamental power), at the distance $z = 0.65 L_{\mathrm{NL}}$ ($L_{\mathrm{NL}} = 1/\gamma_1' E_2$) the second and third harmonic energies are approximately equal: $\eta_{I_2} = \eta_{I_3} = 30\%$. For $\eta_I{}^{\mathrm{e}} = 0.4$ practically 100% conversion to the third harmonic can be achieved.

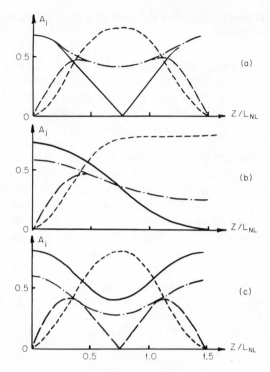

Fig. 40. Simultaneous phase-matched generation of the second (A_2) and third (A_3) harmonics in a medium with quadratic polarization. In the figure the amplitudes A_1^o (—), A_1^e (—·—), A_3 (———), and A_3 (···) are plotted as functions of distance. (a) $P_{1e}/P_1 > 0.4$; (b) $P_{1e}/P_1 = 0.4$; (c) $P_{1e}/P_1 < 0.4$.

The direct fourth and fifth harmonic generation in crystals has been studied theoretically by Akhmanov *et al.* (1974). They estimate the values of $\chi^{(4)}$ and $\chi^{(5)}$ in crystals approximately as $\chi^{(4)} \simeq 10^{-21}$ esu and $\chi^{(5)} \simeq 10^{-26}$ esu. Akhmanov *et al.* have discussed in detail the competition of the direct process of Nth harmonic generation with the cascade processes on the lower order nonlinearities.

It is interesting that, if the direct process is phase matched, cascade processes are also phase matched (see also Yablonovitch *et al.*, 1972). As a result, in several cases it is practically impossible to select the contributions from the direct and cascade processes. This is a very significant complication if direct higher-order harmonic generation was used as a tool to measure the higher-order optical nonlinearities. It is evident, however, that for very large optical fields, the higher-order polarizations dominate. In this case, the direct higher-order processes may be of practical interest for the technique of optical multiplication.

In a recent paper Harris (1973) discussed this problem in connection with optical harmonic generation in the vacuum ultraviolet for gases. It is shown for many practical systems (where the electronic transition frequencies to the ground state are greater than the frequency of the applied laser fields) that at incident power densities which approach the multiphoton ionization limit, the higher-order polarizations may exceed lower-order polarizations. As a first experimental test of this idea Harris (1973) demonstrated fifth harmonic generation in low-pressure xenon ($\lambda_1 = 5320$ Å, $\lambda_5 = 1064$ Å). In this case the direct process ($\omega_1 + \omega_1 + \omega_1 + \omega_1 + \omega_1 = 5\omega_1$), which is connected with the nonlinear susceptibility $\chi^{(5)}$, and two cascade processes ($\omega_1 + \omega_1 + \omega_1 = 3\omega_1$ and $\omega_1 + \omega_1 + 3\omega_1 = 5\omega_1$) which are connected with the nonlinear susceptibility $\chi^{(3)}$, can contribute.

III. TRANSIENT EFFECTS IN OPTICAL MULTIPLIERS

A. Introduction. Harmonic Generation of Ultrashort Laser Pulses

The quasi-stationary theory of harmonic generation which was used in Section II is not valid if the phase or amplitude modulation of the fundamental wave is very rapid, i.e., if the modulation time τ_M is comparable to or less than the characteristic times of group delay $\tau_v = vL$ ($v = u_N^{-1} - u_1^{-1}$ is the group velocity mismatch parameter) and dispersion spreading $\tau_{dis} = (2L \, \partial^2 k/\partial\omega^2)^{1/2}$. In several cases it is more convenient to compare the interaction length L to the characteristic lengths

$$L_\tau = \tau v, \qquad L_{dis} = \tau^2 (2 \, \partial^2 k/\partial\omega^2)^{-1}$$

If $L > L_\tau$ or $L > L_{dis}$, the harmonic generation process is transient in nature. If $\lambda_1 = 1.06 \, \mu m$, $\tau = 2 \cdot 10^{-12}$ sec in a KDP crystal with $L_\tau = 15$ cm, but in LiNbO$_3$, L_τ is only 2 cm. Very important transients connected with the group velocity mismatch in KDP and ADP crystals are in the ultraviolet range; if $\lambda_1 = 0.53 \, \mu m$ and $\tau = 2 \cdot 10^{-12}$ sec, $L_\tau \simeq 1$ cm. Transients connected with dispersion spreading are not essential in typical nonlinear crystals up to the pulse duration 10^{-13} sec, because in the transparency region the typical value $\partial^2 k/\partial\omega^2 = 10^{-27}$ sec cm^{-1}. It should be mentioned that in resonance harmonic generation (as in gases or vapors) dispersion spreading effects should be taken into account. Generally speaking, the theoretical analysis of harmonic generation of short laser pulses should be based on Eqs. (62)–(64) in which both spatial and temporal modulations of the interacting beams are taken into account. Such a general theory is developed in Section III, C. Here, we start with the analysis of the pure temporal transient problem namely the harmonic generation of plane wave packets.

B. Group Velocity Mismatch Effects

When double refraction and diffraction are negligible ($L \ll L_\beta, R_d$) all rays can be considered parallel to the Z-axis and the transverse coordinates x and y appear in the equations as parameters. In the first approximation of the dispersion theory taking into account only group velocity mismatching, the equations for the Nth harmonic generation of a plane wave packet for the O_1O_1–E_N interaction can be written as

$$\partial A_1/\partial z = -i(\gamma_N/n_1) A_N A_1^{*(N-1)} e^{-i\Delta_N z} \qquad (160)$$

$$(\partial A_N/\partial z) + v \, \partial A_N/\partial \eta = -i(\gamma_N/n_N) A_1^N e^{i\Delta_N z} \qquad (161)$$

where the local time $\eta = t - (z/u_1)$ refer to the fundamental wave. It can be seen from (161) that the nonlinear effect at a fixed spacetime point depends on the values of the light fields at another instant of time, because $u_1 \neq u_N$ and $v \neq 0$. If $A_N \ll A_1$ [parametric approximation is valid and $A_1 = A_1(\eta)$], Eq. (161) can be integrated as

$$A_N = -i(\gamma_N/n_N) \widetilde{\int_0^L} A_1{}^N [t - (L/u_N) + vz] e^{i\Delta_N z} \, dz \qquad (162)$$

From (162) one can see clearly the transient character of the harmonic generation of wave packets in a medium with group velocity mismatch. According to (162), a characteristic bandwidth of nonlinear medium $\Omega_{coh} \simeq (vL)^{-1}$ may be introduced. Naturally the transient effects in optical harmonic generation can be studied in the same fashion as the effects of spatial modulation which were considered in Section II. The space–time analogy pointed out in Section II simplifies greatly this analysis, because the problem of

Table I

Analogous Parameters in the Theory of Second Harmonic Generation of Modulated Light Waves

Beam parameter		Plane wave packet parameter	
Beam width	a	Pulse duration	τ
Angular divergence	α	Spectral width	$\Omega\tau$
Walk-off angle	β	Group velocities mismatch parameter	v
Aperture length	$L_\beta = 2/k_1\beta\alpha$	Quasi-static length	$L_v = 2/v\Omega_1$
Aperture length for the diffraction limited beam	$L_a = a/\beta$	Quasi-static length for the pulse without phase modulation	$L_\tau = \tau/v$
Phase-matching angular width	$\theta_{coh} = \pi/k_2\beta L$	Phase-matching frequency width	$\Omega_{coh} = \pi/2vL$

harmonic generation of wave packets is the temporal analog of harmonic generation of two-dimensional beams. Thus many results of Section II can be used directly in the theory of harmonic generation of short laser pulses. In Table I the analogous parameters of spatial and temporal problems are summarized. Using this table and the results of Section II, consider the main questions of the theory of harmonic generation of short pulses.

1. GAUSSIAN PULSE WITH FREQUENCY CHIRP. THE PARAMETRIC APPROXIMATION

Let the fundamental pulse be Gaussian with phase modulation

$$A_1(\eta) = A_1(0) \exp\{-(\eta/\tau)^2(1 - i(\Omega_0 \tau/2))\} \tag{163}$$

Introducing the spectrum of the complex amplitudes

$$S_N(\Omega, z) = (1/2\pi) \int_{-\infty}^{\infty} A_N(\eta, z) e^{-i\Omega \eta} \, d\eta, \qquad \Omega = \omega - N\omega_1$$

from (162), for the spectrum of the second harmonic amplitude we have

$$|S_N|^2 = \frac{\gamma_N^2 A_1^{2N}(0) L^2}{\pi N n_N^2 \Omega_1^2} \operatorname{sinc}^2 \left\{ \frac{\Delta_N(\Omega) L}{2} \right\} \exp \left\{ -\frac{2\Omega^2}{N\Omega_1^2} \right\} \tag{164}$$

where $\Omega_1^2 = \Omega_0^2 + 4\tau^{-2}$, $\Delta_N(\Omega) = \Delta_N + v\Omega$. Using (164) or the results of Section II and the space–time analogy, the spectral intensity of Nth harmonic can be obtained. From (117) the normalized density of energy is

$$\tilde{W}_N = L^2 \zeta^{-2} \{\sqrt{\pi \zeta} \operatorname{erf} \zeta + e^{-\zeta^2} - 1\} \tag{165}$$

where $\zeta = L/L_v$; $L_v = 2\sqrt{2}/\sqrt{N} v\Omega_1$. The above formulas permit one to investigate the transformation of the spectrum and the form of the harmonic pulse.

When $L < L_v$ (the quasi-static regime), the harmonic pulse is Gaussian with the duration $\sqrt{N}$ times less than that of the fundamental, and the harmonic spectrum is correspondingly $\sqrt{N}$ times broader. The harmonic energy grows as L^2 ($\tilde{W}_N = L^2$). When $L > L_v$, the width of the harmonic spectrum is defined by the frequency width of phase matching. Now the harmonic spectrum narrows during propagation in the nonlinear medium:

$$\Omega_N \simeq \Omega_{\text{coh}} = \pi/2vL \tag{166}$$

The narrow harmonic spectrum can be moved within the limits of the fundamental spectrum by means of changing the phase mismatch, Δ_N, for example, by rotation of the crystal.* In this case the harmonic energy equals $\tilde{W}_N = \sqrt{\pi}\, LL_v$. The form of the harmonic pulse depends on the frequency chirp parameter of the fundamental pulse $D_\Omega = \Omega_0 \tau$. In crystals with $L > L_v$, if the frequency chirp is absent, the lengthening of the harmonic pulse occurs during the propagation. The harmonic pulse duration is equal to

$$\tau_N \simeq vL \tag{167}$$

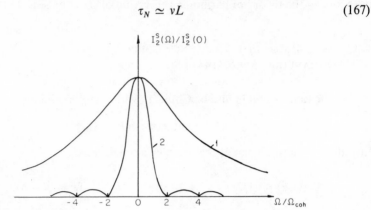

Fig. 41. Frequency spectra in the process of nonquasi-static frequency doubling. Curve 1 corresponds to quasi-static $(L < L_v)$ phase-matched frequency doubling (for Gaussian spectra second harmonics spectral width $\Omega_2 = \sqrt{2}\,\Omega_1$). For curve 2, $L/L_v = 5$; the SH spectrum is significantly narrower than the spectrum of the fundamental wave (from Akhmanov *et al.*, 1968b).

Formulas (164) and (165) are illustrated in Figs. 41–43 along with the theoretical curves characterizing the spectral distribution of the harmonic; the experimental results on harmonic generation of picosecond pulses from Nd-glass lasers in the crystals KDP and LiNBO$_3$ are shown. The possibility of a modulation of the harmonic pulse duration by means of changing the crystal length should be mentioned. Akhmanov *et al.* (1972) observed the variation of the Nd-glass laser fourth harmonic $(\lambda_4 = 2660\,\text{Å})$ pulse duration from $0.5 \cdot 10^{-12}$ to $5 \cdot 10^{-12}$ sec by changing the length of the KDP crystal from 0.2 to 4 cm.

Fast phase modulation $(D_\Omega \gg 1)$ of the fundamental wave leads to the modulation of the envelope of the harmonic pulse. Particularly if $D_\Omega \gg L_\tau/L$, the harmonic pulse has the form

$$I_N \simeq \mathrm{sinc}^2\left\{\frac{\Delta_N L}{2} + \frac{\pi\eta}{2\tau_{\mathrm{coh}}}\right\}\exp\left\{-\frac{2N\eta^2}{\tau^2}\right\} \tag{168}$$

---

* A tunable generator of this type was described by Carman *et al.* (1967). They obtained radiation with a broad spectrum by means of focusing and self-modulation of short and intense laser pulses in liquids and then doubling it in a KDP crystal. Frequency tuning of the second harmonic radiation was achieved by rotation of the nonlinear crystal.

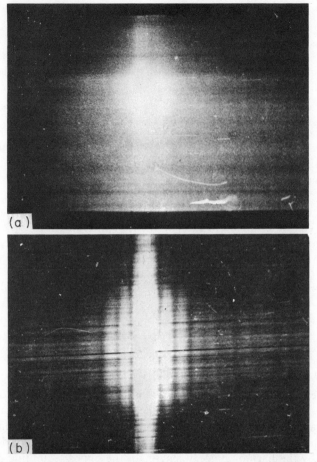

Fig. 42. Spectra of the second harmonic excited with picosecond pulses of a Nd-glass laser in a KDP crystal. (a) $L < L_v$; (b) $L > L_v$ (compare to Fig. 41). In this experiment $L_v = 0.9$ cm.

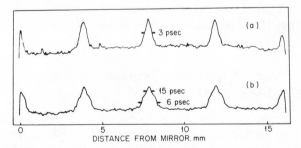

Fig. 43. Second harmonic pulse lengthening due to group velocity mismatch in nonlinear media. (a) Two-photon fluorescence (TPF) track for the SH excited in KDP by picosecond pulses of a Nd-glass laser (group velocity mismatching effects are negligible); (b) TPF track for the SH excited in LiNbO$_3$ with the same laser ($L > L_t$ and sufficient pulse lengthening take place) (from Shapiro, 1968).

where $\tau_{coh} = \tau\Omega_{coh}/N\Omega_0$. By changing the phase mismatch Δ_N, the tuned retardation of the harmonic pulse (of course, over the limits of the fundamental pulse duration) can be obtained. The above mentioned dependence of the harmonic envelope from the frequency chirp parameter D_Ω can be used in principle for the measurement of this significant parameter.

(a) *Lossy Medium. The Quasi-static Harmonic Generation in a Medium with Large Group Velocity Mismatch.*   All the results of the previous sections were obtained for a lossless medium. In the case of equal losses for the fundamental and the second harmonic ($\delta_1 = \delta_2 = \delta$), the substitutions $A_{1,2}(z) = \tilde{A}_{1,2}(z)e^{-\delta z}$ and $z = [1 - e^{-\delta z}]\delta^{-1}$ lead to the equation for $\tilde{A}_{1,2}$, which is exactly the same as in the case of a lossless medium. If $\delta_1 \neq \delta_N$, analytic solutions can be obtained in the parametric approximation ($A_1 \gg A_N$). The fundamental amplitude may be written as $A_1(t,z) = A_1(t-(z/u_1))e^{-\delta_1 z}$. From (161) taking into account the decay of second harmonic, we obtain

$$A_N = -i(\gamma_N/n_N)e^{-\delta_N L}\int_0^L A_1{}^N(t-(L/u_N)+vz)\exp[-(N\delta_1-\delta_N)z]\,dz \tag{169}$$

If $\delta_N > N\delta_1$ and

$$L_\tau \gg (\delta_N - N\delta_1)^{-1} \tag{170}$$

Eq. (169) reduces to the simple algebraic formula

$$A_N = -i(\gamma_N/n_N)A_1{}^2(t-(L/u_1))[e^{-N\delta_1 L}-e^{-\delta_N L}](\delta_N-N\delta_1)^{-1} \tag{171}$$

It is interesting that in the case under consideration harmonic generation for arbitrary group velocity mismatch becomes quasi-static. The physical explanation is that the part of the harmonic pulse lagging behind the fundamental pulse is absorbed very quickly and the doubler efficiency is determined critically by the harmonic decay length. In the opposite limiting case when $N\delta_1 > \delta_N$ and $L_\tau \gg (N\delta_1 - \delta_N)^{-1}$, it follows from (160) that:

$$A_N = -i(\gamma_N/n_N)A_1{}^2(t-(L/u_N))[e^{-\delta_N L}-e^{-N\delta_1 L}](N\delta_1-\delta_N)^{-1} \tag{172}$$

i.e., the second harmonic is excited only near the entrance plane of the nonlinear medium. The process is again quasi-static for arbitrary group velocity mismatch.[*]

2. LIMITING EFFICIENCY OF OPTICAL DOUBLERS EXCITED WITH SHORT LASER PULSES

The analytic theory of the optical doubler excited with short laser pulses, taking into account both the group velocity mismatch effects and the reaction of the harmonic field on the fundamental wave, may be developed in the same manner as in Section II, E, if $D_\Omega = 0$ and $\Delta_2 = 0$. In this case for the complex

[*] Similar effects take place in the interaction of optical beams in a lossy medium; remember the space–time analogy.

amplitude of the fundamental wave we obtain [compare (126)]

$$\partial^2 A_{01}^{-1}/\partial z^2 - (\gamma_2/n_1)^2 G(\eta - vz) A_{01}^{-1} = 0 \tag{173}$$

in which the function G is determined by the boundary conditions

$$G(t) = A_{01}^2(0,t) + A_{02}^2(0,t) - v(n_1/\gamma_2) \, \partial A_{02}(0,t)/\partial t \tag{174}$$

Using solution (127) the modification of the second harmonic and the fundamental pulses can be investigated as a limiting conversion efficiency of the doubler [the parameter g_β in (137) should be replaced by $g_v = (\gamma_2/n_1)(\tau/v) A_{01}(0,0)$; see Fig. 32]. It should be noted that the parameter $g_v \propto (w_1/\tau)^{1/2}$ can be varied by altering the fundamental pulse duration while conserving its energy. Figure 44 illustrates the influence of the group velocity mismatch effect on the limiting efficiency of the optical doubler excited with ultrashort pulses.

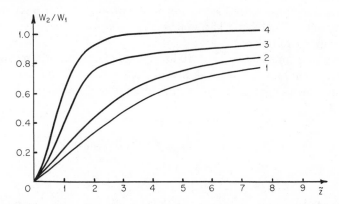

Fig. 44. Efficiency of the optical doubler ($\eta_{W_2} = W_2/W_1$) excited with a short light pulse as a function of the dimensionless distance $\tilde{z} = z(L_\tau^{-1} + L_{NL}^{-1})$ for different values of $g_v = L_\tau/L_{NL}$. Curve 1 shows $g_v^2 = 0.5$; curve 2 shows $g_v^2 = 2$; curve 3 shows $g_v^2 = 10$. Curve 4 corresponds to a plane monochromatic wave (from Akhmanov et al., 1968b).

3. Second Harmonic Pulse Formation.
Interaction of Pulses with Different Durations

One interesting aspect of nonlinear light pulse interaction is related to the possibility of shaping them and in particular to decrease their duration. In the case of quasistatic harmonic generation ($L < L_\tau$) of Gaussian plane wave pockets, the duration of the harmonic pulse contracts $\sqrt{N}$ times. If $L > L_\tau$, lengthening of the harmonic pulse occurs. Thus, for effective doubling of short pulses, simultaneous phase and group velocity matching should be achieved:

Phase-matching condition in the simplest case is

$$\Delta_N = 0, \qquad v_N = v_1, \qquad n_N = n_1$$

Group velocity matching condition is

$$v = 0; \quad u_N = u_1, \quad n_N + \omega_N \, \partial n_N / \partial \omega_N = n_1 + \omega_1 \, \partial n_1 / \partial \omega_1$$

These conditions can be fulfilled simultaneously in inhomogeneous media. There is, however, another mode of operation of nonlinear pulse formation for which the group velocity mismatch is conducive for second harmonic pulse shortening.

Let us consider a relatively narrow second harmonic pulse obtained from an external source, and let this pulse interact with a quasi-fundamental radiation. When a reaction of the harmonic on the fundamental radiation occurs and there is a group velocity mismatch, the character of the energy exchange between the fundamental wave and harmonic is different for the leading edge and the tail of the pulse. If, for example, $v < 0$, $u_2 > u_1$, the leading edge of the second harmonic pulse "runs into" the fundamental wave and extracts a considerable fraction of the energy; less energy is delivered to the tail end of the second harmonic pulse. As a result, narrowing of the second harmonic pulse occurs.

Akhmanov *et al.* (1968b) have studied the amplification of a rectangular second harmonic pulse in the field of a quasicontinuous fundamental wave. The group velocity mismatch leads to the formation of the second harmonic pulse:

$$A_{02} = A_0 \tanh\{\psi + (z - (\eta/v))L_{NL}^{-1}\} \tag{175}$$

$$L_{NL} = (n_1/\gamma_2)A_0^{-1}, \quad A_0 = [A_{01}^2(0) + A_{02}^2(0)]^{1/2}$$

$$\tanh \psi = \frac{A_{02}(0)}{A_0} + \frac{A_{01}(0)}{A_0} \tanh \frac{\eta}{vL_{NL}}$$

On the front of the second harmonic pulse ($t = z/u_2$; i.e., $z = \eta/v$) the amplitude has a jump which equals $A_{02}(0)$, i.e., the same as at the point $z = 0$. According to (175), the width of the wave packet near the front equals

$$\tau_2 = vL_{NL} \tag{176}$$

This formation process was illustrated in Fig. 45. In conclusion it should be noted that the process considered here is an example of transient nonlinear wave interaction where the amplified pulse extracts energy from the quasi-continuous wave in the volume, which has a linear dimension $L \gg \tau_2 u_2$.

C. Aperture Effects in Optical Multipliers Excited with Ultrashort Laser Pulses

In the general case of the nonlinear interaction of waves modulated simultaneously in time and space, the harmonic temporal and angular spectra are strongly coupled to each other. Let us consider these effects for the $O_1 O_1$–E_2

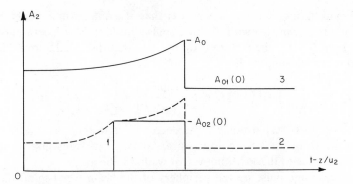

Fig. 45. Second harmonic pulse formation due to the nonlinear interaction of a short SH pulse with quasi-cw fundamental radiation. Curve 1 shows SH input rectangular pulse; curve 2 shows SH pulse in the formation stage; curve 3 shows steady-state harmonic pulse.

phase-matched second harmonic generation process. Now all derivatives in Eqs. (62) should be taken into account. Let us consider the second harmonic excited with a Gaussian beam, which possesses simultaneously Gaussian amplitude modulation in time; frequency chirping would also be taken into account. Thus, for the complex amplitude of the fundamental wave, we can write ($z = 0$)

$$A_1 = A_1(0,0) \exp\left\{ -\frac{x^2+y^2}{a^2}\left(1-\frac{i\alpha_0}{\alpha_d}\right) - \frac{\eta^2}{\tau^2}\left(1-\frac{i\Omega_0\tau}{2}\right)\right\} \quad (177)$$

In the geometrical optics approximation in (62), $\mathscr{L}_j(x,y) = \beta_j(\partial/\partial x)$; $\mathscr{L}_j(\eta) = v_j(\partial/\partial\eta)$, the dependence of the harmonic energy on the crystal length is the same as (117), but now the parameter $\zeta = L/L_{\beta,v}$, where the new characteristic length

$$L_{\beta,v} = (L_\beta^{-2} + L_v^{-2})^{-1/2} \quad (178)$$

characterizes the simultaneous influence of aperture and dispersion effects. In the absence of phase modulation ($\alpha_0 = 0$, $\Omega_0 = 0$) the second harmonic amplitude equals

$$A_2 = -i\left(\frac{\gamma_2}{n_2}\right)A_1^2(0,0)L_{a,\tau}\exp\left\{-\frac{2(vx-\beta\eta)^2}{\beta^2\tau^2+v^2a^2}\right\}$$

$$\times\left\{\mathrm{erf}\left[\frac{L}{L_{a,\tau}}+L_{a,\tau}\left(\frac{\beta x}{a^2}+\frac{v\eta}{\tau^2}\right)\right]-\mathrm{erf}\left[L_{a,\tau}\left(\frac{\beta x}{a^2}+\frac{v\eta}{\tau^2}\right)\right]\right\} \quad (179)$$

where $\mathrm{erf}\, x = (2/\sqrt{\pi})\int_0^x e^{-\xi^2}\,d\xi$.

If $L_a > L > L_\tau$ (the dispersion effect dominates), (179) reduces to

$$A_2 = -i\left(\frac{\gamma_2}{n_2}\right)A_1^2(0,0)L_\tau\exp\left(-\frac{2x^2}{a^2}\right)\left\{\mathrm{erf}\left[\frac{L}{L_\tau}+\frac{\eta}{\tau}\right]-\mathrm{erf}\left(\frac{\eta}{\tau}\right)\right\} \quad (180)$$

The second harmonic beam is narrower than the fundamental beam, but the harmonic pulse spreads with distance (pulse duration is proportional to vL). When $L_\tau > L > L_a$, the picture is opposite to that discussed above. Now the second harmonic amplitude

$$A_2 = -i\left(\frac{\gamma_2}{n_2}\right)A_1^2(0,0)L_a\exp\left(-\frac{2\eta^2}{\tau^2}\right)\left\{\text{erf}\left[\frac{L}{L_a}+\frac{x}{a}\right]-\text{erf}\left(\frac{x}{a}\right)\right\} \quad (181)$$

The duration of the harmonic pulse decreases as in the quasi-static process but the width of the harmonic beam expands, so its transversel dimension grows as βL due to the Poynting vector walk-off effect.

The harmonic pulse on the periphery of the beam lags behind the pulse on the axis. The retardation time is

$$t_r(x) = xv/\beta = \tau(x/a)(L_a/L_\tau)$$

Naturally, spatial sweep of the harmonic pulse occurs. Hence changing the ratio L_a/L_τ leads to a change in the harmonic pulse duration. In the case of a divergent beam ($D_\alpha \gg L/L_a$) and rapid frequency modulation ($D_\Omega \gg L/L_\tau$) the amplitude A_2 can be written as [compare (168)]

$$|A_2|^2 \propto \text{sinc}^2\left(\frac{\Delta_2 L}{2}+\frac{\pi x}{2a_{\text{coh}}}+\frac{\pi\eta}{2\tau_{\text{coh}}}\right)\exp\left[-4\left(\frac{\eta^2}{\tau^2}+\frac{x^2}{a^2}\right)\right] \quad (182)$$

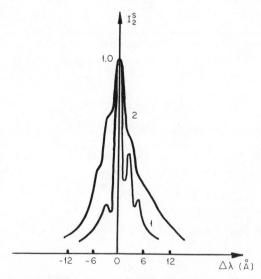

Fig. 46. Experimental SH spectra in a LiNbO$_3$ crystal excited with divergent beams from a Nd-glass picosecond laser. Curve 1 shows small divergence; curve 2 shows strong divergence. It is evident that the spectral width increases with the fundamental beam divergence (from Orlov *et al.*, 1969).

Here the harmonic pulse has a duration $\tau_{coh} = \tau\Omega_{coh}/2\Omega_0$ and the retardation time t_r depends on the transverse coordinate $t_r = \tau_{coh} x/a_{coh}$; $a_{coh} = a\theta_{coh}/2\alpha_0$. Orlov *et al.* (1969) studied these effects experimentally. In Fig. 46, the spectral width of the harmonic versus the beam divergence is shown. It is evident that the spectral width and hence the second harmonic pulse duration depend strongly on the beam divergence.

D. Harmonic Generation in the Presence of Dispersion Spreading of the Wave Packets

For large dispersion or very short pulses, the second-order term in the dispersion theory (the spreading of wave pockets) should be taken into account. The characteristic length of pulse spreading or compression is $L_\Omega = (\tau/\Omega_j)(\partial^2 k/\partial\omega^2)_j$ (its spatial analog is $L_a = a/\alpha$). The theory of the harmonic generation of short pulses in the presence of dispersion spreading may be developed with the help of the space–time analogy. The results of the theory of harmonic generation of the stationary laser beam focused by a cylindrical lens can be used; this is the spatial analog of the problem under consideration. It should be pointed out, however, that the simple analogy is valid only if $\partial^2 k_1/\partial\omega_1^2 = 2 \partial^2 k_2/\partial\omega_2^2$ (in the spatial problem, the diffusion coefficients $(2ik_j)^{-1}$ are coupled with the condition $k_2 = 2k_1$. For a nonlinear medium with such dispersion, the optimum duration of the fundamental pulse, which corresponds to a maximum doubling efficiency, can be estimated (it is a direct analog of optimum focusing) $\tau_{opt} \simeq (L \, \partial^2 k/\partial\omega^2)^{1/2}$ and $L_{dis, opt} \simeq L$.

The dispersion of group velocities in a transparent crystal is very small $\partial^2 k/\partial\omega^2 = 10^{-27} \sec^2 \mathrm{cm}^{-1}$ and the characteristic dispersion length L_{dis} equals approximately 1 cm for pulses with subnanosecond structure ($\Omega_1 \simeq 10^{-13} \sec^{-1}$). On the other hand, the dispersion spreading effects can be significant in resonant frequency multiplication in gases and vapors.

E. Statistical Effects in Optical Harmonic Generation

The models of regular optical beams and pulses, which we have used above, described very well the radiation of single mode or mode-locked lasers. For multimode lasers with random phases of the modes and superluminescent lasers the model of optical noise is much more suitable. The harmonic generation of optical noise has some interesting features. The detailed account of these can be found in the book by Akhmanov and Chirkin (1971). Here, we will discuss briefly only the influence of temporal coherence of a fundamental beam on the harmonic generation process.

1. SECOND HARMONIC GENERATION OF OPTICAL NOISE.
COHERENT AND INCOHERENT INTERACTIONS

The most important new effect which arises when the randomly modulated waves interacting in a dispersive* medium is the decay of phase correlations during the propagation. As a result, after a characteristic length L_{coh} the initially coherent interaction (the phase relations between the interacting waves are fixed) becomes incoherent; now the stochastic interaction of weakly correlated waves takes place. To illustrate this transition from coherent to incoherent interaction, let us consider the harmonic generation of quasi-monochromatic Gaussian noise. Let the fundamental field on entrance into a nonlinear medium be

$$E_1(z, t) = \tfrac{1}{2}A_1(z, t) \exp\{i(\omega_1 t - k_1 z)\} + \text{c.c.} \tag{183}$$

where $A_1(t, z)$ is a random function, which is slowly varying in comparison with $\exp\{i(\omega_1 t - k_1 z)\}$. In the parametric approximation for the second harmonic amplitude, we have

$$\frac{\partial A_2}{\partial z} - \frac{1}{u_2}\frac{\partial A_2}{\partial t} = -i\left(\frac{\gamma_2}{n_2}\right)A_1^{\,2}\left(t - \frac{z}{u_1}\right)e^{i\Delta_2 z} \tag{184}$$

where $u_{1,2}$ is the group velocity of the fundamental and second harmonic waves and $\Delta_2 = k_2 - 2k_1$. From (184) it is evident that the process of harmonic generation depends critically on the relation of the correlation time τ_c of the fundamental wave and the group retardation time $\tau_v = vL$, $v = u_2^{-1} - u_1^{-1}$. If $\tau_c > \tau_v$ in (184), we can put $u_1 = u_2$ and neglect the temporal derivative. In this case, the relation between the fundamental and harmonic intensities is algebraic, $I_2 = (8\pi\gamma_2^{\,2}/cn_1^{\,3})I_1^{\,2}$. For Gaussian noise: $\omega_1(I_1) = (\bar{I}_1)^{-1}\exp(-I_1/\bar{I}_1)$:

$$\bar{I}_2 = (16\pi\gamma_2^{\,2}/cn_1^{\,3})(\bar{I}_1)^2$$

For the arbitrary Nth harmonic, excited with Gaussian optical noise

$$\bar{I}_N = (\gamma_N^{\,2}/n_1^{\,2})(8\pi/cn_1)^{N-1} \cdot N!(\bar{I}_1)^N \tag{185}$$

This regime of frequency multiplication may be termed "coherent"; phase relations between the fundamental and harmonic fields are fixed. It should be emphasized that in the coherent nonlinear interaction process optical noise acts much more effectively than a monochromatic wave with the same intensity. The reason for this enhancement is the sensitivity of the nonlinear process to the peaks of the Gaussian noise.

---

* In this section only linear dispersion (dispersion which is connected with the linear response of the medium) is taken into account.

In the opposite limiting case $\tau_c < \tau_v$, the situation is quite different: the efficiency of the optical multiplier excited with optical noise is less than that excited with a monochromatic wave with the same intensity. In this case the interaction becomes inertial and a nonlinear medium plays the role of a narrow band filter with the bandwidth $\Omega_{coh} = \pi/(2vL)$. The phase correlation between the fundamental and the harmonic waves decays and the harmonic generation process becomes incoherent.* The intensity and correlation function of the harmonic field can be estimated with the help of the general solution (162) (Akhmanov and Chirkin, 1966, 1971). From (162) the correlation function of the second harmonic can be written as

$$B_2(z,\tau) = 2(\gamma_2/n_2)^2 \int\int_0^L B_1{}^2[\tau - v(z_1 - z_2)] \cos[\Delta_2(z_1 - z_2)]\, dz_1\, dz_2 \tag{186}$$

where B_1 is the correlation function of the fundamental wave. From (186), all the results mentioned above may be directly obtained.

It is evident that the characteristic length which divided the regions of coherent and incoherent interactions is, indeed, the coherent length $L_{\tau,\,coh} = \tau_c/v$. Accordingly $L < L_{\tau,\,coh}$ corresponds to coherent interaction, and $L > L_{\tau,\,coh}$ to incoherent interaction. In the frequency domain, if the fundamental bandwidth

$$\Omega_1 \ll \Omega_{coh} \tag{187}$$

coherent interaction takes place. If

$$\Omega_1 \gg \Omega_{coh} \tag{188}$$

incoherent interaction occurs. From (186), for the coherent interaction

$$\bar{I}_2 = (16\pi\gamma_2{}^2/cn_2 n_1{}^2)\bar{I}_1{}^2 L^2 \operatorname{sinc}^2(\Delta_2 L) \tag{189}$$

(see Fig. 47). In this case the harmonic spectrum is somewhat wider than the spectrum of the fundamental wave. If the interaction is incoherent,

$$\bar{I}_2 = 4(\gamma_2/n_2)^2 LL_{coh}\bar{I}_1 \int_0^\infty B_1{}^2(\xi) \cos(\Delta_2 L_{coh}\xi)\, d\xi \tag{190}$$

i.e., $\bar{I}_2 \propto L$ (see Fig. 47).

Now the width of the harmonic spectrum $\Omega_2 \ll \Omega_1$ and $\Omega_2 \simeq \Omega_{coh}$ is independent of the fundamental bandwidth. It is interesting that for the incoherent interaction, Eq. (84) reduces to the equation for the average intensity (rate equation)

$$d\bar{I}_2/dz = \bar{\gamma}_2(\bar{I}_1)^2 \tag{191}$$

---

* In several aspects this process is similar to the transient harmonic generation of short pulses considered in Sections III, A–C.

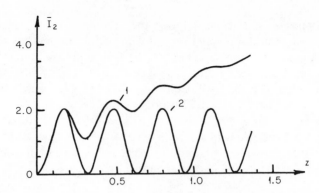

Fig. 47. Average intensity of the second harmonic excited with Gaussian optical noise versus distance. Curve 1 shows incoherent interaction; curve 2 shows coherent interaction. For both cases $\Delta_2 \neq 0$.

The last equation describes fully incoherent harmonic generation. It should be noted here that the nonlinear interaction of regular waves can become incoherent only if during the interaction light waves have excited the internal motions of the medium (as, for example, in stimulated Raman scattering).

(a) *Spectrum of the Second Harmonic.* If the correlation function of the fundamental radiation is known, the spectrum of the second harmonic can be directly estimated from (186). In many aspects the results (Akhmanov and Chirkin, 1966, 1971) are similar to those described in Section III,B (i.e., harmonic spectral broadening for coherent interaction, monochromatization for incoherent interaction, etc.). All these results may be obtained from Fourier transformation of the correlation (186), or directly after averaging of the amplitude spectrum (164). The purpose of this subsection, however, is to describe briefly another approach to the theory of nonlinear optical effects excited with random waves, an approach which is termed the "spectral method"; the method used above in which the truncated equations in partial derivatives were used can be termed the "envelope method." In the spectral method, which is correct only in the parametric approximation, truncated equations for the spectral components are used. Let us consider the Fourier transform of the fundamental amplitude:

$$S_1(\omega) = (1/2\pi) \int_{-\infty}^{\infty} A_1(t) \exp[i(\omega_1 - \omega)t] \, dt \tag{192}$$

In the parametric approximation, S_1 is only a function of the frequency. In a dispersive medium the Fourier component of the frequency ω propagates with phase velocity $v_1 = \omega/k_1(\omega)$. In a nonlinear medium the Fourier component of the second harmonic $S_2(\omega, z)$ will be a slowly varying function of the distance. The fact, that the spectral components of the harmonic arise from doubling and mixing of the fundamental spectral components yields the equation.

$$
dS_2/dz = -i \int\!\!\int_{-\infty}^{\infty} \gamma_2(\omega, \omega', \omega'') n_2^{-1}(\omega) S_1(\omega') S_2(\omega'')
$$

$$
\times \exp\{i\Delta_2(\omega, \omega', \omega'')z\}\, \delta(\omega - \omega' - \omega'')\, d\omega'\, d\omega'' \qquad (193)
$$

where $\Delta_2 = k_2(\omega) - k_1(\omega') - k_1(\omega'')$. In the case of monochromatic waves $S_1 = A_1 \delta(\omega - \omega_1)$ and (193) reduces to (160) if in the last equations temporal derivatives are excluded.

For quasi-monochromatic waves, the wave numbers $k_j(\omega)$ can be expanded in the vicinity of the center frequencies:

$$
k_j(\omega) = k_j(\omega_j) + (\partial k_j/\partial\omega)(\omega - \omega_j) = k_j(\omega_j) + u_j^{-1}(\omega - \omega_j) \qquad (194)
$$

The phase mismatch parameter now equals

$$
\Delta_2(\omega, \omega', \omega'') = \Delta_2 + u_1^{-1}(2\omega_1 - \omega' - \omega'') - u_2^{-1}(2\omega_1 - \omega) \qquad (195)
$$

where Δ_2 is the phase mismatch for the center frequency. Inserting (195) into (193) and neglecting the dispersion of the nonlinearity in the fundamental bandwidth, we can write:

$$
dS_2/dz = -i(\gamma_2/n_2) \exp[i\Delta_2(\Omega)z] \int_{-\infty}^{+\infty} S_1(\omega - \omega') S_1(\omega')\, d\omega' \qquad (196)
$$

where $\Delta_2(\Omega) = \Delta_2 + v\Omega$ and $\Omega = \omega - 2\omega_1$. Equation (196) is really the Fourier transform of Eq. (161) for a fixed amplitude $A_1 = A_1(t - (z/u_1))$. The solution of (196) with zero boundary conditions is

$$
S_2 = -i(\gamma_2/n_2) L \operatorname{sinc}\{\Delta_2(\Omega)L/2\} \exp[i\Delta_2(\Omega)L/2]
$$

$$
\int_{-\infty}^{+\infty} S_1(\omega - \omega') S_1(\omega')\, d\omega' \qquad (197)
$$

Now for the spectral intensity of the second harmonic we obtain:

$$
\overline{S_2 S_2^*} = (\gamma_2/n_2)^2 L^2 \operatorname{sinc}^2\{\Delta_2(\Omega)L/2\}
$$

$$
\times \int\!\!\int_{-\infty}^{+\infty} \overline{S_1(\omega - \omega') S_1(\omega') S_1^*(\omega - \omega'') S_1^*(\omega'')}\, d\omega'\, d\omega'' \qquad (198)
$$

In (198) the integrand depends on the statistics of the fundamental radiation. For stationary Gaussian noise, (198) reduces to

$$
\bar{I}_2(\omega, L) = (16\pi\gamma_2{}^2/cn_1{}^3) L^2 \operatorname{sinc}^2\{\Delta_2(\Omega)L/2\} \int_{-\infty}^{+\infty} I_1(\omega - \omega') I_1(\omega')\, d\omega'
$$

$$
(199)
$$

The spectral method for the study of second harmonic generation of non-monochromatic waves in the quasi-static regime first was used by Strizevsky (1966).

2. THE DISTRIBUTION LAWS.
PHOTON-COUNTING STATISTICS OF SECOND HARMONICS

In several cases not only are the spectra of interest, but the field correlations are also. As an example, in experiments with gas lasers, one-dimensional distribution laws of the harmonic can be measured. Thus we obtain additional information about the light fields and the nonlinear process also. The one-dimensional distribution law of the light field can be obtained by measurement of the statistical distribution of the photocounts (see, for example, Klauder and Sudarshan, 1968).

The probability of n photocounts during the time T follows from the well-known Mandel formula

$$p_j(n, T) = \int_0^\infty [(\alpha_j V_j)^n/n!] \, e^{-\alpha_j v_j} w(V_j) \, dV_j \qquad (200)$$

where α_j is the quantum efficiency of the photodetector and

$$V_j(t, T) = \int_t^{t+T} I_j(t') \, dt', \qquad j = 1, 2 \qquad (201)$$

and $w(V_j)$ is the distribution of the random value V_j. The relative fluctuations of V_j are connected with the momenta of the photocount numbers

$$d_j = [\overline{V_j^2} - (\overline{V_j})^2]^{1/2} (\overline{V_j})^{-1} = [\overline{n_j^2} - (\bar{n}_j)^2 - n_j]^{1/2} (\bar{n}_j)^{-1} \qquad (202)$$

Let us now consider the optical harmonic photocount distributions. Below we consider mainly quasi-static harmonic generation. In this case the relation between the instant harmonic and fundamental intensities for $\Delta_2 = 0$ is

$$I_2(L, t) = (8\pi\gamma_2^2/cn_1^3) L^2 I_1^2(t) \qquad (203)$$

and the photocounting distribution of the second harmonic will be the same as for two-quantum photocounts (see Akhmanov et al., 1969). Let us consider the second harmonic photocount distributions for two models of fundamental radiation: Gaussian noise and multimode radiation.

(a) *Second Harmonic Excited with Thermal Radiation.* The photo-count distribution for thermal radiation, if the observation time T is much smaller than the correlation time τ_c, $T \ll \tau_c$, is described by the well-known Bose–Einstein distribution:

$$p_1(n) = (1 + \bar{n}_1)^{-1} (1 + \bar{n}_1^{-1})^n \simeq (\bar{n}_1)^{-1} \exp(-n/\bar{n}_1) \qquad (204)$$

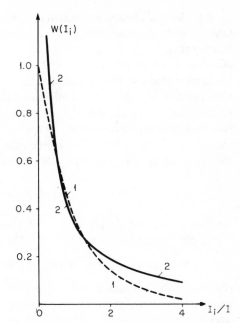

Fig. 48. Intensity distributions of the thermal radiation (curve 1) and its second harmonic (curve 2).

The photocount distribution for the second harmonic, if the group velocity mismatch is negligible ($L \ll L_{\mathrm{coh}}$), can be estimated from (200). For this coherent process (see Fig. 48)

$$p_2(n) = \frac{(2n)!}{2^n \bar{n}_2^{1/2} n!} e^{1/4\bar{n}_2} D_{-(1+2n)}\left((\bar{n}_2)^{-1/2}\right) \tag{205}$$

where $D_{-n}(x)$ is a parabolic function, $\bar{n}_2 = \alpha_2 T \bar{I}_2$. It should be emphasized that $p_2(n)$ differs significantly from $p_1(n)$, so the relative fluctuations

$$d_2 = \sqrt{5} \tag{205a}$$

differ markedly from the Bose–Einstein case ($d_1 = 1$). The theory of second harmonic photocount distribution for incoherent harmonic generation ($L \simeq L_{\mathrm{coh}}$; $L \gg L_{\mathrm{coh}}$) has been given by Akhmanov *et al.* (1969). They found that if $L \gtrsim L_{\mathrm{coh}}$

$$d_2 = \{5 - 4(L/L_{\mathrm{coh}})^2\}^{1/2}$$

For the sufficiently incoherent process ($L \gg L_{\mathrm{coh}}$)

$$d_2 = \{1 + 10(L_{\mathrm{coh}}/L)^2\}^{1/2}$$

One can see that for $L \to \infty$, $d_2 \to 1$. Thus, for $L \gg L_{\mathrm{coh}}$, the second harmonic statistics will be Gaussian; during incoherent nonlinear interaction normalization of the second harmonic field takes place.

(b)   *Second Harmonic Excited with Multimode Laser Radiation*   The interest in the distribution law of the second harmonic excited with multimode lasers began in 1963 and 1964, when Ducuing and Bloembergen (1964) and Akhmanov *et al.* (1963) observed the "excess" fluctuations of the second harmonic intensity. They found that the presence of several equidistant modes with random phases lead to both excess fluctuations in the second harmonic intensity and to an error in the measurement of the corresponding nonlinear susceptibility. Transformation of phase fluctuations of the fundamental radiation into second harmonic intensity fluctuations is of special interest. Representing the fundamental radiation in the form

$$E_1(t) = \sum_{j=1}^{N} \tfrac{1}{2}A_{1j} \exp\{i(\omega_{1j}t+\psi_{1j})\} + \text{c.c.} \tag{206}$$

and supposing for simplicity that $A_{11} = A_{12} = \cdots = A_{1N} = \text{const}$ and that the N dimensional distribution law of the phases $\psi_{11}, ..., \psi_{1N}$ is

$$w_N(\psi_{11}, ..., \psi_{1N}) = w_1(\psi_{11}) \cdots w_1(\psi_{1N}), \qquad w_1(\psi_{1j}) = 1/2\pi$$

(the phase fluctuations in various modes can be assumed statistically independent), we can estimate the corresponding relative second harmonic intensity fluctuations

$$d_2 = [\overline{(I_2-\bar{I}_2)^2}]^{1/2}/\bar{I}_2$$

(see Ducuing and Bloembergen, 1964; Akhmanov and Khokhlov, 1964). It can be shown that for $N = 3$, $d_2 = 0.18$ and for $N = 4$, $d_2 = 0.25$, as $N \to \infty$, $d_2 \to N^{-1/2}$. Akhmanov *et al.* (1969) and Akhmanov and Chirkin (1971) discussed the photocount distributions for these cases. The first experimental check of these formulas was given by Ducuing and Bloembergen (1964). Akhmanov *et al.* (1969) measured the photocount distribution of the second harmonic excited with a multimode gas laser. They found that for $N > 3$, the

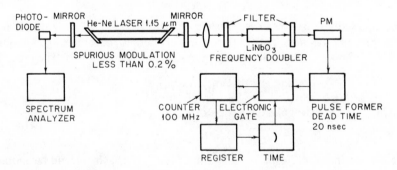

Fig. 49. Experimental setup for measuring second harmonic radiation statistics for a sampling time of 19 μsec with one sample in 2 sec (from Akhmanov *et al.*, 1969).

second harmonics photocount distribution differs markedly from Poisson statistics; d_2 was 0.12. This value differs from the theoretical value $d_2 = 0.25$; Akhmanov et al. attributed this discrepancy to the partial mode locking. The experimental technique is pictured in Fig. 49.

3. Transformation of Space Statistics of the Laser Field
   in Optical Frequency Multipliers

The theory of harmonic generation of spatially incoherent fields was presented in the book by Akhmanov and Chirkin (1971). The key parameter now is the coherence length

$$L_{a, coh} = a_c/\beta \tag{207}$$

where a_c is a correlation radius of fundamental radiation and β is the Poynting vector walf-off angle. The harmonic spatial statistics depend critically on the relation between L and L_{coh}. The situation is very similar to those discussed in Section III, E, 1 in connection with the temporal statistics (with reference to the space–time analogy mentioned above). Arutunyan et al. (1973) have developed an effective polarization interferometer and have made careful measurements of the space statistics of the nonlinear transformed light. In particular, they have found that the space statistics of the second harmonic at the output of a birefringent crystal is sufficiently nonuniform. The correlation radius in the direction of the Poynting vector walk-off grows with the distance $a_c \propto z$, but at the same time the correlation radius in the rectangular direction remains constant (these processes are, of course, simple spatial analogs of the processes discussed above in connection with the transformation of temporal statistics).

IV. OPTICAL MULTIPLIERS.
    DESIGN, CHARACTERISTICS, AND APPLICATIONS

A. Harmonic Generation in the Visible and the Near Infrared.
   Nonlinear Materials

The nonlinear media suitable for harmonic generation can be characterized by the region of transparency, the value of the nonlinearity, the possibility of phase matching (especially 90° phase matching), and damage thresholds for fundamental and harmonic radiations. In Fig. 50 the transmission range for the most commonly used nonlinear optical materials with quadratic nonlinearity $\chi^{(2)}$ are shown (the boundaries in Fig. 50 correspond to 10% transparency for a crystal sample several millimeters long).

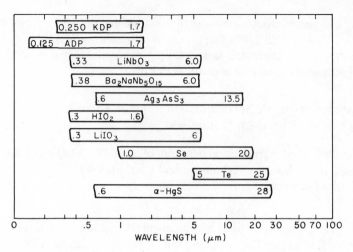

Fig. 50. Optical transmission range for primary nonlinear optical materials (from Kurtz, 1972).

Table II

Absolute and Relative Nonlinear Optical Coefficients for Primary Nonlinear Optical Materials [a]

Name	Formula	Symmetry	mks $(10^{-12}$ m/V$)$	Relative [b] $d_{i\mu,\,r}$ $= d_{i\mu}/d_{36}(KDP)$
KDP	KH_2PO_4	$\bar{4}2m$	$d_{36} = 0.63$	1
ADP	$NH_4H_2PO_4$	$\bar{4}2m$	$d_{36} = 0.76$	1.21
Lithium niobate	$LiNbO_3$	$3m$	$d_{31} = -6.3$	10
			$d_{22} = +3.6$	5.8
			$d_{33} = -47$	75
Barium sodium niobate	$Ba_2NaNb_5O_{15}$	$mm2$	$d_{31} = -20$	31
			$d_{32} = -20$	31
			$d_{33} = -26$	42
Proustite (synthetic)	Ag_3AsS_3	$3m$	$d_{31} = 12.6$	20.0
			$d_{22} = 13.4$	21.3
Iodic acid	HIO_3	222	$d_{14} = 6.6$	10.5
Lithium iodate	$LiIO_3$	6	$d_{31} = -6.6$	11.9
			$d_{33} = -7.8$	12.4
Selenium	Se	32	$d_{11} = 150$	240
Tellurium	Te	32	$d_{11} = 920(4300)$	1460
Cinnabar (synthetic)	HgS	32	$d_{11} = 57$	90

[a] From Kurtz (1972).
[b] $d_{36}(KDP) = (1.5 \pm 20\%) \times 10^{-9}$ esu; $\chi_{ijk} = 2d_{ijk}$.

In the visible and near infrared there are the crystals KDP, ADP, CDA, $LiNbO_3$, $Ba_2NaNb_5O_{15}$, $LiIO_3$, and HIO_3. All these crystals possess relatively high nonlinear susceptibilities, $\chi^{(2)} = 10^{-8}$–10^{-9} esu (precise values of $\chi^{(2)}$ for several nonlinear materials are listed in Table II). The large birefringence of these crystals provides phase matching in a wide spectral range (see Table III for the fundamental wavelength ranges in which phase-matched frequency doubling is possible, presented for several useful nonlinear

Table III

Fundamental Wavelength Range in Which Phase-Matched Doubling Is Possible for Several Nonlinear Materials at Room Temperature

Crystal	$\lambda_1^{(max)}$ μm	$\lambda_2^{(min)}$ μm
KDP	1.7	0.517
ADP	1.7	0.524
$LiNbO_3$	3.75	1.05
$Ba_2NaNb_5O_{15}$	6.0	1.02
HIO_3	1.4	0.602
$LiIO_3$	6.0	0.6

materials). Below, the interaction O_1O_1–E_2 in negative, and E_1E_1–O_2 in positive crystals (the fundamental waves have the same polarization) will be termed "type I" interactions; interactions in which polarizations of the fundamental waves are normal to one another will be termed "type II" interactions (O_1E_1–E_2 and O_1E_1–O_2). In Table IV, the phase-matching properties of several nonlinear materials for second harmonic generation are shown. One can see from these data that with these crystals, phase-matched second harmonic generation could be obtained practically from all useful lasers, operating in the visible and infrared ranges, such as, for example the He–Ne laser ($\lambda_1 = 3.39$; 1.15; 0.63 μm), the Nd laser ($\lambda_1 = 1.06$ μm), the GaAs semiconducter laser ($\lambda_1 = 0.8$ μm), the ruby laser ($\lambda_1 = 0.69$ μm), and the dye lasers ($\lambda_1 = 0.5$–0.9 μm).

The second harmonic of the argon–ion laser ($\lambda_1 = 0.514$ μm) can be obtained in slightly cooled KDP and ADP crystals (Dowley, 1968). The KDP and ADP crystals can also be used for the third and fourth cascade harmonic generation of the Nd laser. Therefore, coherent radiation up to the vacuum ultraviolet can be obtained. The direct phase-matched third harmonic generation (dependent on the cubic nonlinearity $\chi^{(3)}$) of ruby and Nd lasers can be obtained in the $CaCO_3$ crystal and also in KDP and ADP crystals. In the last case, however, cascade third harmonic, which is connected with a two-step processes in the quadratic nonlinearity, can compete with the direct

Table IV

Phase-Matching Properties of Primary SHG Materials[a] ·

Material	PM type	Fundamental wavelength (μm)	PM angle θ_m (deg)	Temp. (°C)	Comments
KDP	I	0.5174 (min)	90		calc
	I	0.6943	50.4 ± 1		meas
	I	1.058	40.3 ± 1		meas
	I	1.70 (max)	58.5		calc
	I	0.5145	90.0	-13.7	meas
	II	0.732 (min)	90.0		calc
	II	0.890	63.6		calc
	II	1.064	59.1		calc
	II	1.70 (max)	83.2		calc
ADP	I	0.5245 (min)	90		calc
	I	0.6943	51.9 ± 1		meas
	I	1.0582	41.9 ± 1		meas
	I	1.064	41.7		calc
	I	1.1523	42.4		
	I	1.70 (max)	59.7		calc
	I	0.5145	90.0	~ -10	meas
	II	0.750 (min)	90		calc
	II	0.890	65.9		calc
	II	1.064	61.6		calc
	II	1.599 (max)	90		calc
LiNbO$_3$	I	1.056 (min)	90	23	calc
	I	1.064	83.6	23	calc
	I	1.1523	67.6 ± 0.3	23	meas
	I	3.756 (max)	90	23	calc
	I	1.1523	90.0	210	meas
	II	1.685 (max)	90	23	calc
	II	2.420 (min)	90	23	calc
Hot "linobate"[b]	I	1.064	90.0	~ 160	meas
Ba$_2$NaNb$_5$O$_{15}$	I	1.024 (min)	0		calc
($n_z = n_b$, $n_r = n_c$	I	1.0642	15 ± 2		meas
for $c_0 < a_0 < b_0$)	I	6.0 (max)	75		calc
	I	1.0642	$0 (\phi = 0)$	101	meas
	I	1.0642	$0 (\phi = 90)$	89	meas
	II	1.434 (min)	0		calc
		6.0 (max)	75		calc
Ag$_3$AsS$_3$	I	1.156 (min)	90		calc
	I	10.590	20.5 ± 0.5		meas
	I	13.5 (max)	30.2		calc
	I	1.152	90.0	12	meas

Table IV (*cont.*)

Material	PM type	Fundamental wavelength (μm)	PM angle θ_m (deg)	Temp. (°C)	Comments
Ag_3AsS_3	II	1.493 (min)	90		calc
	II	10.6	30.9		calc
	II	13.0 (max)	40		calc
HIO_3	I	0.602 (min)	$0(\phi = 90)$		calc
	I	0.659	$0(\phi = 0)$		calc
$(n_z = n_b, n_x = n_a$	I	0.890	$52.5(\phi = 29.2)$		meas
for $a_0 < b_0 < c_0$)	I	1.064	$41.5(\phi = 0)$		meas
	I	1.064	$60.0(\phi = 90)$		meas
	I	1.4 (max)	$53.6(\phi = 0)$		calc
	II	0.837 (min)	$0(\phi = 90)$		calc
	II	0.914	$0(\phi = 0)$		calc
	II	1.064	$24(\phi = 0)$		meas
	II	1.064	$38(\phi = 90)$		meas
	II	1.4	$53.6(\phi = 90)$		calc
	II	1.4 (max)	$76.5(\phi = 90)$		calc
$LiIO_3$	I	0.600 (min)	62.8		calc
	I	0.6943	50.3		calc
	I	0.800	36.6		meas
	I	1.065	29.7		calc
	I	1.1523	27.2 ± 0.3		meas
	I	6.0 (max)	5.0		calc
	II	0.707 (min)	90		calc
	II	0.890	54.2		calc
	II	1.065	43.1		calc
	II	6.0 (max)	7.0		calc
Se	I	1.060	6.5 ± 0.5		meas
	I	10.600	~ 10		meas
Te	I	10.0 (min)	15.3		calc
	I	10.5915	14.8 ± 0.25		meas
	I	25.0 (max)	5.7		calc
	II	10.0 (min)	15.0		calc
	II	25.0 (max)	5.0		calc
HgS	I	1.131 (min)	90		calc
	I	1.1523	77.5		calc
	I	10.6	20.75 ± 0.75		meas
	I	18.2 (max)	90		calc
	II	1.464 (min)	90		calc
	II	10.6	30.0		calc
	II	16.5 (max)	90		calc

[a] From Kurtz (1972).

[b] Hot "linobate" is commercial terminology for stochiometric $LiNbO_3$ grown from a eutectic melt.

process (see also Section II, H). It should be mentioned that the very large birefringence of the $CaCO_3$ crystal makes possible the direct phase-matched fifth harmonic generation of the Nd laser, which is connected with the nonlinear polarization $\mathscr{P}^{NL} = \chi^{(5)} E^5$.

In practice, however, these higher order harmonics in crystals have been obtained with the help of cascade processes; the efficiencies of the direct processes are relatively small. The large list of nonlinear crystals suitable for applications in the visible and near infrared ranges makes it possible to realize different schemes of frequency multiplication. The choice of the scheme and the nonlinear crystal are mainly determined by the laser operating characteristics. Below, we describe separately the harmonic generation of Q-switched laser pulses, cw and quasi-cw laser radiation, and picosecond pulses, which are obtained from mode-locked lasers.

1. Q-Switched Lasers

The high intensities of giant laser pulses provides efficient generation of second, third, fourth, and even fifth harmonics. Harmonic generators of Ndglass and YAG : Nd lasers are the most widely used at the present time. Such systems provide for the construction of efficient and powerful coherent sources in the near infrared ($\lambda_1 = 1.06 \ \mu m$), visible ($\lambda_2 = 0.53 \ \mu m$), and ultraviolet regions ($\lambda_3 = 0.35 \ \mu m$, $\lambda_4 = 0.265 \ \mu m$, and $\lambda_5 = 0.212 \ \mu m$). Tunable radiation in the range 0.28–$0.35 \ \mu m$ can be obtained by second harmonic generation of dye lasers operated in the visible range.

As was shown in Section II, the power conversion efficiency of the doubler which is excited with a plane monochromatic wave depends only on one parameter, $g_\gamma = L/L_{NL}$ the relation of the length of nonlinear crystal L to the characteristic nonlinear length $L_{NL} = (n_1 n_2)^{1/2} \gamma_2^{-1} A_{01}^{-1}(0)$ [see Eq. (80c)]. Even for $L = 1$ cm, $g_\gamma \gtrsim 1$ for relatively low Q-switched laser intensities; there are no problems in obtaining $g_\gamma \gg 1$. For example, for $L = 1$ cm, $g_\gamma = 1$ if $I_1 = 350$ MW cm^{-2} for a KDP crystal; for a $LiNbO_3$ crystal the corresponding value of the fundamental intensity is $I_1 = 10$ MW cm^{-2}. Such intensities can be obtained not only for focused laser beams, but also for unfocused beams. It should be mentioned, however, that the achievement of conversion efficiency of the order of 100% in real laser beams is a problem; in practice, the energy conversion efficiency of an optical doubler excited with a powerful Q-switched laser often does not exceed $\eta_{W_2} \simeq 20$–30%. The main reasons for such a situation are the angular divergence of laser beam, its nonmonochromaticity and frequency instabilities, namely, aperture and transient effects which limit the conversion efficiency. Therefore, in designing efficient frequency multipliers, not only the effective nonlinearity of the crystal $\chi_{2,\,\text{eff}} = \chi_2/n_1^{3/2}$ should be taken into account,* but also such important parameters as angular width of phase matching (see Section II)

$$\theta_{\text{coh}} = \frac{\lambda_1}{4L} \left| \frac{\partial n_1}{\partial \theta} - \frac{\partial n_2}{\partial \theta} \right|^{-1}$$

and the frequency width of phase matching (see Section III)

$$\Delta\lambda_{\text{m}} = \frac{\Omega_{\text{coh}}\lambda_1^{\,2}}{2\pi c} = \frac{\lambda_1}{4L} \left| \frac{\partial n_1}{\partial \lambda_1} - \frac{1}{2}\frac{\partial n_2}{\partial \lambda_2} \right|^{-1}$$

Along with the parameter $\Delta\lambda_{\text{m}}$ is the very important derivative

$$\frac{\partial\theta_{\text{m}}}{\partial\lambda_1} = -\left(\frac{\partial n_1}{\partial \lambda_1} - \frac{1}{2}\frac{\partial n_2}{\partial \lambda_2}\right)\Big/\left(\frac{\partial n_1}{\partial \theta} - \frac{\partial n_2}{\partial \theta}\right)$$

which determines the sensitivity of the phase matching direction to the fundamental wavelength (see Fig. 51). The derivative $\partial\theta_{\text{m}}/\partial\lambda_1$ is of special interest. if the SH intensity fluctuations due to laser instabilities are considered.

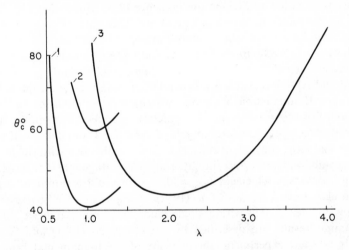

Fig. 51. Frequency dependence of the SHG phase-matching angles for a KDP crystal (curve 1 shows O_1O_1–E_2 interaction; curve 2 shows O_1E_1–E_2 interaction) and for a LiNbO$_3$ crystal (curve 3 shows O_1O_1–E_2 interaction). In the graphs, the phase-matching angle is presented as a function of the fundamental wavelength.

* The effective nonlinearity of the crystal $\chi_{2,\,\text{eff}}$ determines the characteristic nonlinear length L_{NL}. Thus for a plane monochromatic wave the parameter $\chi_2/n_1^{3/2}$ may be considered as a figure of merit of the nonlinear crystal from the point of view of frequency multiplication. Harris (1969) introduced as a figure of merit the ratio $\chi_2^{\,2}/n_1^{\,3}$ which determines the SHG efficiency. It should be mentioned that for multipliers excited with real laser beams another parameter should be taken into account.

Table V

Characteristic Parameters Which Determine the Efficiency of Frequency Doublers Excited with Real Laser Beams[a]

Crystal	θ_m (deg)	$L \times \theta_{coh}$ (cm · mrad)	$L\Delta\lambda_m$ (cm · Å)	$\partial\theta_m/\partial\lambda_1$ (mrad/Å)
KDP (type I interaction)	41.5	0.95	130	$5 \cdot 10^{-3}$
KDP (type II interaction)	59.5	1.85	31	$3 \cdot 10^{-2}$
CDA	83	24	55	$5 \cdot 10^{-2}$
$LiNbO_3$	83.6	24	1.0	1.05
$LiIO_3$	30	0.35	3.4	$9 \cdot 10^{-2}$

[a] The fundamental wavelength $\lambda_1 = 1.06\,\mu m$.

The value θ_{coh} has a maximum when $\theta_m = 90°$. Using 90° phase matching is a very effective method to exclude the influence of aperture effects (both diaphragm and angular) on harmonic generation. Therefore, the crystals having 90° phase matching are of special interest in applied nonlinear optics. For Nd lasers these are $LiNbO_3$, $Ba_2NaNb_5O_{15}$, and CDA. It should be mentioned, however, that the values $\Delta\lambda_m$ for these crystals are significantly less than that for KDP and ADP (see Table V). For KDP and ADP crystals, the value of θ_{coh} for type II interaction is almost two times larger than for the type I interaction. For this reason type II interaction is preferable when the fundamental radiation possesses a large angular divergence. From data listed above, it follows that large conversion efficiency could be obtained with single-mode frequency stabilized laser. Accordingly, Hagen and Magnante (1969) obtained an energy conversion efficiency $\eta_{W_2} = 51\%$, with a diffraction-limited beam from Nd-glass laser. Zhdanov et al. (1972) have achieved 60% energy in the frequency doubler excited with a single-mode frequency-stabilized Nd-glass laser. Similar results are described by Eremin et al. (1971) and Andreev (1972). In all these experiments the intensity of the fundamental beam was very high and the second harmonic was generated in a collimated beam.

It should be noted that the power conversion η_{P_2} and especially the intensity conversion in center of the beam η_{I_2} are generally larger than the energy conversion η_{W_2}. This is a consequence of a relatively small conversion in the "wings" of the real laser beam and the pulse (see Section II,B). Therefore, as an example, in the Zhdanov et al. (1972) experiments, the intensity conversion at the center of the beam $\eta_{I_2} = I_2(0)/I_1(0)$ exceeds 75%. Some experimental data on harmonic generation of Nd lasers are summarized in a Table VI. Together with the data obtained with Nd-glass lasers, characteristics of

Table VI

Characteristics of Optical Multipliers Excited by Nd-Glass and Nd:YAG Lasers

Laser	Regime of operation	Fundamental radiation $\lambda_1 = 1.06\ \mu m$ Energy (W_1) Power (P_1)	Harmonics Harmonic number N	$\lambda_N\ \mu m$	W_{N,P_N}^{max}	Efficiency η	Material	Reference
Nd-glass	Q-switched	$W_1 = 30$ J	2	0.53	$W_2 = 15$ J	$\eta_{P_2} = 70\%$	KDP	Hagen and Magnante (1969)
	Q-switched	$W_1 = 3$ J	2	0.53	$W_2 = 1.8$ J		KDP	Zhdanov et al. (1972)
	Q-switched	$W_1 = 3$ J	3^a	0.35	$W_3 = 0.8$ J		KDP	Zhdanov et al. (1972)
	Q-switched	$W_1 = 3$ J	4^a	0.26	$W_4 = 0.2$ J		KDP, ADP	Zhdanov et al. (1972)
	Q-switched	$W_1 = 1,2$ J	5^a	0.212	$W_5 = 10^{-4}$ J		KDP, ADP	Akhmanov et al. (1969)
Nd:YAG	Mode-locked	$W_1 = 10^{-2}$ J (train)	2	0.53	$W_2 = 8 \cdot 10^{-3}$ J	$\eta_{w_2} = 80\%$	KDP	Kung et al. (1972)
	Mode-locked	$W_1 = 10^{-2}$ J (train)	3^a	0.35	$W_3 = 10^{-3}$ J (train)	$\eta_{w_3} = 10\%$	KDP	Kung et al. (1972)
	cw	$P_1 = 1.1$ W	2^b	0.53	$P_2 = 1.1$ W	$\eta_{P_2} = 100\%$	$Ba_2NaNb_5O_{15}$	Geusic et al. (1968)
	cw	$P_1 = 0.9$ W	2^b	0.53	$P_2 = 0.14$ W	$\eta_{P_2} = 16\%$	$LiNbO_3$	Dmitriev et al. (1972)
	Q-switched high repetition rate	—	2	0.53	$P_2 = 10$ W (average)	—	$LiNbO_3$	Yarborouyh and Ammann (1973)
	Mode-locked	—	5^a	0.212	$P_5^{max} = 10^5$ W	—	$CaCO_3$	Tunkin et al. (1972)

^{a} Cascade multiplier.
^{b} Intracavity doubler.

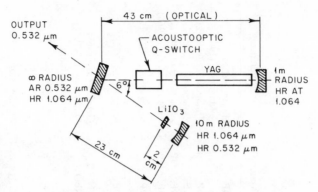

Fig. 52. Schematic diagram of the intracavity second harmonic generator of a Nd : YAG laser. In this scheme of intracavity harmonic generation, absorption of the harmonic power in the laser rod is fully excluded (from Chesler *et al.*, 1970).

the multipliers, excited with Nd : YAG lasers (see Fig. 52) are also presented. Taking into account the fact that these lasers can operate in the regime with a high repetition rate, data on average harmonic powers available are also presented in Table VI.

An important characteristic of nonlinear materials which are useful in powerful optical multipliers is the optical damage threshold I_{thr}. We list in the accompanying tabulation values of I_{thr} for the useful nonlinear materials; the data correspond to optical damage in the field of a single nanosecond laser pulse. It is interesting to note that in several cases these threshold intensities decrease significantly when efficient second harmonic generation takes place in the crystal; this can be explained by self-focusing in the biharmonic light field. The doublers of the giant dye-laser pulses described by Bradley *et al.* (1971) and Bokuth *et al.* (1972), KDP and ADP crystals, were used the second harmonic tuned in the region 0.38–0.28 μm with the power 0.5–1 MW were achieved. The energy conversion efficiency η_{w_2} was 10%. Such tunable ultraviolet lasers are very interesting for nonlinear spectroscopy.

The use of the higher-order nonlinearities (particular cubic) for harmonic generation from Q-switched lasers have been found not to be effective. Although the phase-matching conditions for third harmonic generation of

Material	I_{thr} W cm^{-2}
KDP, ADP	5–$6 \cdot 10^{8}$
LiNbO$_3$	5–$10 \cdot 10^{7}$
LiIO$_3$	10^{8}
HIO$_3$	10^{8}
Ba$_2$NaNbO$_{15}$	10^{7}–10^{8}

Nd and ruby lasers are fulfilled in calcite and in KDP and ADP (Terhune *et al.*, 1962; Sukhorukov and Tomov, 1970a), the cascade harmonic generators are more efficient.

2. THE CW LASERS

Optimum focusing inside the laser resonator provides effective conversion (up to 100%) to the second harmonic of the radiation of cw lasers with powers 0.1–10 W (argon, YAG:Nd^{3+} and dye cw lasers). The 90° phase matching is very important for these devices. The specific effects making cw harmonic generation difficult are optical damage and thermal self-action of the fundamental and harmonic beams. Phase-matched harmonic generation from a cw He–Ne laser was first demonstrated by Ashkin (Ashkin *et al.*, 1963); Geusic *et al.* in 1968 obtained effective second harmonic generation of the cw YAG:Nd^{3+} laser radiation in the crystal $Ba_2Nb_5O_{15}$. In their experiment intracavity harmonic generation was used. Optical damage did not occur in this crystal and 100% conversion efficiency of the laser power to the second harmonic was achieved.

The second harmonic of cw radiation of the argon–ion laser ($\lambda_2 = 0.257$ μm) with power 415 mW and efficiency 50% was obtained by Dowley (1968) in KDP. He also used intracavity harmonic generation. In spite of a small linear absorption of the fundamental and second harmonic waves ($\delta_1 = 10^{-3}$ cm and $\delta_2 = 10^{-2}$ cm^{-1}), Dowley was able to observe strong thermal self-actions in his doubler. Gabel and Hercher (1972) have used the lithium formate crystal to obtain the second harmonic of the cw dye laser. In the intracavity scheme, they obtained a cw radiation, tunable in the range 0.26–0.325 μm with a power of several milliwatt. The power of the fundamental radiation from tunable dye laser was of the order of 250 mW. An intracavity doubler of the cw Nd:YAG laser on the $LiNbO_3$ crystal was described by Dmitriev *et al.* (1972). There are problems in using $LiNbO_3$ crystals for the effective harmonic generation of cw lasers, because typically in such crystals it is impossible to obtain simultaneously 90° phase matching and to exclude reversible optical damage. Dmitriev *et al.* have prepared the $LiNbO_3$ crystals with special stoichiometry to avoid these difficulties; in an intracavity doubler they obtained energy conversion efficiency of the order of 16% with the cw fundamental power $P_1 = 0.9$ W. The problems connected with the effective doubling of the radiation of quasi-cw radiation (free running lasers, millisecond pulses with high repetition rate) are very similar to those considered above. A recent study by Golyaev *et al.* (1974) has shown that self-induced thermal effects may be very important in these devices. Volosov *et al.* (1972) investigated the second harmonic generation of the Nd-glass laser operated in the free running regime with a large pulse energy (several hundred joules per pulse); the threshold for breakdown of the KDP crystal in the field of such a pulse was 10^3 J cm^{-2}. The energy conversion to the second harmonic was very small, less than 1%.

3. MODE-LOCKED LASERS

The added problem arising from the harmonic generation of picosecond pulses concerns the group velocity mismatch of the fundamental and harmonic waves. Now for efficient harmonic generation it is necessary to have $L_{NL} < L < L_\tau = \tau/v$ [see formulas (162)]. The data on group velocity mismatch for the usual nonlinear crystals are listed in Table VII. The effect of group velocity mismatch is strongly pronounced in LiNbO$_3$ and LiIO$_3$ for second harmonic generation of picosecond Nd laser pulses, and is especially strong for ultraviolet frequency multipliers.

Table VII

Characteristic Group Delay Length in Second Harmonic Generation of a 1 psec Pulse in Several Nonlinear Materials[a]

The fundamental radiation parameters	L_τ cm			
	KDP	LiNbO$_3$	LiIO$_3$	CDA
$\lambda_1 = 1.06\,\mu$m $\tau = 10^{-12}$ sec	3.7	0.2	0.5	1.4
$\lambda_1 = 0.53\,\mu$m $\tau = 10^{-12}$ sec	0.3	—	—	—

[a] The quasi-static length $L_\tau = \tau |u_2^{-1} - u_1^{-1}|$.

It should be noted that the threshold intensities of optical breakdown in crystals increase for the short pulses as the inverse laser pulse duration $I_{thr} \simeq \tau^{-1}$ as has been shown by Fradin et al. (1973) and usually equals 10^{10}–10^{11} W cm^{-2} for picosecond pulses. The value of the characteristic nonlinear length L_{NL} for such high intensities is less than 0.1 cm for the usual nonlinear materials and consequently the condition $L_{NL} < L$ can be easily fulfilled for second and even for fourth harmonic of Nd lasers. Akhmanov et al. (1972) demonstrated the efficient cascade harmonic generation of the picosecond Nd-glass laser. The schematic diagram of their experimental arrangement is shown in Fig. 53; corresponding experimental data are summarized in Table VIII. Akhmanov et al. were able to change the pulse duration of the fourth harmonic from 0.5 to 5 psec by changing the KDP crystal length in the second doubler.

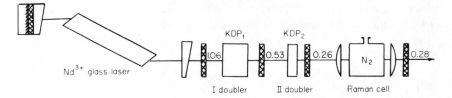

Fig. 53. Schematic diagram of the cascade harmonic generator of picosecond pulses. The fundamental radiation (from a Nd-glass laser) has the pulse duration $\tau = 3 \cdot 10^{-12}$ sec. First doubler is a KDP crystal ($L = 4$ cm; $L_\tau = 12$ cm). Second doubler is also a KDP crystal ($L = 0.2$ cm; $L_\tau = 1$ cm). In liquid nitrogen, stimulated Raman scattering is effectively excited by ultraviolet picosecond pulses with $\lambda_4 = 0.26$ μm. This leads to the discrete tuning of picosecond pulses in ultraviolet range (from Akhmanov *et al.*, 1972).

Along with KDP and ADP crystals, the crystal CDA is very useful for the efficient second harmonic generation of picosecond pulses obtained from Nd lasers operating in the regime with many transverse modes (Akhmanov *et al.*, 1972; Rabson *et al.*, 1972; Orlov *et al.*, 1973). Usually the energy conversion of the doublers excited with trains of picosecond pulses are relatively small, no more than 20–30% (see, for example, Akhmanov *et al.*, 1972, and Rabson *et al.*, 1972). This is probably due to imperfect mode locking. At the same time Kung *et al.* (1972) obtained very high conversion efficiency of the doubler and cascade tripler ($\eta_{W_2} = 80\%$; $\eta_{W_3} = 10\%$), using bandwidth limited picosecond pulses from a Nd:YAG laser ($\tau = 50$ psec; $P_1 = 20$ MW).

The large breakdown threshold of nonlinear materials in the field of picosecond pulses makes it possible to use the higher-order nonlinearities

Table VIII

Experimental Characteristics of the Cascade Frequency Multiplier Excited with Picosecond Pulses from a Nd-Glass Laser[a]

Wavelength	Energy of the train (J)	Spectral width (cm^{-1})	Peak intensity (W cm^{-2}) (unfocused beam)
$\lambda_1 = 1.06$ μm (fundamental)	$5 \cdot 10^{-2}$	8	$4 \cdot 10^{10}$
$\lambda_2 = 0.53$ μm (second harmonic)	$1 \cdot 10^{-2}$	25	$1.7 \cdot 10^{10}$
$\lambda_4 = 0.26$ μm (fourth harmonic)[b]	$1 \cdot 10^{-3}$	50	$5 \cdot 10^9$

[a] From Akhmanov *et al.* (1972).

[b] At the fourth harmonic bandwidth limited pulses with the duration $\tau_4 \simeq 6 \cdot 10^{-13}$ sec were obtained.

for practical higher-order harmonic generation. As an example, Tunkin *et al.*
(1972) obtained the fifth harmonic ($\lambda_5 = 0.21$ μm) of the Nd-glass laser
using two successive four-photon interactions based on the cubic nonlinearity
of the $CaCO_3$ crystal. The fundamental picosecond pulses excite the third
harmonic in the first $CaCO_3$ crystal and then in another $CaCO_3$ crystal the
fifth harmonic is the result of the mixing of third harmonic and the fundamental. The power of the fifth harmonic pulses obtained was ~ 100 kW.

At present, harmonic generation using cw mode-locked lasers (Nd : YAG
laser, dye lasers) is of practical interest. Because the peak power of such a
laser is relatively small, intracavity harmonic generators are widely used.
Several authors have used the same nonlinear crystal simultaneously as a
modulator and a doubler in such devices (Gurski, 1969; Hitz and Osternik,
1971; Rice and Burkhardt, 1971). Gurski has used mode locking of a cw
laser for the enhancement of the efficiency of the intracavity frequency
doubler, which was excited with the Nd : YAG laser; his experimental arrangement is shown in Fig. 54. Rice and Burkhardt have used a similar scheme in
the doubling of the frequency of the $YAlO_3$: Nd laser with a $Ba_2NaNb_5O_{15}$
crystal. They obtained an energy conversion of $\eta_{W_2} \simeq 50\%$; average harmonic
power was about $\langle P_2 \rangle = 300$ mW. It should be emphasized that for
$YAG : Nd^{3+}$, the crystal $YAlO_3 : Nd^{3+}$ is transparent for the second harmonic
radiation ($\lambda_2 = 0.54$ μm) and hence, in a simplest scheme of harmonic
generation inside the laser resonator, it was possible to obtain effective
harmonic output in both directions.

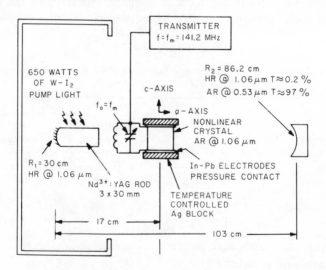

Fig. 54. An experimental laser in which simultaneous mode locking and second harmonic
generation were achieved using the same nonlinear crystal (from Gurski, 1969).

B. Harmonic Generation in the Infrared

In this spectral range the practical interests are centered on the harmonic generation of the CO_2 laser ($\lambda_1 = 10.6 \, \mu m$), the CO laser ($\lambda_1 = 5 \, \mu m$), the He–Ne laser ($\lambda_1 = 3.39 \, \mu m$), the disprosium laser ($\lambda_1 = 2.36 \, \mu m$), and tunable spin-flip lasers with CO_2 and CO laser pumps. The phase-matchable second harmonic was first generated in the far infrared range in the crystals Te, Se, Ag_3AsS_3 and HgS; in the near infrared, phase-matchable harmonic generation could be obtained in $LiNbO_3$, $LiIO_3$, and $Ba_2NaNb_5O_{15}$ (see Fig. 50). As was mentioned above (see Table II), the Te crystal has the largest nonlinear susceptibility ($\chi_{Te}^{(2)} = 1460\chi_{KDP}^{(2)}$). Unfortunately, it could not be used for the design of effective optical multipliers in the far infrared. This is due to the strong absorption of the fundamental and harmonic radiation by light-induced free carriers.

Gandrud and Abrams (1970) and Taynai et al. (1971) demonstrated that the efficiency of the Te crystal doubler excited with the Q-switched CO_2 laser could not exceed several percent due to nonlinear absorption. Taynai et al. have also estimated the optical breakdown threshold of the Te crystal in the field of Q-switched CO_2 lasers. They found $I_{thr} \simeq 2\text{–}4\cdot10^6$ W cm^{-2}. Several authors (Ernest and Witteman, 1972; Henningsen et al., 1971) have used synthetic prousite for frequency doubling of the CO_2 laser; the conversion efficiencies are also small.

Recently, it was discovered that the ternary semiconductors of crystal symmetry $42m$ (the crystals of chalopyrite type) are very effective nonlinear materials for infrared devices (see Boyd. 1972). Such crystals as $ZnGeP_2$, $AgGaS_2$, $CdGeP_2$, $CdGeAs_2$, and others have sufficiently large birefringence for phase matching and nonlinear susceptibility on the order of 10^{-8} esu. Kildal and Mikkelsen (1973) have used the ternary semiconductor $GdGeAs_2$ for intracavity doubling of the CO_2 laser radiation. They obtained an energy conversion efficiency of 15%. The optical breakdown threshold $I_{thr} = 40$ MW cm^{-2} for this crystal was measured at the wavelength $\lambda_1 = 10.6 \, \mu m$.

As was mentioned in Section I inhomogeneous and periodic nonlinear media can be used for phase-matched nonlinear interactions in the infrared. The experimental studies carried out to date, however, have only demonstrated these possibilities (see Section I, C); due to the rapid progress of integrated optics, this is where they will find the real, practical applications in the near future.

C. Harmonic Generation in the Vacuum Ultraviolet

The first successful experiments on harmonic generation in the vacuum ultraviolet range were only carried out in 1972 (Kung et al.). Because the

Table IX

Calculated Parameters for Third Harmonic Generation in Metal Vapors[a]

Parameter	$10{,}640 \to 3547$ Å	$6943 \to 2314$ Å
Metal vapor	Rb	Na
Buffer gas	Xe	Xe
Metal atoms (cm^3)	$2.25 \cdot 10^{17}$	$7.73 \cdot 10^{16}$
$\chi^{(3)}$ per atom (esu)	$7.42 \cdot 10^{-32}$	$5.86 \cdot 10^{-33}$
L_{coh} in vapor (cm)	$9.62 \cdot 10^{-2}$	$2.6 \cdot 10^{-1}$
Phase matching		
Buffer gas concentration	412:1	100:1
$2\delta_1$ (cm^{-1})	$4.17 \cdot 10^{-3}$	$2.13 \cdot 10^{-4}$
$2\delta_3$ (cm^{-1})	$5.7 \cdot 10^{-3}$	$5.7 \cdot 10^{-3}$
I_1 (W cm^{-2})		
For 50% conversion		
with $L = 50$ cm	$7.27 \cdot 10^8$	$1.75 \cdot 10^{10}$
I_{sat} (W cm^{-2})	$1.68 \cdot 10^8$	$3.8 \cdot 10^9$
W_{sat} (J cm^{-2})	5	51.8

[a] From Harris and Miles (1971).

crystals which admit phase-matched second harmonics generation are transparent only to about 2000 Å in vacuum UV range only nonlinear media with the center of inversion can be used.[*] Harris et al. (1971, 1972, 1973) demonstrated the advantages of using inert gases and vapors of alkali metals as a nonlinear medium for phase-matched harmonic generation in the vacuum ultraviolet range. In 1971, Harris and Miles proposed phase-matched third harmonic generation in mixtures of alkali metal vapors and inert gases. Their calculations show that the combination of near resonant nonlinear susceptibilities, the ability to phase match, and the relatively high ultraviolet transparency of vapors should allow high conversion efficiency for picosecond pulses with a peak power of 10^8–10^9 W. Their numerical results are presented in Table IX where the calculated parameters for phase-matched third harmonic generation in Rb and Na vapors are summarized; in both cases, Xe is considered as a buffer gas for a phase matching (see Section I).

The very important characteristics of such resonant nonlinear media as gases and vapors are the saturation power density I_{sat} and saturation energy density w_{sat} for the given operating conditions. The corresponding data are also listed in Table IX. If $I_1 \simeq I_{sat}$ for relatively long pulses, or $w_1 \simeq w_{sat}$, for short pulses, the nonlinearity disappears. For picosecond pulses, only w_{sat} should be taken into account; thus it is evident that the saturation limit is very high.

[*] The exception is the crystal $BeSO_4$, which is transparent to 1600 Å.

As a experimental test of these ideas, Young *et al.* (1971) and Kung *et al.* (1972) demonstrated phase matched third harmonic generation in the Rb–Xe mixture and in the Cd–Ar mixture. In the latter paper, the generation of 1773, 1520, and 1182 Å radiation by frequency tripling and summing was achieved. For the third harmonic process $5390 \rightarrow 1773$ Å phase-matching occurs for Cd and Ar atoms in the ratio 1:25, and the measured cubic nonlinearity per atom was $\chi^{(3)}_{at} = 2 \cdot 10^{-34}$ esu. The energy conversion efficiency to 1773 Å was about 10^{-4}, yielding a peak power of 7 kW in the vacuum ultraviolet range. Kung *et al.* (1973) reported the generation of 1182 Å radiation in a phase-matched mixture of Xe and Ar, i.e., in a mixture where no metal vapor is employed. The large nonlinearity of Xe in the vacuum ultraviolet and the homogeneous mixing of inert gases allows a conversion efficiency of 2.8% at an input power of about 10 MW. These experiments are, of course, only demonstrations of the usefulness of the technique of optical frequency multiplication in design of coherent sources in the vacuum ultraviolet. It is clear now (see also Harris, 1973) that by using the inert gases and vapors, it is possible in principle to obtain coherent radiation with wavelengths down to 100–200 Å; it is very interesting that for such short wavelength harmonic generation not only cascade processes but also direct processes employing the higher-order nonlinearities of single atoms can be used.

Very important developments in the UV frequency multipliers and converters are connected with utilizing the large resonance enhancement in the nonlinear susceptibility $\chi^{(3)}$ that exists when the frequency of the input radiation is tuned to a nonallowed double-quantum transition. This idea was first proposed by Maker *et al.* (1964) and Afanasiev and Manykin (1965). Hodgson *et al.* (1974) used the double-quantum transition in Sr vapor to generate third-harmonic and tunable UV radiation. As a result of the resonant enhancement, the power density of the pumping laser is restricted to a value that is several orders of magnitude lower than would be the case if this resonance had not been employed. So nanosecond pulses from tunable lasers can be effectively used as pump sources.

D. Applications. Concluding Remarks

Optical multipliers are widely used in modern laser physics. For example, the YAG : Nd laser and the frequency doubler (using the $LiNbO_3$, $LiIO_3$, or $BaNaNbO_{15}$ crystals) is a unique powerful source of coherent optical radiation in the green range of the visible spectra. It was mentioned above that the doubling of tunable dye lasers provides the construction of effective tunable sources in the ultraviolet range. Many physical experiments have been conducted with optical multipliers as the sources of powerful coherent radiation. For example, many authors used the multifrequency source constructed on the basis of a single-mode Nd-glass laser and its second, third, and fourth harmonic to measure the dispersion of the Raman and Brillouin gain in liquids and solids. They also obtained the frequency dependence of optical breakdown in solids throughout the visible range; these data are very useful in estimating the physical mechanism of optical breakdown. It should be mentioned that along with these obvious applications, several other applications of frequency multipliers exist, applications in which the harmonic generation process is used for spectroscopic purposes. Let us consider again Fig. 51. It shows that if the harmonic frequency falls near a crystal band gap, the dispersion of the phase-matching angle is very strong. Akmanov et al. (1968) and Volosov (1968) used this circumstance for the frequency discrimination of laser radiation. They showed that the spectral resolution that one may obtain in this way is an order of magnitude greater than the resolution of a good grating spectrograph. Frequency multipliers are widely used also in correlation spectroscopy. Armstrong (1967) demonstrated the usefulness of the frequency doubler in measuring the temporal intensity correlation function

$$G_2(\tau_d) = \langle I_1(t) I_1(t - \tau_d) \rangle$$

In this experiment the $O_1 E_1$–E_2 interaction in a GaAs crystal was used; the dependence of the second harmonic power from the time delay τ_d between the ordinary and extraordinary fundamental beams gives directly the temporal intensity correlation function $G_2(\tau_d)$. Arutunyan et al. (1973) used the optical doubler in measurements of spatial intensity correlations of laser fields. This technique is now widely used in examining the multimode nature (temporal or spatial) of laser fields; using higher-order harmonics, higher-order correlations can be measured. These measurements are of special interest for the problem of picosecond pulse formation. In spite of the fact that the first experiment on optical harmonic generation was done in 1961, new ideas are still being contributed in this field. Very interesting physical problems and applications are connected with the higher-order nonlinearities and harmonic generation in the far ultraviolet and soft x-ray range. Many important problems still remain in the theory, for example, the nonlinear theory of parametric processes in Gaussian beams [some very recent results in this field are contained in the Karamzin and Sukhomnov (1974) paper] and the theory of harmonic generation in integrated optics devices.

REFERENCES

Afanasiev, A. M., Manykin, E. A. (1965). *Sov. Phys. JETP* **Y8**, Y83.
Akhmanov, S. A., and Chirkin, A. S. (1966). *Radiotekh. Electron.* **11**, 1915.
Akhmanov, S. A., and Chirkin, A. S. (1971). "Statistical Effects in Nonlinear Optics", Izd. Moscow State Univ., Moscow. English Translation (1973). Plenum, New York.
Akhmanov, S. A., and Khokhlov, R. V. (1964). "Problems of Nonlinear Optics." Acad. of Science, Moscow. English Translation (1972). Gordon and Breach, New York.
Akhmanov, S. A., Kovrygin, A. I., Khokhlov, R. V., and Chunaev, O. N. (1963). *Sov. Phys. JETP* **45**, 1336.
Akhmanov, S. A., Kovrygin, A. I., Piskarskas, A. S., and Khokhlov, R. V. (1965a). *Sov. Phys. JETP Lett.* **2**, 141.
Akhmanov, S. A., Kovrygin, A. I., and Kulakova, N. K. (1965b). *Sov. Phys. JETP* **48**, 1445.
Akhmanov, S. A., Dmitriev, V. G., and Modenov, V. P. (1965c). *Radiotekh. Elektron.* **10**, 649.
Akhmanov, S. A., Dmitriev, V. G., Kovrygin, A. I., and Khokhlov, R. V. (1965d). *In* "Physics of Quantum Electronics" (P. L. Kelley, B. Lax, and P. E. Tannewald, ed.). McGraw-Hill, New York.
Akhmanov, S. A., Kovrygin, A. I., Strukov, M. M., and Khokhlov, R. V. (1965e). *Sov. Phys. JETP Lett.* **1**, 42.
Akhmanov, S. A., Kovrygin, A. I., Kulakova, N. K., and Khokhlov, R. V. (1965f). *Sov. Phys. JETP* **48**, 1202.
Akhmanov, S. A., Sukhorukov, A. P., and Khokhlov, R. V. (1966a). *Sov. Phys. JETP* **50**, 474.
Akhmanov, S. A., Kovrygin, A. I., Chirkin, A. S., and Chunaev, O. N. (1966c). *Sov. Phys. JETP* **50**, 829.
Akhmanov, S. A., Sukhorukov, A. P., and Khokhlov, R. V. (1967a). *Sov. Phys. Usp.* **93**, 19. English translation **10**, 809.
Akhmanov, S. A., Sukhorukov, A. P., and Chirkin, A. S. (1967b). Izv. Vyssh. *Ucheb. Zaved. Radiofiz.* **10**, 1639.
Akhmanov, S. A., Chirkin, A. S., Drabovich, K. N., Kovrygin, A. I., Sukhorukov, A. P., and Khokhlov, R. V. (1968a). *IEEE J. Quant. Electron* **4**, 591.
Akhmanov, S. A., Sukhorukov, A. P., and Chirkin, A. S. (1968b). *Sov. Phys. JETP* **55**, 1480.
Akhmanov, S. A., Chirkin, A. S., and Tunkin, V. G. (1969). *Opto-electronics* **1**, 196.
Akhmanov, S. A., Skidan, I. B., Orlov, R. Yu., and Telegin, L. I. (1972). *Sov. Phys. JETP Lett.* **16**, 471.
Akhmanov, S. A., Dubovik, A. N., Saltiel, S. M., Tomov, N. B., and Tunkin, V. G. (1974). *Sov. Phys. JETP Lett.* **20**, 264.
Akmanov, A. G., Kovrygin, A. I., and Sukhorukov, A. P. (1967). Paper presented at *Sov. Conf. Nonlinear Opt. Erevan, 1967* **3**, 142. Moscow State Univ., Moscow.
Akmanov, A. G., Akhmanov, S. A., Kovrygin, A. I., Khokhlov, R. V., Piskarskas, A. S., and Sukhorukov, A. P. (1968). *IEEE J. Quant. Electron.* **4**, 828.
Akmanov, A. G., Akhmanov, S. A., Kovrygin, A. I., Zdanov, B. V., and Khokhlov, (1969). *Sov. Phys. JETP Lett.* **10**, 244.
Anderson, D. B., and Boyd, G. D. (1970). *Appl. Phys. Lett.* **17**, 388.
Andreev, R. B., Volosov, V. D., and Kalinsev, A. G. (1972). *Kvant. Electron.* **6**, 44.
Armstrong, J. A., (1967). *Appl. Phys. Lett.* **10**, 16.
Armstrong, J. A., Bloembergen, N., Ducuing, J., and Pershan, P. (1962). *Phys. Rev.* **127**, 1918.

Arutunyan, A. G., Akhmanov, S. A., Golayev Yu. D., Tunkin, V. G., and Chirkin, A. S. (1973). *Sov. Phys. JETP* **64**, 1511.

Ashkin, A., Boyd, G. D., and Dziedzic, J. M. (1963). *Phys. Rev. Lett.* **11**, 14.

Ashkin, A., Boyd, G. D., and Dziedzic, J. M. (1966). *J. Quant. Electron.* **2**, 109.

Bey, P. P., Giuliani, J. F., and Rabin, H. (1967). *Phys. Rev. Lett.* **19**, 819.

Bey, P. P., Giuliani, J. F., and Rabin, H. (1968). *IEEE J. Quant. Electron.* **4**, 932.

Bey, P. P., Giuliani, J. F., and Rabin, H. (1971). *IEEE J. Quant Electron.* **7**, 86.

Bjorkholm, J. E. (1966). *Phys. Rev.* **142**, 126.

Bjorkholm, J. E., and Siegman, A. E. (1967). *Phys. Rev. Lett.* **19**, 835.

Bloembergen, N. (1965). "Nonlinear Optics." Benjamin, New York.

Bloembergen, N, (1966). *Opt. Acta* **13**, 311.

Bloembergen, N., and Pershan, P. S. (1962). *Phys. Rev.* **128**, 606.

Bloembergen, N., Simon, H. J., and Lee, C. H. (1969). *Phys. Rev.* **181**, 261.

Bokuth, B. V., and Khatkevich, A. G. (1964). *Zh. Prikl. Spectrosk.* **1**, 971.

Bokuth, B. V., Kasak, N. S., Mashenko, A. G., Mostovikov, V. A., and Rubinov, A. N. (1972). *Sov. Phys. JETP Lett.* **15**, 26.

Boyd, G. D. (1972). Paper presented to *Intern. Quant. Electron. Conf. Montreal May 1972,* 7.

Boyd, G. D., and Kleinman, D. A. (1968). *J. Appl. Phys.* **39**, 3597.

Boyd, G. D., and Patel, C. K. N. (1966). *Appl. Phys. Lett.* **8**, 313.

Boyd, G. D., Ashkin, A., Dzidzic, J. M., and Kleinman, D. A. (1965). *Phys. Rev.* **137**, 1305.

Boyd, G. D., Nash, F. R., and Nelson, D. F. (1970). *Phys. Rev. Lett.* **24**, 1298.

Bradley, D. J., Nicholas, J. V., and Shaw, J. R. (1971). *Appl. Phys. Lett.* **19**, 172.

Carman, P. I., Hanus, J., and Weinberg, D. I. (1967). *Appl. Phys. Lett.* **11**, 250.

Chesler, R. B., Karr, M. A., and Geusic, J. E. (1970). *J. Appl. Phys.* **41**, 4125.

Comly, J., and Garmire, E. (1968). *Appl. Phys. Lett.* **12**, 7.

De Maria, A. J., Glenn, W. H., Brienza, M. J., and Mack, M. E. (1969). *Proc. IEEE* **57**, 2.

Dmitriev, V. G., Ershov, A. G., Kovrygin, A. I., Kushnir, V. R., Rustamov, R. S., and Schkunov, N. V. (1972). *Kvant. Electron. no.* **5**, 133.

Dowley, M. W. (1968). *Appl. Phys. Lett.* **13**, 395.

Ducuing, J. Blomebergen, N. (1964). *Phys. Rev.* **133A**, 1493.

Eremin, V. I., Kolosov, V. A., and Norinski, L. V. (1971). *Prib. Tech. Eksp.* **5**, 196.

Ernest, G. J., and Witterman, W. I. (1972). *IEEE J. Quant. Electron.* **8**, 382.

Eznst, G. J., and Witteman, W. I. (1972). *IEEE J. Quant. Electron.* **8**, 382.

Fradin, D. F., Bloembergen, N., and Letellier, J. (1973). *Appl. Phys. Lett.* **22**, 635.

Francois, G. E., and Siegman, A. E. (1965). *Phys. Rev.* **139A**, 1965.

Franken, P. A., and Ward, J. F. (1963). *Rev. Mod. Phys.* **35**, 23.

Franken, P., Hill, A., Peters, C., and Weinreich, G. (1961). *Phys. Rev. Lett.* **7**, 118.

Freund, I. (1968). *Phys. Rev. Lett.* **21**, 1404.

Gabel, G., and Hercher, M. (1972). *Int. Quant. Electron. Conf. Dig. Tech. papers* **7**, 6.

Gandrud, W. B., and Abrams, R. L. (1970). *Appl. Phys. Lett.* **17**, 302.

Geusic, J. E., Levinstein, H. J., Singh, S., Smith, R. C., and Van Uitert, L. (1968). *Appl. Phys. Lett.* **12**, 306.

Giordmaine, J. A. (1962). *Phys. Rev. Lett.* **8**, 19.

Giordmaine, J. A., and Miller, R. C. (1965). *Phys. Rev. Lett.* **14**, 973.

Golyaev, Yu. D., Gryaznova, T. G., and Sukhorukov, A. P. (1974), *Sov. Conf. Nonlinear Opt. Tashkent,* 1974, **7**, 416.

Gorokhov, Yu. A., Krindach, D. P., Nikogosyan, D. N., and Sukhorukov, A. P. (1974). *Kvant. Electron.* **1**, 679.

Grütter, A. A., Dändliker, H. P., and Dändliker, R. (1969). *Phys. Rev.* **185**, 629.

Gursky, D. (1969). *Appl. Phys. Lett.* **15**, 5.

Hagen, W. F., and Magnante, P. C. (1969). *J. Appl. Phys.* **40**, 219.
Harris, S. E. (1969). *Proc. IEEE* **57**, 2096.
Harris, S. E. (1973). *Phys. Rev. Lett.* **31**, 341.
Harris, S. E., and Miles, H. (1971). *Appl. Phys. Lett.* **19**, 385.
Henningsen, T., Feichtner, G. D., Melamed, N. T., (1971) *IEEE* **QE-7**, 482.
Hitz, C. B., and Osternik, L. M. (1971). *Appl. Phys. Lett.* **18**, 378.
Hobden, M. V. (1967). *J. Appl. Phys.* **38**, 4365.
Hodgson, R. T., Sozokin, P. P., Wynn, J. J. (1974). Int. Conf. Quantum Electr. Digest of Tech. Papers, **8**. M4, p.*49*.
Javan, A., and Kelley, P. L. (1966). *IEEE* **QE-2**, 470.
Johnson, F. M. (1964). *Nature* **204**, 985.
Karamzin, Yu, N., and Sukhorukov, A. P. (1974). *Sov. Phys. JETP Lett.* **20**, 734.
Klauder, J. R., and Sudarshan, E. C. G. (1968). "Fundamentals of Quantum Optics." Benjamin, New York.
Khokhlov, R. V. (1961). *Radiotek. Electron.* **6**, 1116.
Kildal, H., and Mikkelsen, A. (1973). Paper presented at *Conf. Laser Eng. Appl., Washington, May 1973.*
Kleinman, D. A. (1962). *Phys. Rev.* **128**, 1761.
Kleinman, D. A., and Miller, R. C. (1966). *Phys. Rev.* **148**, 302.
Kleinman, D. A., Ashkin, A., and Boyd, G. D. (1966). *Phys. Rev.* **145**, 338.
Kovrygin, A. I., Podsotskaya, N. K., and Sukhorukov, A. P. (1967). *Opt. Spectrosc.* **23**, 766.
Kovrygin, A. I., Podsotskaya, N. K., and Sukhorukov, A. P. (1969). *Opt. Spectrosc.* **26**, 393,
Krivoshchekov, G. V., Nikulin, N. G., and Sokolovskii, G. I. (1971). *Opt. Spectrosc.* **31**. 448.
Kung, A., Young, J. F., Bjorklund, G., and Harris, S. E. (1972). *Phys. Rev. Lett.* **29**, 985.
Kung, A., Young, J. F., and Harris, S. E. (1973). *Appl. Phys. Lett.* **22**, 301.
Kurtz, S. (1972). "Laser Handbook" vol. 1, p. 942 North-Holland, Amsterdam.
Labuda, E. F., and Johnson, A. M. (1967). *IEEE J. Quant. Electron.* **3**, 164.
Lokhov, Yu. N., Mospanov, V. S., and Fiveyskii, Yu. D. (1972). *Kvan. Electron.* **no. 8**, 103.
Maier, M., Kaiser, W., and Goirdmaine, J. A. (1966). *Phys. Rev. Lett.* **17**, 1275.
Maker, P. D., Terhune, R. W., Nisenoff, M., and Savage, C. M. (1962). *Phys. Rev. Lett.* **8**, 21.
Maker, P. D., Terhune, R. W., and Savage, C. M. (1964). *Quant. Electron. Proc. Third Int. Congr. Paris.* **2**, 1559
McFee, J. H., Boyd, G. D., and Schmidt, P. H. (1970). *Appl. Phys. Lett.* **17**, 57.
McGeoch, M. W., and Smith, R. C. (1970). *IEEE J. Quant. Electron.* **6**, 203.
Mikhina, T. V., Sukhorukov, A. P., and Tomov, I. V. (1971). *Zh. Prikl. Spektrosk.* **15**, 1001.
Miles, R., and Harris, S. E. (1973). *IEEE J. Quant. Electron.* **9**, 470.
Miller, R. C. (1964a). *Appl. Phys. Lett.* **5**, 17.
Miller, R. C. (1964b). *Phys. Rev.* **134A**, 1313.
Miller, R. C. (1968). *Phys. Lett.* **26A**, 177.
Miller, R. C., Boyd, G. D., and Savage, A. (1965). *Appl. Phys. Lett.* **6**, 77.
Moldavskaya, V. M. (1967). *Izv. Vyssh. Ucheb. Zaved. Radiofi.* **10**, 876.
Murray, J. E., and Harris, S. E. (1970). *J. Appl. Phys.* **41**, 609.
Okada, M., and Ieiri, Sh. (1971). *IEEE J. Quant. Electron.* **7**, 469.
Orlov, R. Yu., Chirkin, A. S., and Usmanov, T. (1969). *Sov. Phys. JETP* **57**, 1069.
Orlov, R. Yu., Skidan, I. B., Telegin, L. I., and Rez, I. S. (1973). *Zh. Prikl. Spektrosk.* **19**, 719.
Ostrovsky, L. A. (1967). *Sov. Phys. JETP Lett.* **5**, 331.
Patel, C. K. N. (1965). *Phys. Rev. Lett.* **15**, 1027.

Patel, C. K. N., and Van Tran, N. (1969). *Appl. Phys. Lett.* **15**, 189.
Polloni, R., and Svelto, O. (1968). *IEEE J. Quant. Electron.* **4**, 528.
Rabin, H., and Bey, P. P. (1967). *Phys. Rev.* **156**, 1010.
Rabson, T. A., Raiz, H., Shen, P., and Tittel, F. (1972). *Appl. Phys. Lett.* **20**, 203.
Rice, R. R., and Burkhardt, G. H. (1971). *Appl. Phys. Lett.* **19**, 225.
Shapiro, S. M. (1968). *Appl. Phys. Lett.* **13**, 19.
Shelton, J. W., and Shen, Y. R. (1970). *Phys. Rev. Lett.* **25**, 23.
Shelton, J. W., and Shen, Y. R. (1972). *Phys. Rev.* **5A**, 1867.
Simon, H. J., and Bloembergen, N. (1968). *Phys. Rev.* **171**, 1104.
Smith, R. G. (1970). *IEEE J. Quant. Electron.* **6**, 215.
Strizhevskii, V. L. (1966). *Opt. Spectrosc.* **20**, 516.
Strizhevskii, V. L., Karpenko, S. G., and Bugaev, A. V. (1970). *Opt. Spectrosc.* **29**, 953.
Sukhorukov, A. P. (1966). Izv. *Vyssh. Ucheb. Zaved. Radiofiz.* **9**, 765.
Sukhorukov, A. P., and Tomov, I. V. (1969). *Opt. Spectrosc.* **27**, 119.
Sukhorukov, A. P., and Tomov, I. V. (1970a). *Sov. Phys. JETP* **58**, 1626.
Sukhorukov, A. P., and Tomov, I. V. (1970b). Izv. *Vyssh. Ucheb. Zaved. Radiofizi.* **13**, 266.
Sukhorukov, A. P., and Tomov, I. V. (1970c). *Opt. Spectrosc.* **28**, 1211.
Taynai, J., Targ, R., and Tiffany, W. (1971). *IEEE J. Quant. Electron.* **7**, 412.
Teich, M. C., Abrams, R. L., and Gandrud, W. B. (1970). *Opt. Commun.* **2**, 206.
Terhune, R. W., Maker, P. D., and Savage, C. M. (1962). *Phys. Rev. Lett.* **8**, 404.
Terhune, R. W., Maker, P. D., and Savage, C. M. (1963). *Appl. Phys. Lett.* **2**, 54.
Terhune, R. W., Maker, P. D., Savage, C. M. (1965). *Phys. Rev. Lett.* **14**, 681.
Tien, P. K., Ulrich, R., Martin, R. J. (1970). *Appl. Phys. Lett.* **17**, 447.·
Tunkin, V. G., Usmanov, T., and Shakirov, V. A. (1972). *Kvant. Electron.* **5**, 118.
Van der Ziel, J. P. (1967). *Appl. Phys. Lett.* **5**, 27.
Venkin, G. V., Dneprovskii, V. S., Protasov, V. P., Smirnov, N. D., and Sukhorukov, A. P. (1971). *Kvant. Electron.* **6**, 97.
Volosov, V. D. (1968). *Sov. J. Tech. Phys.* **38**, 1769.
Volosov, V. D. (1969). *Sov. J. Tech. Phys.* **39**, 2188.
Volosov, V. D., and Rashchektaeva, M. I. (1970). *Opt. Spectrosc.* **28**, 105.
Volosov, V. D., Dukhovny, A. M., Krylov, V. H., Sokolova, T. V. (1972). *Kvant. Electron.* **8**, 1972.
Ward, J. F., and New, G. H. (1969). *Phys. Rev.* **185**, 57.
Yablonovitch, E., Flytzanis, C., and Bloembergen, N. (1972). *Phys. Rev. Lett.* **29**, 865.
Yarborough, J. M., and Ammann, E. O. (1973). Paper presented at *Conf. Laser Eng. Appl. Washington, May 1973*.
Young, J. F., Bjorklund, G. C., Kung, A. H., Miles, R. B., and Harris, S. E. (1971). *Phys. Rev. Lett.* **27**, 1551.
Zeldovich, B. Ya. (1966). *Sov. Phys. JETP* **50**, 680.
Zernike, F., and Berman, P. R. (1965). *Phys. Rev. Lett.* **15**, 999. Erratum: *ibid.* (1966), **16**, 117.
Zhdanov, B. V., Kovrygin, A. I., and Pershin, S. M. (1972). *Prib. Tekh. Eksp.* **3**, 206.

9

Optical Parametric Oscillators

ROBERT L. BYER

Department of Applied Physics
Stanford University
Stanford, California

I. INTRODUCTION

A. Brief Historical Review

The field of nonlinear optics has developed rapidly since the demonstration of second harmonic generation by Franken et al. (1961). This experimental demonstration of the nonlinear response of a medium to intense optical fields initiated the rapidly expanding field of nonlinear optics. The progress in non-linear optics since 1961 has been described in monographs (Akhmanov and Khoklov, 1964; Bloembergen, 1965; Butcher, 1965; Franken and Ward, 1963) and review articles (Pershan, 1966; Ovander, 1965; Bonch Bruevich and Khodovoy, 1965; Minck et al., 1966; Bloembergen, 1967; Rank, 1970; Akhmanov et al., 1967a; Terhune and Maker, 1968; Starunov and Fabelinskii, 1969; Kielich, 1970). Theoretical aspects of nonlinear optics have also been investigated (Armstrong et al., 1962; Bloembergen and Pershan, 1962; Kleinman, 1962a; Pershan, 1963).

The development of parametric devices (Yariv and Pearson, 1969) in the microwave region (Louisell, 1960) was well under way when Kingston (1962), Kroll (1962), Akhmanov and Khoklov (1962), and Armstrong et al. (1962) proposed extending parametric interactions to optical frequencies. Three years later in 1965 Wang and Racette (1965) observed significant parametric gain and Giordmaine and Miller (1965) demonstrated the first optical para-metric oscillator in $LiNbO_3$. The success of the first parametric oscillator leaned heavily on earlier work in second harmonic generation. The concept of phase matching (Kleinman, 1962b; Giordmaine, 1962; Maker et al., 1962; Akhmanov et al., 1963), effects of focusing (Bjorkholm, 1966; Kleinman et al., 1966), double refraction (Bloembergen and Pershan, 1962; Kleinman, 1962b; Boyd et al., 1965), and operation with the nonlinear element inside the laser cavity (Wright, 1963; Smith et al., 1965) and within an external resonator (Ashkin et al., 1966a; Akhmanov et al., 1965a) had been previously suggested and studied. The importance of these effects, demonstrated by second harmonic generation, later was extended to the three frequency case.

Soon after Giordmaine reported the first $LiNbO_3$ parametric oscillator, Akhmanov et al. (1966) achieved oscillation in KDP (KH_2PO_4). The majority of reported parametric oscillators which followed were pumped by high peak power pulsed lasers and utilized either KDP or $LiNbO_3$ as nonlinear crystals. In addition, the tuning range remained limited to the visible and near infrared region.

In 1966, Giordmaine and Miller (1966a) extended the $LiNbO_3$ oscillator tuning range and reached the visible by tuning from 0.73 to 1.93 μm, (Giordmaine and Miller, 1966b). The following year Miller and Nordlund (1967) obtained oscillation with an external resonator, and Kreuzer (1967) demonstrated electric field tuning over a limited range. Bjorkholm (1968a, b) studied the singly resonant oscillator operation using a ruby laser pumped LiNbO$_3$ oscillator. By 1969 LiNbO$_3$ oscillator work included a noncollinear ruby pumped oscillator by Falk and Murray (1969) frequency control by radiation injection (Bjorkholm and Danielmayer, 1969) and 1.06-μm Nd:YAG pumped degenerate operation at 2.12 μm (Ammann et al., 1969). In addition Yarborough et al. (1969) achieved stimulated polariton scattering in LiNbO$_3$ with output in the 50–238 μm region.

In Russia work progressed on KDP parametric oscillators and generators. Akhmanov et al. (1965b, c) constructed superradiant parametric generators using KDP and ADP. Possible applications of superradiant parametric generators were reviewed by Akhmanov et al. (1967, 1968c). In addition, Russian work also demonstrated electric field tuning (Krivoshchekov et al., 1968) and singly resonant oscillation (Belyaev et al., 1969) in KDP.

In 1966 Boyd and Ashkin (1966) published a theoretical calculation showing the possibility of cw parametric oscillation in LiNbO$_3$. Three years later Smith et al. (1968a, b) reported low threshold operation and finally cw oscillation in Ba$_2$NaNbO$_{15}$. Simultaneously Byer et al. (1968) obtained cw operation in the visible using LiNbO$_3$. Following theoretical calculations of increased doubly resonant oscillator efficiency without pump reflections by Bjorkholm (1969b) cw oscillation was extended to a ring cavity configuration (Byer et al., 1969b).

This summary represents the progress in optical parametric oscillators through 1969. Oscillation was limited to the three materials KDP, LiNbO$_3$, and Ba$_2$NaNb$_5$O$_{15}$ and covered only a spectral region from 0.53 to 2.6 μm with bandwidths of approximately 1 cm^{-1}. In spite of the lack of progress in device development, the theory of parametric interactions including efficiency for DRO (doubly resonant oscillator), and SRO (singly resonant oscillator) operation frequency stability problems for the DRO, operation with optical parametric oscillation inside the laser cavity, noise and bandwidth consideration, and plane wave and Gaussian beam focusing had been considered in detail. An important paper by Harris (1969a) reviewed both the theory and device aspects of optical parametric oscillators.

This chapter stresses the progress made in the understanding and operation of optical parametric oscillators since the review by Harris. The aim is to present the theory of optical parametric oscillators in enough depth to describe their important operation characteristics. The discussion considers factors such as efficiency, bandwidth, frequency stability, and gain which are important in the applications of optical parametric oscillators. Following this discussion, particular attention is paid to recent optical parametric oscillator devices and their operation characteristics. Finally, present research efforts are reviewed, particularly in new nonlinear materials with the aim of predicting future optical parametric oscillator device characteristics.

B. Nonlinear Susceptibility

The application of radiation to a material leads to the generation of an induced polarization wave. The linear response of the medium is described by the constitutive relation $\mathbf{P} = \varepsilon_0 \chi \mathbf{E}$, where ε_0 is the free space permittivity, χ the linear susceptibility, and $\mathbf{E}$ the applied field. For intense fields appreciable nonlinear response of the media may occur. The susceptibility can be expanded by perturbation approach to include second-, third-, and higher-order terms. This expansion has been carried out and discussed in detail (Armstrong *et al.*, 1962; Terhune and Maker, 1968; Butcher, 1965; Pershan, 1966).

We are interested in the second-order response which by symmetry arguments occurs in noncentrosymmetric materials. In general, the second-order nonlinear polarization can be written as

$$\mathbf{P}_i = \varepsilon_0 \sum_{jk} \chi_{ijk} : \mathbf{E}_j \mathbf{E}_k \tag{1}$$

where χ_{ijk} is the nonlinear polarizability tensor. The nonlinear polarization is driven by the square of the electric field so that dc and second harmonic terms appear. If more than one frequency is present, the polarization contains various sum and difference frequency terms.

The generated polarization acts as a driving term in the wave equation

$$\mathbf{V} \times (\mathbf{V} \times \mathbf{E}_i) + \frac{\partial}{\partial t}(\sigma \mu_0 \mathbf{E}_i) + \mu_0 \varepsilon \frac{\partial^2 \mathbf{E}_i}{\partial t^2} = -\mu_0 \frac{\partial^2 \mathbf{P}_i}{\partial t^2} \tag{2}$$

for each frequency component present.

The instantaneous electric field and polarization can be written in terms of their Fourier frequency components as

$$\mathbf{E}(\mathbf{r}, t) = \tfrac{1}{2}[\mathbf{E}(\mathbf{r}. \omega) \exp i(\mathbf{k} \cdot \mathbf{r} - \omega t) + \text{c.c.}]$$

and $\tag{3}$

$$\mathbf{P}(\mathbf{r}, t) = \tfrac{1}{2}[\mathbf{P}(\mathbf{r}, \omega) \exp i(\mathbf{k} \cdot \mathbf{r} - \omega t) + \text{c.c.}]$$

It is convention to define the susceptibility $\chi_{ijk}(-\omega_3, \omega_2, \omega_1)$ in terms of the Fourier amplitude relations

$$\mathbf{P}_i(\mathbf{r}, \omega_3) = \varepsilon_0 \sum_{jk} \chi_{ijk}(-\omega_3, \omega_2, \omega_1) : \mathbf{E}_j(\mathbf{r}, \omega_2) \mathbf{E}_k(\mathbf{r}, \omega_1) \exp[i(\mathbf{k}_2 + \mathbf{k}_1) \cdot \mathbf{r}] \tag{4}$$

where $\omega_3 = \omega_2 + \omega_1$.

The nonlinear susceptibility satisfies two important symmetry relations. The first is permutation symmetry (Armstrong *et al.*, 1962) such that

$$\chi_{ijk}(-\omega_3, \omega_2, \omega_1)$$

$$= \chi_{jik}(\omega_2, -\omega_3, \omega_1) = \chi_{kji}(\omega_1, \omega_2, -\omega_3) = \chi_{ikj}(-\omega_3, \omega_1, \omega_2) \tag{5}$$

This symmetry ensures that the nonlinear coefficient is the same regardless of the three frequency process. In addition the relation

$$\chi_{ijk}(-\omega_3, \omega_2, \omega_1) = \chi_{ijk}^*(\omega_3, -\omega_2, -\omega_1)$$

holds which allows $E(r, t)$ and $P(r, t)$ to remain real. The second symmetry relation, called "Kleinman's symmetry" (Kleinman, 1962a), states that $\chi(-\omega_3, \omega_2, \omega_1)$ should be independent of frequency as long as all three frequencies are in a lossless frequency region of the material. Thus χ is real and invariant to any permutation of ijk.

In addition to the above considerations, the form of the χ_{ijk} tensor must satisfy the material symmetry requirements. The χ_{ijk} tensor has the same symmetry as the piezoelectric d tensors which have been tabulated (Nye, 1960). For example, the tensor for $\bar{4}2m$ point group symmetry to which KDP and chalcopyrite crystals belong has the components

$$P_x = 2d_{14}E_yE_z, \qquad P_y = 2d_{14}E_zE_x, \qquad P_z = 2d_{36}E_xE_y \qquad (6)$$

Here x, y, and z refer to the crystal axes and $d_{i(jk)} = d_{im}$ has been written in the condensed notation

$$(jk) = (11) \quad (22) \quad (33) \quad (23) \quad (13) \quad (12)$$

$$m = \quad 1 \qquad 2 \qquad 3 \qquad 4 \qquad 5 \qquad 6$$

The 3×6 d_{im} matrix operates on a column vector $(EE)_m$, where

$$(EE)_1 = E_x^2, \qquad (EE)_2 = E_y^2, \qquad (EE)_3 = E_z^2$$

$$(EE)_4 = 2E_yE_z, \qquad (EE)_5 = 2E_xE_z, \qquad (EE)_6 = 2E_xE_y$$

to generate the polarization components. In this notation the polarization for second harmonic generation is

$$P(r, \omega) = \varepsilon_0 \sum_{m=1}^{6} d_{im}(EE)_m$$

with

$$\chi_{im}(-2\omega, \omega, \omega) = 2d(-2\omega, \omega, \omega) \qquad (7)$$

as the relation between the nonlinear susceptibility and the d coefficients used to describe second harmonic generation. For three frequency interactions

$$P(r, \omega_1) = \varepsilon_0 2d : E^*(r, \omega_2)E(r, \omega_3)$$

$$P(r, \omega_2) = \varepsilon_0 2d : E^*(r, \omega_1)E(r, \omega_3) \qquad (8)$$

$$P(r, \omega_3) = \varepsilon_0 2d : E(r, \omega_1)E(r, \omega_2)$$

where the condensed notation is implied.

For propagation along a direction not parallel to a crystal axis the component of the field effective in generating the polarization must be used. These

Table I

Properties of Selected Nonlinear Crystal

Crystal	Index	$d(m/V) \times 10^{12}$	$d^2/n^3 \times 10^{24}$	$d^2/n^3\lambda_3^3 \times 10^6$	Transparency (μm)
KDP	1.51	$d_{36} = 0.50^a$	0.08	4.25	0.22–1.1
	1.47			($\lambda_3 = 0.266 \, \mu$m)	
ADP	1.53	$d_{36} = 0.57^{b,c}$	0.10	5.31	0.20–1.1
	1.48			($\lambda_3 = 0.266 \, \mu$m)	
LiNbO$_3$	2.24	$d_{31} = 6.25^d$	3.86	25.7	0.35–4.5
	2.16			($\lambda_3 = 0.532 \, \mu$m)	
LiIO$_3$	1.85	$d_{31} = 7.5^e$	1.68	11.2	0.31–5.5
	1.72			($\lambda_3 = 0.694 \, \mu$m)	
Ag$_3$AsS$_3$	2.76	$d_+ = 11.6^f$	8.2	6.9	0.60–13
(proustite)	2.54			($\lambda_3 = 1.06 \, \mu$m)	
CdSe	2.45	$d_{31} = 19^{g,h}$	24.5	4.0	0.75–25
	2.47			($\lambda_3 = 1.83 \, \mu$m)	

[a] Jerphanon and Kurtz (1970).   [e] Campillo and Tang (1970).
[b] Francois (1966).   [f] Levine and Bethea (1972).
[c] Bjorkholm and Siegman (1967).   [g] Boyd et al. (1971).
[d] Byer and Harris (1968).   [h] Herbst (1972).

effective nonlinear coefficients have been tabulated for both uniaxial (Boyd and Kleinman, 1968) crystal classes and biaxial crystal classes (Midwinter and Warner, 1965; Hobden, 1967).

Values of d for some important nonlinear crystals are listed in Table I. The values are given in mks units to be consistent with the text. Note that d(mks) has the units of meters per volt and is related to d(cgs) by $[4\pi/(3 \times 10^4)] \, d(cgs) = d(mks)$.*

The coupled equations for the three frequency parametric interaction follow from the wave equation [Eq. (2)]. Using Eq. (8) for the driving polarization we find

$$\frac{d}{dr}E_1 + \alpha_1 E_1 = i\kappa_1 E_3 E_2^* \exp i \, \Delta k \, r \qquad (9a)$$

$$\frac{d}{dr}E_2 + \alpha_2 E_2 = i\kappa_2 E_3 E_1^* \exp i \, \Delta k \, r \qquad (9b)$$

$$\frac{d}{dr}E_3 = i\kappa_3 E_2 E_1 \exp(-i \, \Delta k \, r) \qquad (9c)$$

---

* The use of d in meters per volt is becoming standard practice. In the past $d = d(m/V)/\varepsilon_0$ has been used often [see Harris (1969a)].

where $\kappa_i = \omega_i d/n_i c$ and we have let $\alpha_i = \frac{1}{2}\mu_0 \sigma_i c$ be the round trip electric field loss. Here

$$\omega_3 = \omega_2 + \omega_1 \tag{10}$$

is the energy conservation condition and

$$\mathbf{k}_3 = \mathbf{k}_2 + \mathbf{k}_1 + \Delta\mathbf{k} \tag{11}$$

is the momentum conservation or phase-matching condition. The momentum-matching condition is a vector relation, but if we assume collinear waves and $\Delta\mathbf{k} = 0$ for the moment, it can be written in terms of the frequency ω_2 as

$$\omega_2 = \omega_3 (n_3 - n_1)/(n_2 - n_1)$$

which explicitly illustrates that a change in the index of refraction at, say n_3, forces the signal frequency ω_2 to tune. Since the index of refraction can be altered by a number of means including crystal rotation, temperature, and electric field these are all possible tuning techniques for optical parametric oscillators.

Since the 1969 review of optical parametric oscillators by Harris (1969a) and a review shortly afterwards by Sushchik *et al.* (1970) in Russia, considerable development of optical parametric devices has been achieved*. Most of the work has, however, made use of the existing nonlinear crystals KDP and LiNbO$_3$.

Oscillation has been achieved recently in four new materials: ADP, LiIO$_3$, Ag$_3$AsS$_3$ (proustite), and CdSe. These materials have extended the tuning range available to parametric oscillators. The development of nonlinear devices, however, remains severely materials limited. As a consequence, the rapid pace that characterized initial discoveries and demonstrations of parametric oscillators has slowed recently, reflecting the time needed for the development of new materials.

II. THEORY OF OPERATION

A. Introduction

Parametric oscillators have been operated in a number of configurations. To avoid confusion these devices are briefly described here. The particular operation characteristics are considered in more detail in later sections.

Figure 1a shows a schematic of an optical parametric oscillator (OPO). The incident laser pump beam is properly focused into the nonlinear crystal.

* Parametric oscillators are the subject of a recent review article by Smith (1973).

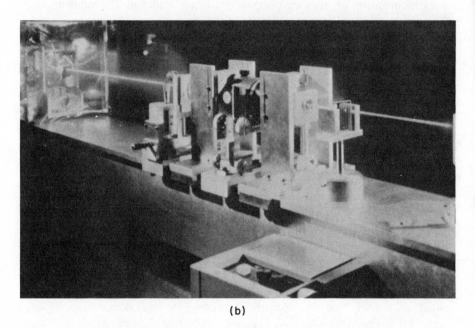

(a)

(b)

Fig. 1. (a) Schematic of an optical parametric oscillator. (b) Photograph of a LiNbO$_3$ singly resonant parametric oscillator pumped with 0.473 μm and operating at 0.59 and 2.5 μm.

When the gain of the nonlinear interaction exceeds the cavity loss, the device exceeds threshold and breaks into oscillation in a manner very similar to a laser. Like a laser oscillator, the OPO generates coherent light at a marked increase in efficiency once above threshold. Figure 1b shows a photograph of a LiNbO$_3$ parametric oscillator operating in the visible and near infrared.

An immediate distinction can be made between a parametric generator and oscillator. The former operates without a resonator and due to its very high gain amplifies spontaneous noise to incident pump power levels. Its operation is similar to superfluorescent laser emission. The parametric oscillator operates with resonant feedback at both the signal and idler frequencies (ω_2 and ω_1) in the doubly resonant oscillator configuration (DRO) or with feedback at only the single or the idler frequency in the singly resonant oscillator (SRO) configuration. In addition, the oscillator may operate either in a continuous mode (cw) or pulse manner depending on the operation of the pump laser. Finally, a distinction must be made between operation external or internal to the laser cavity. The theoretical and practical aspects of the various parametric oscillator configurations are discussed below.

Equations (9a)–(9c) are the basic equations for the following discussion. They have been solved exactly by Armstrong *et al.* (1962) for various input conditions. We first note that, neglecting loss, the equations lead to a photon conservation relation

$$\frac{1}{\omega_1}\frac{dI_1}{dz} = \frac{1}{\omega_2}\frac{dI_2}{dz} = -\frac{1}{\omega_3}\frac{dI_3}{dz} \tag{12}$$

where

$$I(\mathbf{r}, \omega) = \tfrac{1}{2}nc\varepsilon_0 |E(\mathbf{r}, \omega)|^2$$

is the intensity, n the index of refraction, c the velocity of light in vacuum, and z the direction of propagation. For every photon converted from the pump field a signal and a corresponding idler photon are generated.

Equation (9) leads to solutions for up-conversion or sum generation, mixing or difference frequency generation, and for parametric generation. For up-conversion (Kleinman and Boyd, 1969; Andrews, 1970; Warner, 1971) the phase of the coupled equations assuming $dE_2/dz = 0$ leads to solutions of the form $E_1 \propto \cos \Gamma z$ and $E_3 \propto \sin \Gamma z$ which show periodic conversion with no net gain. The lack of gain and therefore the lack of generated parametric noise emission (Byer and Harris, 1968; Kleinman, 1968; Giallorenzi and Tang, 1968) is advantageous in some applications. For difference frequency generation, the phase of the coupled equations assuming $dE_3/dz = 0$ leads to solutions of the form $E_1 \propto \cosh \Gamma z$ and $E_2 \propto \sinh \Gamma z$ which have an exponential increase and therefore net gain. Parametric mixing has been used to measure generated gain (Wang and Racette, 1965; Boyd and Ashkin, 1966; Herbst and Byer, 1971), as well as to generate tunable infrared radiation by mixing visible frequency tunable sources (Zernike and Berman, 1965; Martin and Thomas, 1966; Yajima and Inave, 1968; Faries *et al.*, 1969; Dewey and Hocker, 1971).

B. Gain and Bandwidth of a Parametric Amplifier

For parametric generation in a crystal of length l we neglect pump depletion and assume solutions of the form

$$E_1{}^* = E_1{}^* \exp[(\Gamma' - \tfrac{1}{2}i\,\Delta k)z] \qquad \text{and} \qquad E_2 = E_2 \exp[(\Gamma' + \tfrac{1}{2}i\,\Delta k)z]$$

where the complex conjugate of Eq. (9a) is used. The solution of the resulting characteristic equations

$$E_2\left(\Gamma' + \alpha_s + \frac{i\,\Delta k}{2}\right) + E_1{}^*\left(-\frac{i\omega_2\,dE_3}{n_2\,c}\right) = 0$$

$$E_2\left(\frac{i\omega_2\,dE_3{}^*}{n_2\,c}\right) + E_1{}^*\left(\Gamma' + \alpha_i - \frac{i\,\Delta k}{2}\right) = 0$$

gives

$$\Gamma_{\pm}{}' = \frac{-(\alpha_1 + \alpha_2)}{2}$$

$$\pm \frac{1}{2}\left[(\alpha_1 - \alpha_2)^2 + \frac{i\,\Delta k}{2}(\alpha_1 - \alpha_2) - 4\left(\frac{\Delta k}{2}\right)^2 + \frac{4\omega_1\omega_2\,|d|^2\,|E_3|^2}{n_1 n_2 c^2}\right]^{1/2}$$

We now define the parametric gain coefficient by

$$\Gamma^2 = \omega_1\omega_2\,|d|^2\,|E_3|^2/n_1 n_2 c^2 = 2\omega_1\omega_2\,|d|^2 I_3/n_1 n_2 n_3 \varepsilon_0 c^3 \tag{13}$$

where I_3 is the pump intensity. The general solution is therefore

$$E_1{}^*(z) = (E_{1+}^* e^{gz} + E_{1-}^* e^{-gz})\,e^{-\alpha z}\exp(-i\,\Delta k\,z/2) \tag{14a}$$

$$E_2(z) = (E_{2+} e^{gz} + E_{2-} e^{-gz})\,e^{-\alpha z}\exp(+i\,\Delta k\,z/2) \tag{14b}$$

where we have set $\alpha_1 = \alpha_2 = \alpha$ and have defined a total gain coefficient

$$g = [\Gamma^2 - (\tfrac{1}{2}\,\Delta k)^2]^{1/2} \tag{15}$$

For parametric amplification we assume input fields of $E_1{}^*(0)$ and $E_2(0)$ at $z = 0$. These input fields determine $E_{1\pm}$ and $E_{2\pm}$ and give for the field amplitudes at l

$$E_1(l)\,e^{\alpha l} = E_1(0)\,e^{i\Delta k l/2}\left[\cosh gl - \frac{i\,\Delta k}{2g}\sinh gl\right]$$

$$+ i\frac{\kappa_1 E_3}{g}E_2{}^*(0)\,e^{i\Delta k l/2}\,[\sinh gl] \tag{16a}$$

and

$$E_2(l)\,e^{\alpha l} = E_2(0)\,e^{i\Delta k l/2}\left[\cosh gl - \frac{i\,\Delta k}{2g}\sinh gl\right]$$

$$+ i\frac{\kappa_2 E_3}{g}E_1{}^*(0)\,e^{i\Delta k l/2}\,[\sinh gl] \tag{16b}$$

where $\kappa_1 = \omega_1 d/n_1 c$, $\kappa_2 = \omega_2 d/n_2 c$, and $\Gamma^2 = \kappa_1\kappa_2\,|E_3|^2$. These expressions are identical under and interchange of subscripts $1 \leftrightarrow 2$. Thus we may arbitrarily call one the signal field and the other the idler field.

When a single frequency at ω_2 is incident on the parametric amplifier, the single-pass power gain is

$$G_2(l) = \frac{|E_2(l)|^2}{|E_2(0)|^2} - 1 = \Gamma^2 l^2 \frac{\sinh^2 gl}{(gl)^2} \qquad (17)$$

In this case the gain is independent of the phase of the input field $E_2(0)$ relative to the pump field since the generated field $E_1(0)$ assumes the proper phase for maximum gain. If, however, two fields $E_1(0)$ and $E_2(0)$ are incident on the amplifier, the gain depends on their phases relative to the pump field.

Superfluorescent parametric emission is an extension of the parametric amplifier to the case of extremely high gain such that the input noise fields are amplified to an output of the order of the pump field.

In the high gain limit Eq. (17) becomes

$$G_2(l) = [1 + (\Delta k/2g)^2] \sinh^2 gl$$

which can be written as

$$G_2(l) = \tfrac{1}{4} \exp 2\Gamma l \qquad (18)$$

when $\Delta k/2 < g$.

The noise input per mode for a parametric amplifier is equivalent to one photon in either the signal or idler channel (Byer and Harris, 1968; Kleinman, 1968; Giallorenzi and Tang, 1968) or one-half photon in each. It is interesting to note that $\tfrac{1}{2}\hbar\omega$ input into a laser amplifier also leads to the correct amplified spontaneous emission output (Wagner and Hellwarth, 1964). For efficient superradiant operation the input noise field must be amplified by approximately 10^{16} so that $\Gamma l \approx 20$.

In the low gain limit where $\Gamma^2 l^2 < (\Delta k/2)^2$ the gain of a parametric amplifier is

$$G_2(l) = \Gamma^2 l^2 \operatorname{sinc}^2 \{[(\tfrac{1}{2}\Delta k)^2 - \Gamma^2]^{1/2} l\} \qquad (19)$$

where $\operatorname{sinc} x \equiv (\sin x)/x$. Equations (17) and (19) show that as the gain varies from $\Gamma^2 l^2 > (\Delta k/2)^2$ to $\Gamma^2 l^2 < (\Delta k/2)^2$ the $(\sinh x)/x$ function continuously goes to the $\operatorname{sinc} x$ function (cf. Fig. 2, Harris, 1969a).

We define the gain bandwidth by

$$|(\tfrac{1}{2}\Delta k)^2 - \Gamma^2|^{1/2} l = \pi \qquad (20)$$

For low gain such that $\Gamma^2 l^2 \ll \pi^2$ the bandwidth expression reduces to

$$\tfrac{1}{2}\Delta k\, l = \pi \qquad (21)$$

Taking the ratio of the bandwidth (BW) in the high gain limit to the bandwidth at low gain we find

$$\frac{\mathrm{BW\,(high\ gain)}}{\mathrm{BW\,(low\ gain)}} = [1 + (\Gamma^2 l^2/\pi^2)]^{1/2} \qquad (22)$$

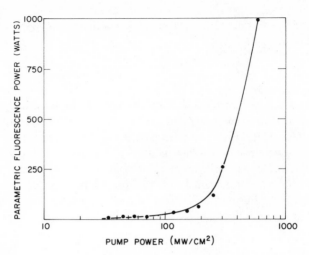

Fig. 2. Superradiant output power versus pump intensity for a $Ba_2NaNb_5O_{15}$ crystal pumped by a mode-lock doubled Nd : glass laser [after Rabson *et al.* (1972)].

Therefore, until $\Gamma^2 l^2 \simeq \pi^2$ the bandwidth broadening is small. For superfluorescent operation where $\Gamma^2 l^2 \simeq 400$ the bandwidth increases by a factor of six over the low gain case.

It is convenient at this point to expand Δk in Eq. (21) as a function of frequency and write an expression for the bandwidth in the low gain limit. To second order

$$k_n = k_{n\,0} + u_n^{-1}\delta\omega_n + \tfrac{1}{2}g_n\delta^2\omega_n \tag{23}$$

where

$$u_n = (\partial k_n/\partial\omega_n)^{-1} \tag{24a}$$

is the group velocity and

$$g_n = \partial^2 k_n/\partial\omega_n{}^2 \tag{24b}$$

is the group velocity dispersion.

Since $\Delta k = k_3 - k_2 - k_1$ and $k_{30} - k_{20} - k_{10} = 0$ at phase matching, we have

$$\Delta k = \left(\frac{1}{u_1} - \frac{1}{u_2}\right)\delta\omega_2 + \frac{1}{2}(g_1 + g_2)\delta\omega_2{}^2$$

$$= \beta_{12}\delta\omega_2 + \frac{1}{2}\gamma_{12}\delta\omega_2{}^2 \tag{25}$$

where $\beta_{12} = (1/u_1) - (1/u_2)$ and $\gamma_{12} = g_1 + g_2$. Here we have used $\delta\omega_2 = -\delta\omega_1$.

Equation (21) defines the bandwidth in the low gain limit. Except for operation close to degeneracy ($\omega_2 = \omega_1$) or near a turning point in the dispersion curves, $|4\pi\gamma_{12}/l\beta_{12}^2| < 1$ so that

$$\delta\omega_2 = \frac{1}{2}\left|\left(\frac{\beta_{12}}{\gamma_{12}}\right)\left\{\left(1 + \frac{4\pi\gamma_{12}}{l\beta_{12}^2}\right)^{1/2} - \left(1 - \frac{4\pi\gamma_{12}}{l\beta_{12}^2}\right)^{1/2}\right\}\right| \approx \frac{2\pi}{l\beta_{12}} \quad (26)$$

which is the usual expression for the gain bandwidth [cf. Harris, 1969a; Eq. (28)]. For the special case when $\beta_{12} \approx 0$ (group velocity matching) and $|4\pi\gamma_{12}/l\beta_{12}^2| > 1$, the bandwidth is

$$\delta\omega_2 = \left(\frac{\beta_{12}}{\gamma_{12}}\right)\left(1 + \left|\frac{4\pi\gamma_{12}}{l\beta_{12}^2}\right|\right)^{1/2} \approx \left(\left|\frac{4\pi}{l\gamma_{12}}\right|\right)^{1/2} \quad (27)$$

In practice the bandwidth can be written in terms of whatever parameter is important in varying the refractive indices. This is discussed further in Section II, H.

C. Superfluorescent Parametric Amplification

Experimental observation of superfluorescent parametric emission was first reported by Akhmanov et al. (1967a, 1968c). The early experiments in KDP achieved a gain of $\Gamma^2 l^2 = 10$ with low conversion efficiency. Later a traveling wave oscillator (TWO) was constructed using ADP in a multipass configuration. The interaction length was 30–35 cm with 6 or 7 passes. The device operated with a 1–2-Å bandwidth and for a 5320-Å pump intensity of 70 MW/cm² gave an output power of 100 kW. For this device the estimated gain is approximately $\Gamma^2 l^2 \simeq 85$ using the parameters shown in Table II for ADP. Akhmanov et al. (1967) point out that the gain and thus conversion efficiency are exponentially increasing which leads to the possibility of higher conversion efficiencies.

Table II

Parametric Gain Coefficients and Superfluorescent Power Density

Crystal	λ_p (μm)	l (cm)	$\Gamma^2 l^2$ at 1 MW/cm²	P/A ($\Gamma^2 l^2 = 400$) (MW/cm²)	Damage power density[a] (MW/cm²)	Minimum length for superfluorescent operation (cm)
ADP	0.266	5	0.13	2960	> 1000	8.5
LiNbO₃	0.532	5	1.3	267	50–140	7–11
CdSe	1.833	5	0.68	588	60	15.6
CdGeAs₂	5.3	1	0.04	10,000	20–40	15

[a] Damage intensity for 10 to 200 nsec pulse lengths. See Section IV, A for further discussion.

Recent interest in the generation of picosecond pulses has led to further work in high gain parametric devices. Burneika *et al.* (1972) reported parametric amplification in KDP with a mode-locked pump source at 0.532 μm. They obtained 10^{-2} J output in the 1-μm region in three passes through 18 cm of KDP. Using a mode-locked frequency-doubled Nd:glass pump source, Rabson *et al.* (1972) obtained greater than 300 W of 10 psec pulses in $Ba_2NaNb_5O_{15}$. Figure 2 shows the exponential increase of the measured fluorescence power versus pump power density for that experiment. Harris *et al.* (1972) recently reported a 1% energy conversion efficiency in a 4-cm $LiNbO_3$ amplifier pumped with a mode-locked frequency-doubled Nd:YAG laser source. The signal output was in the visible near 6700 Å. This experiment demonstrated the importance suppressing feedback to prevent oscillation. The 20-psec pump pulse length had an interaction length with the generated signal wave of less than the 4-cm crystal length due to group velocity mismatch. When the laser did not properly mode-lock, the longer pump pulse lengths led to observed coherent oscillation in the resonator formed by the crystal surfaces. Similar high gain oscillation had been observed in ADP pumped at 0.2660 μm by Yarborough and Massey (1971).

D. Picosecond Pulse Propagation and Pulse Narrowing

The possibility of significant pulse sharpening by parametric amplification was first discussed by Glenn (1967). The early work on both the theoretical (Glenn, 1969) and experimental aspects (Comly and Garmire, 1968; Miller, 1968; Shapiro, 1968; Hagen and Magnante, 1969) of picosecond pulse propagation in a nonlinear crystal concerned second harmonic generation. It was recognized that group velocity mismatch limited the interaction length and thus the SHG conversion efficiency. In addition, group velocity mismatch led to broadening of the SHG pulse width.

Akhmanov *et al.* (1968a) analyzed the dynamics of pulse shortening in a parametric amplifier. The results were later extended to degenerate parametric amplification with a quasi-continuous pump source and a pulsed pump source (Akhmanov *et al.*, 1968b). In more recent work, Akhmanov *et al.* (1969) and Sukhorukov and Shchednova (1967) found general solutions to the problem of picosecond pulse amplification by utilizing the analogy between temporal pulse modulation and spatial modulation. This analogy is applied in the discussion of short pulse amplification given below.

The problem of steady-state propagation of short pulses in a parametric amplifier was also considered by Akhmanov *et al.* (1968b, 1969) and by Armstrong *et al.* (1970). Armstrong's analysis shows that steady-state pulses exist under proper conditions and are analogous to pi pulses propagating in a two-level system. Because of the extremely short relaxation time of the electronic nonlinearity ($\sim 10^{-15}$ sec), pulse narrowing in high gain parametric amplifiers may approach 10^{-14} sec limited by group velocity dispersion.

For parametric amplification of short pulses Eqs. (9a)–(9c) must be extended to include group velocity effects. The driving polarization in Eq. (1) becomes

$$P_1(\mathbf{r}, t) = \int\limits_0^\infty\!\!\int dt_1\, dt_2 \int\!\!\int \varepsilon_0 \chi(\mathbf{r}_1 \mathbf{r}_2\, t_1\, t_2)\, \mathbf{E}(\mathbf{r}-\mathbf{r}_1,\, t-t_1)\, \mathbf{E}(\mathbf{r}-\mathbf{r}_2,\, t-t_2)\, d\mathbf{r}_1\, d\mathbf{r}_2 \tag{28}$$

If we look for solutions in the slowly varying envelope approximation of the form

$$\mathbf{E}(\mathbf{r}, t) = A_n(\mathbf{r}, t)\, \exp[i(\omega t - \mathbf{k} \cdot \mathbf{r})] + \text{c.c.} \tag{29}$$

Eq. (2) reduces to

$$\left(\beta_{n3}\frac{\partial}{\partial \eta} + s_n \frac{\partial}{\partial x}\right) A_n - \frac{i}{2}\left(\frac{\nabla_\perp}{2k_n} + g_n \frac{\partial^2}{\partial \eta^2}\right) A_n = -\kappa_n A_m \tag{30}$$

where the variable $\eta = t - (z/u_3)$, $u_n^{-1} = \partial k_n/\partial \omega_n$ the group velocity, $\beta_{n3} = u_n^{-1} - u_3^{-1}$, $g_n = \partial^2 k_n/\partial \omega_n^2$, $\nabla_\perp$ the transverse Laplacian, and s_n a unit vector parallel to the ray vector.

By separating the spatial and temporal modulation of the wave, the propagation is governed by the pair of equations

$$\left(\beta_{n3}\frac{\partial}{\partial \eta} - \frac{i}{2} g_n \frac{\partial^2}{\partial \eta^2}\right) A_n = -\kappa_n A_m \tag{31a}$$

$$\left(s_n \frac{\partial}{\partial x} - \frac{i}{2}\frac{\nabla_\perp}{k_n}\right) A_n = -\kappa_n A_m \tag{31b}$$

The space–time analogy is immediately obvious with

$$x \leftrightarrow \eta, \qquad \rho \leftrightarrow \beta, \qquad k_n^{-1} \leftrightarrow g_n \tag{32}$$

where $\rho = s_n \cdot s_m$ is the anisotropy angle.

Solving Eq. (31b) for an input Gaussian beam of spatial profile

$$A_n(x) = A_0 \exp(-x^2/w^2)$$

shows that the profile changes significantly after traveling a length

$$z > z_R = \tfrac{1}{2} w^2 k_n$$

Here z_R is the Rayleigh range equal to one-half the confocal distance b given by

$$b = k_n w^2 \tag{33}$$

where $k_n = 2\pi/\lambda_n$ and w is the Gaussian beam radius. If $\rho \neq 0$, due to a non-zero double refraction angle in an anisotropic crystal, for example, we can also define the usual aperture length (Boyd et al., 1965)

$$l_a = \sqrt{\pi}\, w/\rho \tag{34}$$

By the space–time analogy the "confocal" distance for a Gaussian pulse of width τ is

$$b_\tau = \tau^2/g \tag{35}$$

and the group velocity aperture length is

$$l_\tau = \sqrt{\pi}\, \tau/\beta \tag{36}$$

Equations (30) have been solved for second harmonic generation and degenerate parametric amplification. For SHG the equations in the time domain are

$$\frac{\partial}{\partial z} A_1 = -i\kappa A_2 A_1^*, \qquad \left(\frac{\partial}{\partial z} + \beta \frac{\partial}{\partial \eta}\right) A_2 = -i\kappa A_1^2 \tag{37}$$

where $\beta = (1/u_1) - (1/u_2)$ is the difference in group velocities and $2\omega_1 = \omega_2$ are the fundamental and second harmonic frequencies. The solution of these equations for a specified input pump shape is given by Akhmanov et al. (1968b, 1969). The solutions are complicated functions involving cosh and sinh but reduce to the familiar $A_1(\eta, z) \propto$ sech Γz and $A_2(\eta, z) \propto$ tanh Γz in the limit of negligible pulse walk-off $l_\tau \gg l_{\text{NL}}$, where

$$l_{\text{NL}}^{-1} = \kappa A_{10} = \Gamma \tag{38}$$

is the nonlinear interaction length.

Integrating over z and solving for $A_2(t, z)$ in Eq. (37) we find

$$A_2(t, z) = -i\kappa \int_0^z A_{10}^2 [t - (z/u_2) + \beta z']\, e^{-i\Delta k z'}\, dz' \tag{39}$$

The solution in the frequency domain follows if we define the Fourier transform

$$A(\omega, z) = (1/2\pi) \int_{-\infty}^{\infty} A(t, z)\, e^{-i\omega t}\, dt \tag{40}$$

so that

$$A_2(\omega, z) = -2i\kappa e^{i\Delta k z} \frac{\sin[\Delta k z/2]}{\Delta k} \int_{-\infty}^{\infty} A_{10}(\omega - \omega')\, A_{10}(\omega')\, d\omega' \tag{41}$$

The second harmonic intensity $I_2(2\omega - \omega_0) \propto A_2(\omega) A_2^*(\omega)$ increases according to

$$I_2(2\omega - \omega_0) \propto z^2 \operatorname{sinc}^2 [\beta(2\omega - \omega_0) - \Delta k_0](z/2) \tag{42}$$

where we have expanded Δk, and let $\Delta k_0 = k_{2\omega,0} - k_{\omega,0}$. The usual coherence length $\Delta k z/2 = \pi$ is now generalized to include group velocity walk-off. Glenn (1965) has discussed further generalizations of this solution.

The parametric amplifier provides the possibility of considerable pulse narrowing. There are two cases of interest; the amplitude-modulated pump case and the quasi-static pump case where $l_\tau \gg l_{NL}$ or equivalently $\tau_p \gg \tau_{cr}$ where $\tau_{cr} = \beta l_{NL}$ is the minimum pulse length at which 70% conversion efficiency occurs.

For degenerate parametric amplification Eq. (37) modifies to

$$\frac{\partial A_s}{\partial z} + \alpha_s A_s = i\kappa A_p A_s, \qquad \left(\frac{\partial}{\partial z} + \beta \frac{\partial}{\partial \eta}\right) A_p = -i\kappa A_s^{\;2} \qquad (43)$$

where $\eta = t - (z/u_s)$ and $\beta = (1/u_s) - (1/u_p)$. Here we have let $1 \leftrightarrow s$ and $2 \leftrightarrow p$ to denote the signal and pump fields. For low conversion efficiency such that $A_p \gg A_s$ the intensity of the signal wave is

$$I_s(t,z) = I_{s,0}[t - (z/u_s)] \exp\left\{2\kappa \int_0^z A_{p0}(\eta - \beta z')\,dz' - 2\alpha_s z\right\} \qquad (44)$$

One immediate consequence of Eq. (44) is that for short pulses there exists a threshold energy instead of intensity. This is discussed further in Section II, F where we consider the rise time of a parametric oscillator.

When the gain is such that $\Gamma z \gg 1$, the signal intensity given by Eq. (44) becomes Gaussian with a pulse width

$$\tau_s \approx \sqrt{2}\,\tau_p(\Gamma z)^{-1/2} \qquad (45)$$

Thus pulse narrowing can be achieved by either high gain operation in reasonably short crystal length or by multipass operation synchronously with the pump pulse. Pulse narrowing occurs until limited by the group velocity walk-off length l_τ or ultimately by the dispersion length b_τ. The pulse begins to broaden for $z > l_\tau$ or for $z > b_\tau$ whichever is shorter.

Armstrong et al. (1970) and Akhmanov et al. (1968b) discussed the existence of a steady-state signal pulse. The pulse is formed by parametric amplification of the leading edge and attenuation of the trailing edge by second harmonic generation. The energy density of the steady-state pulse is

$$W_s = 2I_p(\tau + \tau_{cr}) \qquad (46)$$

where τ is the pulse length and $\tau_{cr} = \beta l_{NL}$. If the stationary pulse length is longer than τ_{cr}, then the pulse amplitude equals that of the pump and

$$W_s = 2I_p\tau, \qquad \tau \gg \tau_{cr}$$

When the pulse length shortens such that $\tau \ll \tau_{cr}$, its amplitude increases and the energy density reaches the limit

$$W_{s_{lim}} = 2I_p \tau_{cr} \tag{47}$$

Armstrong *et al.* (1970) discuss the steady-state solution in more detail including nondegenerate operation of the parametric amplifier and amplification of a chirped pulse.

The general solution of Eq. (43) for a quasi-continuous pump and an arbitrary input pulse shape is difficult to solve. In the limit $A_s \ll A_p$, however, Akhmanov *et al.* (1968b) showed that the coupled equations reduce to a single Ricatti equation with a solution given by

$$A_s(\beta,z) = A_s(\beta)\left\{ \exp(-\Gamma z) + \frac{\sinh \Gamma z}{\tau_{cr}A_0{}^2}\int_0^\infty A_s{}^2(\eta-t')\exp\left(-\frac{2t'}{\tau_{cr}}\right)dt'\right\}^{-1} \tag{48}$$

For a Gaussian input pulse $A_s(t) = A_{s0}\exp(-t^2/2\tau^2)$, the pulse length reduces according to Eq. (45). In addition, the pulse energy density increases until it reaches $W_{s_{lim}}$ given by Eq. (47). After reaching this energy density the pulse shortening causes the amplitude to increase by $A_s \simeq A_{s0}(\tau_{cr}/\tau)^{1/2}$.

Figure 3 shows the pulse formation for a pulse with a sharp boundary of the form $A_s(t) = A_{s0}(t/\tau)^m$. The amplification process forms a quasistationary

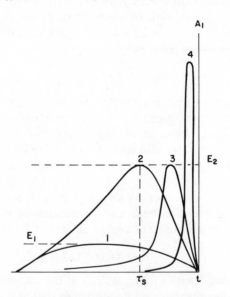

Fig. 3. Parametric pulse narrowing in a quasi-cw pump field: Curve (1) shows the input pulse with $\tau > \tau_{cr}$; curve 2 the amplified pulse with $A_s \simeq E_p$; curve 3 the quasi-stationary pulse with $\tau_f \simeq \tau_{cr}$, curve 4 the pulse showing exponential amplitude growth at constant energy density $W_{s_{lim}}$ [after Akhmanov *et al.* (1968a)].

pulse of energy density $W_{s_{lim}}$ with a front edge characterized by $\tau_f \simeq$ $(\tau, \tau_{cr})_{min}$. The pulse then narrows at constant energy density with a resultant amplitude increase. The limit to the pulse narrowing is finally set by dispersion spreading and the equivalent "confocal" length given by Eq. (35).

Table III gives calculated values for l_τ and b_τ at $\tau = 1$ psec for four representative nonlinear crystals. In addition, the table gives estimates for the limiting pulse duration τ_{lim} assuming group velocity matching where

$$\tau_{lim} \approx \sqrt{2}\,\tau/(\Gamma z)^{1/2}, \qquad z = b_\tau \qquad (49)$$

The values of Γ_{max} used in estimating τ_{cr} and τ_{lim} are based on the burn power density estimates given in Table II. Thus the Γ_{max} values are conservative for short pulses where the burn density is expected to be much higher (for further discussion see Section IV).

Table III

Properties of Nonlinear Crystals for Picosecond Pulse Amplification

Crystal	Γ_{max} (1/cm)	$\beta \times 10^{12}$ (sec/cm)	$g \times 10^{27}$ (sec^2/cm)	l_τ (cm) ($\tau = 1$ psec)	b_τ (cm) ($\tau = 1$ psec)	τ_{cr} (psec)	τ_{lim} (psec)
ADP							
$\lambda_p = 0.266\,\mu m$	2.3	0.37	1.35	4.8	740	0.16	0.034
LiNbO$_3$							
$\lambda_p = 0.532\,\mu m$	2.9	1.04	5.65	1.7	177	0.36	0.063
CdSe							
$\lambda_p = 1.833\,\mu m$	1.3	0.77	4.2	2.30	238	0.59	0.080
CdGeAs$_2$							
$\lambda_p = 5.3\,\mu m$	1.2	0.30	16	5.9	62	0.25	0.16

Table III shows that parametric amplifiers may provide a source of tunable subpicosecond pulses. The limiting pulse narrowing by parametric amplifiers is not surprising since the parametric gain bandwidth depends on crystal dispersion and is independent of amplifier gain. Thus, unlike lasers, wide bandwidths with high gains are possible. In addition, crystals allow both phase matching and group velocity matching so that pulse narrowing is possible to the limit set by group velocity dispersion.

E. Parametric Oscillator Threshold

In a typical parametric oscillator shown schematically in Fig. 1a, the pump makes a single pass through the nonlinear crystal. The generated idler and signal waves $E_1(l)$ and $E_2(l)$ grow according to Eqs. (16a) and (16b) during the single pass in the pump wave direction. Following reflection and the backward trip in the cavity, the waves are again traveling with the pump and are amplified. The gain is thus single pass and the corresponding loss is the round trip electric field loss α.

For the small gain case and $\Delta k = 0$, the threshold condition is

$$E_1(l = 0) e^{\alpha_1 l} = E_1(0) \cosh \Gamma l + i(\kappa_1/\Gamma) E_3 E_2^*(0) \sinh \Gamma l \qquad (50a)$$

and

$$E_2(l = 0) e^{\alpha_2 l} = E_2(0) \cosh \Gamma l + i(\kappa_2/\Gamma) E_3 E_1^*(0) \sinh \Gamma l \qquad (50b)$$

Letting $e^{\alpha_1 l} \simeq 1 + \alpha_1 l$ and $e^{\alpha_2 l} \simeq 1 + \alpha_2 l$ and taking the complex conjugate of Eq. (50b), the solution of the two simultaneous equations is

$$\cosh \Gamma l = 1 + \frac{\alpha_1 l \cdot \alpha_2 l}{2 + \alpha_1 l + \alpha_2 l} \qquad (51)$$

For low loss at both waves, or the doubly resonate oscillator case (DRO), Eq. (51) reduces to

$$\Gamma^2 l^2 \simeq \alpha_1 l \cdot \alpha_2 l = a_1 a_2 \qquad (52)$$

where for small losses the round trip electric field loss αl equals the single pass power loss a.

If only one wave is resonant, the singly resonant oscillator (SRO) condition, Eq. (51) with $\alpha_1 l \gg \alpha_2 l$ reduces to

$$\Gamma^2 l^2 \simeq 2\alpha_2 l = 2a_2 \qquad (53)$$

A comparison of Eqs. (52) and (53) shows that the singly resonant oscillator has a threshold that is $2/a_1$ times that of the doubly resonant oscillator. For a single-pass power loss of 1% the SRO threshold is thus 200 times the DRO threshold. This of course assumes that both waves of the DRO are resonant simultaneously. In practice, this requires a single-frequency pump source and careful cavity control to assure resonance at both parametric waves. Lack of cavity length control or pump frequency control leads to instabilities in the threshold condition resulting in amplitude fluctuations of the oscillator. For these reasons, the lower threshold advantage of the DRO is not utilized except where necessary such as for cw operation. For example, if $a_1 = a_2 = 2\%$, then $\Gamma^2 l^2 = 4 \times 10^{-4}$ for a DRO. For a 5-cm $LiNbO_3$ crystal the required pump power density at $\lambda_p = 0.532 \, \mu m$ to achieve threshold is only 3.13×10^{-4} MW/cm^2. If the pump beam is focused into the crystal with a confocal parameter b equal to the crystal length, the effective pump beam area is $A = l\lambda_p/2n = 1.22 \times 10^{-4}$ cm^2. The threshold pump power is therefore 38 mW which is easily obtained on a cw basis. The corresponding SRO threshold for the same 2% single-pass loss is 100 times higher or approximately 3.8 W.

The possibility of cw operation was first pointed out by Boyd and Ashkin (1966). It was not until 1968 that Smith *et al.* (1968b) using a 5-mm crystal of $Ba_2NaNb_5O_{15}$ pumped at 0.532 μm achieved cw operation at degeneracy. The threshold for that experiment was 45 mW. Later, Smith demonstrated a cw pumped oscillator with only 3-mW threshold.

Using an idea proposed by Harris (1966) that the full power of a multiaxial mode laser may be used to pump a parametric oscillator if the DRO axial mode interval equals that of the pump, Byer *et al.* (1968) constructed a visible cw DRO using a 1.7-cm $LiNbO_3$ crystal. This oscillator tuned from 0.68 to 0.71 μm and 1.9 to 2.1 μm for a 0.5145-μm argon–ion laser pump. It achieved a threshold of less than 500 mW in a long cavity that included an internal lens to achieve proper focusing.

The low gains and resultant difficulty of constructing cw oscillators has held back work in this area. Recently, however, Laurence and Tittel (1971a) and Weller and Andrews (1972) extended the work on cw argon–ion laser pumped $Ba_2NaNb_5O_{15}$ parametric oscillators. Weller and Andrews achieved 60 mW of output at a total efficiency of 15% in a tuning range from 0.93 to 1.15 μm.

Although a cw SRO parametric oscillator has not yet been demonstrated, it should be possible with existing $LiNbO_3$ crystals and high-power argon–ion lasers. In this case, the full multiaxial mode power of the laser could be used to achieve threshold. In addition, since only one frequency is resonant, the SRO stability should be improved relative to the DRO. A possible aid in achieving threshold for a cw SRO would be to use a round trip pump wave as opposed to the usual single-pass configuration.

The resultant net gain of the round trip SRO has been considered by Harris (1969b). The analysis is a straightforward extension of Eqs. (50a) and (50b) to include the net gain after an arbitrary phase shift suffered upon reflection from the end mirror of the SRO cavity. The importance of phase on the gain of an amplifier and DRO has also been considered by Smith (1970) and Bjorkholm *et al.* (1970).

At $\Delta k = 0$ the net round trip SRO gain for a zero phase shift is $\Gamma^2(2l)^2$. Thus the threshold is reduced by a factor of four relative to the single pass gain case. For a phase shift upon reflection of $\Delta(\phi_p - \phi_s - \phi l) = \pi/2$ the gain of the first pass is exactly cancelled by the return pass so that the net gain is zero. As shown by Harris (1969b), however, the gain is not zero at $\Delta k \neq 0$ and has a maximum of $0.4\Gamma^2(2l)^2$ at $|\Delta k|(2l) \gtrsim 2\pi$. Therefore, the round trip SRO reduces threshold compared to the single-pass SRO by 1.6 to 4 times. Here we have assumed that all three frequencies are perfectly reflected—a situation that may be difficult to achieve in practice.

Harris (1969b) extended the above analysis to include a gas cell between two successive nonlinear crystals, one inverted with respect to the other. This scheme allows for the possibility of locking a parametric oscillator to an absorbing or phase-shifting transition in a gas. The net gain of the two inverted crystals is nonzero at $\Delta k = 0$ if the gas cell blocks the improperly phased non-resonant wave generated by the first crystal. The second crystal then generates gain over the gas cell absorption frequency range which may be much less than the oscillator's gain bandwidth.

Initial experiments (Miles, 1972) have shown that locking does occur. Cavity instabilities and the resultant frequency fluctuations made interpretation of the results difficult, however. This type of locking is unique to parametric devices where the gain is dependent on the relative phase of the interacting waves.

The threshold conditions, Eqs. (52) and (53), for a SRO and DRO have been investigated in detail by Falk (1971). He considered the effects of some feedback on the stability and threshold of a nearly singly resonant oscillator. Starting with Eqs. (16a) and (16b), we find following the substitution $E_i = \mathscr{E}_i e^{i\phi_i}$

$$\mathscr{E}_1(l)\, e^{\alpha l} e^{-i\Delta k l/2} = \mathscr{E}_1(0)\left[\cosh gl - \frac{i\,\Delta k}{2g}\sinh gl\right]$$

$$+ (i\kappa_1\mathscr{E}_3)\frac{\mathscr{E}_2(0)}{g}\, e^{i\Delta\phi}\sinh gl \qquad (54a)$$

and

$$\mathscr{E}_2(l)\, e^{\alpha l} e^{-i\Delta k l/2} = \mathscr{E}_2(0)\left[\cosh gl - \frac{i\,\Delta k}{2g}\sinh gl\right]$$

$$+ (i\kappa_2\mathscr{E}_3)\frac{\mathscr{E}_1(0)}{g}\, e^{i\Delta\phi}\sinh gl \qquad (54b)$$

where $\Delta k = k_3 - k_2 - k_1$ and $\Delta\phi = \phi_3 - \phi_2 - \phi_1$. If we solve for $|\mathscr{E}_1(l)|^2$, the idler power as a function of l, and maximize with respect to $\Delta\phi$, we find for the optimum phase $\Delta\phi_m$ that

$$\tan\Delta\phi_m = 2g/(\Delta k \tanh gl) \qquad (55)$$

which in the limit of low gain reduces to

$$\cos(\Delta\phi_m + \tfrac{1}{2}\Delta k\, l) = 0 \qquad (56)$$

or

$$\Delta\phi_m + \tfrac{1}{2}\Delta k\, l = \pm\tfrac{1}{2}\pi$$

By inspection of Eqs. (54a) and (54b) or by integrating Eq. (16c), assuming small changes in $\mathscr{E}_1$ and $\mathscr{E}_2$ with z, we see that the phase $\Delta\phi + \tfrac{1}{2}\Delta k\, l = -\tfrac{1}{2}\pi$ converts power from the pump to the signal and idler waves and the opposite phase converts power back to the pump wave (Smith, 1970).

Falk (1971) extended the threshold results of Eqs. (52) and (53) by including the possibility of phase shifts upon reflection and traversal of the optical cavity. Letting $r = Re^{i\psi}$ be the round trip electric field reflectivity of the cavity, the threshold condition becomes

$$R_1 e^{i\psi_1} \mathscr{E}_1(l) = \mathscr{E}_1(0), \qquad R_2 e^{i\psi_2} \mathscr{E}_2(l) = \mathscr{E}_2(0) \tag{57}$$

Using Eq. (55), Eq. (54b) and the complex conjugate of Eq. (54a) can be written in the form

$$\mathscr{E}_1(l) e^{i\Delta kl/2} = \mathscr{E}_1(0) \exp\left(\frac{i\pi}{2} - i\,\Delta\phi_m\right)\left[1 + \frac{\Gamma^2}{g^2}\sinh^2 gl\right]^{1/2}$$

$$+ \frac{\Gamma^2 \mathscr{E}_2(0)}{g\kappa_2 E_p}\exp\left[-i\left(\frac{\pi}{2}\right) - i\,\Delta\phi\right]\sinh gl$$

and

$$\mathscr{E}_2(l) e^{-i\Delta kl/2} = \mathscr{E}_2(0) \exp\left[-\left(\frac{i\pi}{2} + i\,\Delta\phi_m\right)\right]\left[1 + \frac{\Gamma^2}{g^2}\sinh^2 \Gamma l\right]^{1/2}$$

$$+ \frac{\Gamma^2 \mathscr{E}_1(0)}{g\kappa_1 E_p{}^*}\exp\left[i\left(\frac{\pi}{2}\right) + i\,\Delta\phi\right]\sinh gl$$

Substituting these equations into Eq. (57) and setting the determinant equal to zero gives the threshold condition

$$0 = 1 + R_1 R_2 e^{i(\psi_1 + \psi_2)} + i\left(1 + \frac{\Gamma^2}{g^2}\sinh^2 gl\right)^{1/2}$$

$$\times \left\{R_2 \exp\left[i\left(\psi_2 + \Delta\phi_m + \frac{\Delta k\,l}{2}\right)\right] - R_1 \exp\left[-i\left(-\psi_1 + \Delta\phi_m + \frac{\Delta k\,l}{2}\right)\right]\right\} \tag{58}$$

For the singly resonant oscillator with $R_2 = 0$, Eq. (58) simplifies to

$$1 = R_1 \left[1 + (\Gamma^2/g^2)\sinh^2 gl\right]^{1/2} \tag{59}$$

with the phase condition

$$\tfrac{1}{2}\Delta k\,l + \Delta\phi_m - \psi_1 = -\tfrac{1}{2}\pi \tag{60}$$

The phase condition agrees with our previous results for maximum amplifier gain except we have now included the possible cavity phase shift ψ_1. In the limit of small gain Eq. (59) reduces to

$$\Gamma^2 l^2 \simeq [2(1 - R_1)/R_1][\text{sinc}^2(\tfrac{1}{2}\Delta k\,l)]^{-1} \tag{61}$$

which agrees with the previous result [cf. Eq. (53)] for the threshold of a SRO if we make the identification that the single-pass power loss $a_1 = (1 - R_1)/R_1$ for small losses.

The threshold condition, Eq. (58), does not depend on the relative phase $\Delta\phi$. It does, however, depend on the phase shift $\psi_2 - \psi_1$ introduced by the cavity. For a round trip in the optical cavity

$$\Delta\psi = \psi_2 - \psi_1 = \psi_0 - 2\,\Delta k\,l \tag{62}$$

with

$$\psi_0 = (2\omega_p/c)[L + (n_p - 1)l]$$

where L is the cavity length and l the crystal length with index of refraction n_p at the pump wave. Using Eq. (62) and $\phi_m + \frac{1}{2}\,\Delta k\,l = -\frac{1}{2}\pi$ Eq. (58) can be written in the form

$$\frac{\Gamma^2 l^2}{2}\operatorname{sinc}^2\left(\frac{\Delta k\,l}{2}\right) = \frac{B\sin\{2\tan^{-1}[\sin\Delta\psi/(C+\cos\Delta\psi)]-\Delta\psi\}}{\sin\{\tan^{-1}[\sin\Delta\psi/(C+\cos\Delta\psi)]\}} - 1 \tag{63}$$

where, following Falk, we have defined

$$B \equiv R_1(1-R_2)^2/(R_1^2 - R_2^2) \qquad \text{and} \qquad C \equiv R_2(1-R_1^2)/R_1(1-R_2^2)$$

Equation (63) gives the threshold of a parametric oscillator for arbitrary phase shifts and mirror reflectivities. It is, however, unwieldy and difficult to apply. For ease of interpretation, Eq. (63) can be simplified in the limit of small $\Delta\psi$ to

$$\tfrac{1}{2}\Gamma^2 l^2\operatorname{sinc}^2(\tfrac{1}{2}\,\Delta k\,l) \simeq B(1-C)(1-\sin^2\Delta\psi) - 1 \tag{64}$$

which, for $\Delta\psi = 0$, reduces to

$$\tfrac{1}{2}\Gamma^2 l^2\operatorname{sinc}^2(\tfrac{1}{2}\,\Delta k\,l) = (1-R_1)(1-R_2)/(R_2+R_1) \tag{65}$$

or

$$\Gamma^2 l^2\operatorname{sinc}^2(\tfrac{1}{2}\,\Delta k\,l) \approx a_1 a_2$$

in agreement with our previous result for the threshold of a doubly resonant oscillator [cf. Eq. (52)].

Falk also investigated the frequency stability and threshold behavior of a parametric oscillator for a randomly varying $\Delta\psi$ as a function of the signal electric field reflectivity R_2. For $R_2 \ll R_1$ and $\psi \simeq 0$ so that $\Delta\psi \simeq \psi_2$ the frequency stability is dependent only on feedback at the nonresonant wave. For this case Eq. (58) reduces to

$$[\Gamma^2 l^2/2(1-R_1)]\operatorname{sinc}^2(\tfrac{1}{2}\,\Delta k\,l) \simeq 1 - 2R_2\cos\Delta\psi \tag{66}$$

in agreement with Falk (1971). [In Eq. (13) of Falk a factor $1-R_1$ has been omitted due to a printing error.]

Figure 4 shows the maximum frequency instability $|\tfrac{1}{2}\,\Delta k\,l|_{\max}$ and the average frequency instability $|\tfrac{1}{2}\,\Delta k\,l|_{\text{ave}}$ as a function of R_2 for small reflectivities at the nonresonant wave. The average frequency instability is defined as the average oscillation frequency as $\Delta\psi$ varies over the interval 0 to 2π. As R_2 increases, $|\tfrac{1}{2}\,\Delta k\,l|_{\max}$ approaches $\tfrac{1}{4}\pi$ and $|\tfrac{1}{2}\,\Delta k\,l|_{\text{ave}}$ approaches $\tfrac{1}{8}\pi$. The

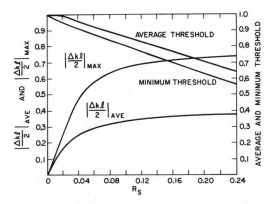

Fig. 4. Frequency instabilities and threshold variations as a function of R_2 for $R_2 \ll R_1$ in a DRO. Also shown is the average and minimum threshold of a DRO for small R_2. The thresholds are normalized to a SRO threshold with $R_1 \simeq 1$ [from Falk (1971)].

frequency instability $\frac{1}{2} \Delta k \, l = \frac{1}{4} \pi$ corresponds to one-half of the frequency spacing between successive dual cavity resonances and is therefore the maximum expected frequency instability. An important result of these calculations is that the frequency stability of a SRO is sensitive to even slight feedback at the nonresonant wave. For example, Fig. 4 shows that $\Delta k \, l \simeq 80\%$ of the average jumping expected for a DRO when $R_2 \simeq 10\%$ corresponding to a round trip power reflectivity of only 1%. Although experiments have not been performed to verify the expected frequency instability, the results compare favorably with previous spectral measurements of a DRO and SRO carried out by Bjorkholm (1968b).

In addition to the DRO and SRO with a single-pass pump wave generating single pass parametric gain, there exist obvious extensions involving a round trip pump wave. The advantage of reflecting the pump wave back through the nonlinear crystal is a reduction in threshold and possibly a higher conversion efficiency.

The round trip DRO has been discussed by Bjorkholm *et al.* (1970). The threshold is

$$\Gamma^2 l^2 \operatorname{sinc}^2 (\tfrac{1}{2} \Delta k \, l) = a_1 a_2 / [1 + R_3{}^2 + 2R_3 \cos(\Delta\psi + \Delta k \, l)] \qquad (67)$$

where R_3 is the electric field reflectivity of the second cavity mirror at the pump frequency. For $R_3 = 0$ this reduces to Eq. (65). The threshold depends on $\Delta\psi$ and $\Delta k \, l$; however, for $R_3{}^2 > 0.12$ the double pass DRO always has a lower threshold. Figure 5 shows the ratio of the minimum threshold of a single-pass DRO to double-pass DRO for various $\Delta\psi$. For the ideal case of $\Delta\psi = 0$ the threshold is reduced by a factor of four.

There are two variations of the round trip SRO; the first where only the pump wave is reflected by the second cavity mirror which is transparent to

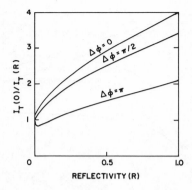

Fig. 5. The ratio of minimum single-pass DRO threshold ($\Delta k = 0$) to minimum double-pass DRO threshold ($\Delta k \neq 0$ in general) for several values of $\Delta\phi$ as a function of R_3 [after Bjorkholm *et al.* (1970)].

the nonresonant wave, and the second where all three waves are reflected by the second cavity mirror.

In the first case Bjorkholm *et al.* (1970) show that the threshold is

$$\Gamma^2 l^2 \operatorname{sinc}^2 (\tfrac{1}{2} \Delta k\ l) = 2a_1/(1 + R_p{}^2) \tag{68}$$

For this case the threshold is independent of the phase and for $R_p \simeq 1$ is a factor of two lower than the single-pass SRO. The only disadvantage of this device is the requirement for mirror coatings that reflect both the pump and the resonant wave.

The SRO where all three waves are reflected from the second cavity mirror has been discussed previously. For this case the threshold is dependent on the relative phase of the reflected waves. In the low gain limit the reduction in threshold is the same as that for the round trip DRO given by Eq. (67) if the reflectance is equal for all three waves at the second mirror. [Here $\cos(\Delta\psi + \Delta k\ l) = \pm 1$ for proper and improper phase upon reflection.) As the gain is increased toward the high gain limit, the net double-pass SRO gain approaches $(\cosh^2 \Gamma l \pm \sinh^2 \Gamma l)^2 - 1$. Thus the net gain varies between zero and $e^{4\Gamma l}$ depending on phase. Recently, Herbst and Byer (1972) used a gold-coated mirror to reflect all three waves and thus reduce the threshold of a SRO infrared CdSe parametric oscillator.

F. Rise Time

Thus far we have discussed the threshold of a parametric oscillator under steady-state conditions. In fact, most optical parametric oscillators are presently constructed using a Q-switched pump source with pump pulse lengths that typically vary from 10 nsec to 1 μsec. When the pump is incident on the nonlinear crystal the initial signal and idler fields are amplified. The oscillator is not "on," however, until the fields are amplified from the initial noise level to a magnitude of the order of the pump field.

The number of round trips in the optical cavity necessary to amplify fully the signal and idler waves multiplied by the cavity round trip time leads to the rise time necessary to achieve threshold. The excess gain which is proportional to the incident pump power, the pump pulse length, and the cavity length are important parameters in the threshold of a pulsed parametric oscillator.

The rise time of a DRO for a step input pump pulse has been discussed by Byer (1968) and Kreuzer (1968). Recently, Pearson et al. (1972b) extended the analysis to include an arbitrary input pulse shape for both the DRO and SRO cases.

Before discussing these results, we give an approximate solution to the rise time of a DRO and SRO based on the model discussed above. In particular, the rise time can be expressed as an additional equivalent loss to be exceeded to reach threshold (Kildal, 1972). For equal signal and idler losses, Eq. (50) reduces to

$$P_1(t) = P_1(t = 0) \exp[2\Gamma(t)l - 2a] \tag{69}$$

where $P_1(t = 0)$ is the initial noise power present at the time the pump pulse is incident on the crystal. The power P_1 grows for the duration of the pump pulse τ_p at a rate proportional to the excess gain. Letting the number of single-pass trips during the pump pulse duration be $n = \tau_p/\tau_{cav} = \tau_p c/L_c$, where L_c is the equivalent length of the cavity plus crystal, we have

$$\ln[P_1(T)/P_1(0)] = (c/L_c) \int_0^{\tau_p} (2\Gamma(t)l - 2a) \, dt$$

$$(\Gamma^2 l^2)^{1/2} = a + (1/2n) \ln[P_1(T)/P_1(0)] \tag{70}$$

where the second term is an equivalent loss due to the finite rise time of the DRO. As an example, if we let $P_1(t)/P_1(0) = 10^{12}$, $\tau_p = 200$ nsec, and $L_c = 6$ cm, the equivalent loss is 1.4%. If, however, $\tau_p = 20$ nsec, the loss increases to 14%.

The exact value of $P_1(t)/P_1(0)$ is not very critical since the ratio appears in a logarithm. We have, however, let $P_1(0)$ equal the noise power present in the cavity for single mode. As discussed previously, the noise is an equivalent single photon so that $P_1(0) \approx h\omega_1/\tau_{cav}$. For definiteness we let $P_1(t) = aP_3$, where P_3 is the peak pump power.

Applying the above procedure to the SRO we have

$$P_1(t) = P_1(0) \exp(\Gamma^2 l^2 - 2a_1) \tag{71}$$

so that

$$\Gamma^2 l^2 = 2a_1 + (1/n) \ln[P_1(t)/P_1(0)] \tag{72}$$

For the SRO with the above parameters the equivalent loss is 2.8 and 28% for $\tau_p = 200$ and 20 nsec, respectively.

A more direct way to determine the rise time of a parametric oscillator is to integrate Eq. (9) with time dependence included. For the DRO we assume that the spatial variations of E_1 and E_2 are negligible so that the coupled equations reduce to

$$\frac{1}{c_1}\frac{\partial E_1}{\partial t} + \alpha_1 E_1 = i\kappa_1 E_3 E_2^* \exp(i\,\Delta k\,r),$$

$$\frac{1}{c_2}\frac{\partial E_2}{\partial t} + \alpha_2 E_2 = i\kappa_2 E_3 E_1^* \exp(i\,\Delta k\,r) \tag{73}$$

where $c_1 = c/n_1$ and $c_2 = c/n_2$ are the phase velocities. Assuming $\Delta k = 0$, $\alpha_1 = \alpha_2 = a/L_c$, where L_c is the effective cavity length, and a step pump source turned on at $t = 0$ the calculated rise time is

$$\tau_R = (L_c/c)\ln[P_1(t)/P_1(0)]/2a[\sqrt{N}-1] \tag{74}$$

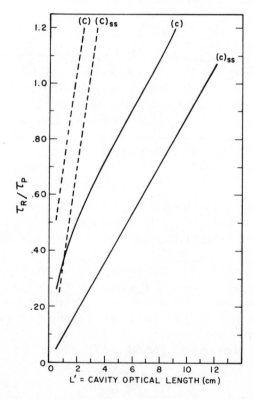

Fig. 6. DRO rise time τ_R per pump pulse width τ_p versus cavity length. Case (c) is exact and (c$_{ss}$) steady-state analysis at $\tau_p = 200$ nsec, and case (C) is exact and (C$_{ss}$) is steady-state analysis at $\tau_p = 50$ nsec. Here $\alpha = 1\%$ [after Pearson *et al.* (1972b)].

where $N = \Gamma^2 l^2/a^2$ is the number of times above threshold. When $\tau_R = \tau_p$ this solution agrees with Eq. (70). At degeneracy Eq. (73) integrates directly for pump pulses of the form $E_p(t) = E_p f(t)$. For example, a Gaussian pump pulse $E_p(t) = E_p \exp(-t^2/\tau^2)$, where $\tau^2 = \tau_p^2/2 \ln 2$ and τ_p is the full width at half maximum of the pump intensity, modifies Eq. (74) by

$$\sqrt{N} \rightarrow \sqrt{N}(1/\tau_R) \int_0^{\tau_R} f(t)\, dt \simeq 0.88\sqrt{N} \tag{75}$$

for $\tau_R < \tau_p$.

Pearson *et al.* (1972b) considered the rise time of parametric oscillators in detail. Figure 6 shows the calculated ratio of rise time to pump pulse length τ_R/τ_p versus cavity length for various values of N_0, where $N_0 = N(100a_1)$ with N the number of times above threshold and a the percent single-pass power loss of the resonant wave. For 1% loss N_0 is the number of times above the cw threshold. For the DRO Pearson *et al.* (1972b) plot, a minimum number of times above cw threshold the oscillator must be pumped in order to achieve a pulsed threshold. This result is shown in Fig. 7. The solid lines are the exact

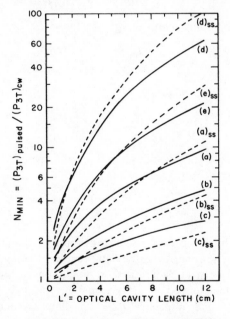

Fig. 7. Minimum number of times the cw DRO threshold to reach threshold for the pulse DRO versus cavity length. Dashed lines are steady-state analysis. [After Pearson *et al.* (1972b)]. For curve (a), $\tau_2 = 200$ nsec, $a = 1\%$; curve (b), $\tau_2 = 200$ nsec, $a = 2\%$; curve (c), $\tau_2 = 200$ nsec, $a = 4\%$; curve (d), $\tau_2 = 50$ nsec, $a = 1\%$; curve (e), $\tau_2 = 50$ nsec, $a = 2\%$.

result corresponding to the use of integral Eq. (75). The dashed results assume a steady-state square pump pulse solution given by Eq. (70). For most practical cases the steady-state and exact results are in good agreement. Figures 6 and 7 illustrate the importance of keeping the cavity length to a minimum in pulsed parametric oscillator operation.

Operation of pulsed parametric oscillators leads to consideration of a minimum threshold energy density rather than power density. The peak gain must remain on long enough to allow the fields to build to the threshold value. For the DRO the minimum threshold energy density follows from Eq. (70):

$$W_{min}^{DRO} > \frac{I_{pk}}{\sqrt{N-1}} \left(\frac{\tau_{cav}}{2a}\right) \ln\left(\frac{P_1(t)}{P_1(0)}\right) \tag{76}$$

When the DRO is pumped at peak intensities many times above threshold such that $\sqrt{N} > 1$, the minimum energy necessary to reach threshold approaches a limiting value proportional to $(I_{pk})^{1/2}$ and independent of the loss given by

$$W_{min}^{DRO} \simeq \frac{I_{pk}}{2\Gamma l} \frac{\tau_{cav}}{2} \ln\left(\frac{P_1(t)}{P_1(0)}\right) \tag{77}$$

For the SRO the minimum pump energy density is

$$W_{min}^{SRO} = \frac{I_{pk}}{N-1} \frac{\tau_{cav}}{2a} \ln\left(\frac{P(t)}{P_0}\right) \tag{78}$$

which in the limit $N > 1$ reduces to

$$W_{min}^{SRO} \simeq \frac{I_{pk}}{\Gamma^2 l^2} \tau_{cav} \ln\left(\frac{P_1(t)}{P_1(0)}\right) \tag{79}$$

It is interesting to note that the minimum threshold energy ratio for the DRO to SRO in the limit of $N > 1$ reduces to

$$\frac{W_{min}^{DRO}}{W_{min}^{SRO}} = \frac{N^{SRO} - 1}{(N^{DRO})^{1/2} - 1} \approx \frac{\Gamma l}{2} \tag{80}$$

which as expected shows the low threshold advantage of the DRO at low gains. For high gain operation, however, the SRO requires the minimum energy to achieve oscillation from consideration of oscillation build-up time. For a 5-cm LiNbO$_3$ crystal pumped at 20 MW/cm^2, a typical operating intensity, Table II shows that $\Gamma^2 l^2 = 25$ or $\Gamma l/2 = 2.5$. Thus we may expect that a partially reflecting DRO using LiNbO$_3$ at this gain will reach threshold and oscillate as a SRO. This may be important in determining the frequency and power stability of the oscillator in view of the expected frequency fluctuations for a low-gain DRO.

Table IV lists the minimum energy density W_{min}^{SRO} for SRO operation and the minimum energy J_{min}^{SRO} assuming optimum focusing in the limit $N > 1$. For the crystals listed, except $LiIO_3$, the threshold energy is less than 1 mJ and therefore easily reached with a low-energy Q-switched Nd : YAG laser source.

Table IV

Minimum Threshold Energy Density and Energy[a]

Crystal	l (cm)	λ_p (μm)	W_{min}^{SRO} (mJ/cm^2)	J_{min}^{SRO} [b] (μJ)
ADP	5	0.266	57	2.62
LiNbO$_3$	5	0.532	5.9	0.37
LiIO$_3$	1	0.694	1380	1360
CdSe	2	1.83	85	5.8
CdGeAs$_2$	1	5.3	190	16

[a] $L_c = 10$ cm optical length.
[b] Minimum pump energy assuming optimum focusing (cf Section I)..

G. Conversion Efficiency

Parametric oscillators are potentially efficient devices for converting fixed frequency pump radiation to tunable signal and idler output. In this section we examine the theoretical conversion efficiency for various parametric oscillator configurations. We proceed by considering a steady-state solution for interacting plane waves. In addition, we assume that the resonant fields vary slowly in passing through the nonlinear medium. This assumption allows solutions to be obtained that are valid even for high conversion efficiency without having to obtain exact solutions for coupled nonlinear equations.

Following the above discussion, modifications to the plane wave results for multimode Gaussian amplitude beams are discussed. In addition, non-steady-state results and exact solutions of the coupled equations are compared to the steady-state and slowly varying resonant wave approximation solutions.

The conversion efficiency for a DRO was first treated by Siegman (1962). His assumption that the spatial variation of the resonated waves is nearly constant is the basis for the following conversion efficiency calculations. For the DRO we assume that dE_2/dz and dE_1/dz are negligible so that Eq. (9) for the remaining forward and backward traveling pump waves becomes

$$d\mathscr{E}_3^+/dz = i\kappa_3\mathscr{E}_{20}\mathscr{E}_{10} \exp[-i(\Delta k\, z+\Delta\phi_+)] \tag{81a}$$

$$d\mathscr{E}_3^-/dz = -i\kappa_3\mathscr{E}_{20}\mathscr{E}_{10} \exp[i(\Delta k\, z-\Delta\phi_-)] \tag{81b}$$

where $E_i = \mathscr{E}_i e^{i\phi_i}$ and $+$ and $-$ refer to the forward and backward directions.

These equations can be integrated under the present assumption that $\mathscr{E}_{20}$ and $\mathscr{E}_{10}$ are independent of z. Integrating Eq. (81a) gives

$$\mathscr{E}_p{}^+(l) = \mathscr{E}_{p0} + i\kappa_3\mathscr{E}_{20}\mathscr{E}_{10}l\,\text{sinc}(\tfrac{1}{2}\,\Delta k\,l)\exp[-i(\tfrac{1}{2}\,\Delta k\,l + \Delta\phi_+)] \quad (82)$$

where $\mathscr{E}_{p0}$ is the incident pump field at $z = 0$. A similar expression follows integration of Eq. (81b) for $\mathscr{E}_3{}^-$ except that $\mathscr{E}_3{}^-(l) = 0$ is the assumed boundary condition. For maximum energy transfer from the pump to the signal and idler $\Delta\phi_+ + \tfrac{1}{2}\,\Delta k\,l = -\tfrac{1}{2}\pi$ which agrees with our previous result in Eq. (55).

We now use an extension of Eq. (12) to relate the signal and idler powers by

$$\mathscr{E}_{20}^2/\mathscr{E}_{10}^2 = \omega_2 n_1 a_1/\omega_1 n_2 a_2 \quad (83)$$

where as before a is the single-pass power loss. In addition, conservation of intensity leads to

$$n_3[\mathscr{E}_{p0}^2 - \mathscr{E}_p^{+2}(l) - \mathscr{E}_p^{-2}(0)] = 2a_2 n_2 \mathscr{E}_{20}^2 + 2a_1 n_1 \mathscr{E}_{10}^2 \quad (84)$$

Solving Eqs. (82)–(84) for the signal and idler output powers gives

$$\frac{\omega_3}{\omega_2}\frac{P_2}{P_{30}} = \frac{\omega_3}{\omega_1}\frac{P_1}{P_{30}} = 2\frac{1}{N}[\sqrt{N}-1] \quad (85)$$

where N is the number of times above threshold defined by $\Gamma^2 l^2 = a_1 a_2$ where Γ^2 is given by Eq. (13).

The conversion efficiency of the DRO is therefore

$$\eta = \frac{P_1 + P_2}{P_{30}} = \frac{2}{N}[\sqrt{N}-1] \quad (86)$$

which reaches a maximum of 50% at four times above threshold. The generated power is divided between the two waves in the ratio ω_2/ω_1 since equal signal and idler photons are generated.

The actual output power transmitted through the cavity mirrors with transmittance $T_1{}^2 = 1 - R_1{}^2$ is

$$\frac{\omega_3}{\omega_1}\frac{P_1}{P_3(\text{TH})} = 2\left(\frac{T_1{}^2}{a_1}\right)[\sqrt{N}-1] \quad (87)$$

where $P_3(\text{TH})$ is the threshold pump power and $a_1 = T_1{}^2 + a_D$ is the sum of the mirror transmittance and dissipative losses.

The transmitted pump power P_3' limits at the threshold power $P_3(\text{TH})$ which forms the basis for the optical limiter first proposed by Siegman (1962). Finally, the reflected pump power is given by

$$P_3(\text{R})/P_3(\text{TH}) = [\sqrt{N}-1]^2 \quad (88)$$

The reflected pump power is not truly "reflected" but is generated by sum generation of the back-traveling signal and idler waves. The back-generated pump acts to reduce the signal and idler powers and simultaneously feed power back into the pumping laser which induces pump laser frequency and power instabilities.

Bjorkholm (1969) analyzed the DRO without the back-generated pump wave and showed that its efficiency increases from 50 to 100% at $N = 4$. One method to ensure no backward pump generation is to construct a ring cavity oscillator. This oscillator is referred to as a "ring resonator oscillator" (RRO) which implies that it is doubly resonant. Recently Fischer (1970, 1971) treated the RRO in detail.

The conversion efficiency calculation for the RRO proceeds as for the DRO except there is only forward interaction. Equating the signal and idler powers coupled out of the cavity to the power lost by the pump and using Eq. (83) gives

$$\frac{\omega_3}{\omega_2}\frac{P_2}{P_{30}} = \frac{\omega_3}{\omega_1}\frac{P_1}{P_{30}} = \frac{4}{N}[\sqrt{N}-1] \tag{89}$$

where the threshold remains the same as for the DRO. Maximizing Eq. (89) with respect to N shows that the conversion efficiency $(P_2 + P_1)/P_{30} = 100\%$ at $N = 4$. Thus, in addition to eliminating feedback into the pump laser, the conversion efficiency is increased from 50 to 100% for the RRO.

The transmitted pump power for the RRO is

$$\frac{P_3'}{P_{30}} = 1 - \frac{4}{N}[\sqrt{N}-1] \tag{90}$$

Figure 8 shows the conversion efficiency and transmitted pump power for the

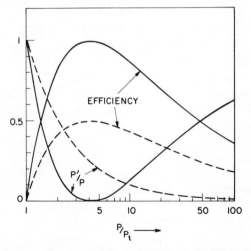

Fig. 8. Transmitted pump power P_3'/P_{30} and efficiency of a DRO and RRO versus the number of times above threshold in the plane wave limit. The dashed curves are for the DRO and the solid curves are for the RRO.

DRO and the RRO. The results are valid in the limit of low gain and interacting plane waves. Modifications of these results for Gaussian beams and
also for high gains are discussed later.

There exists one further form of the DRO in which the pump is reflected
from the second cavity mirror back through the nonlinear crystal. The
threshold for the double-pass DRO is reduced compared to that for the DRO
by between two and four times for $R_3 \simeq 1$ [cf. Eq. (67) and Fig. 5]. In addition,
the conversion efficiency is increased as shown by Bjorkholm et al. (1970) to

$$\eta = \frac{1 + R_3^2 + 2R_3 \cos(\Delta\phi + \Delta k \, l)}{1 + R_3 \cos(\Delta\phi + \Delta k \, l)} \frac{2}{N} [\sqrt{N} - 1] \tag{91}$$

which, for $R_3 = 1$ and the proper phase, reaches 100% at $N = 4$. The double-
pass DRO thus has 100% conversion efficiency similar to the ring resonator
but in addition a decreased threshold. It has a disadvantage, however, in that
the threshold conversion efficiency and optimum Δk depend on the phase
change $\Delta\phi$ induced upon reflection.

All of the DRO configurations suffer from the disadvantage of requiring
simultaneous resonance at the signal and idler waves within a single cavity.
Due to crystal dispersion, the axial mode interval at the signal and idler
frequencies are not the same so that the particular signal and idler cavity modes
that exactly sum to the pump frequency may occur off gain center. Further-
more, slight shifts in crystal index or cavity length cause the modes that align
to shift about rapidly. Typically a few modes within a "cluster" are aligned
or nearly aligned and the next "cluster" of modes is spaced a number of axial
mode intervals away. Giordmaine and Miller (1965, 1966) first discussed the
cluster effect which has since been observed by Bjorkholm (1968a). The
frequency instability due to double resonance is the main reason for sacrificing
the low threshold of the DRO for operation as a single resonant oscillator
whenever possible.

The SRO was first demonstrated by Bjorkholm (1966) and later analyzed
by Kreuzer (1969a). It has since become the most common parametric
oscillator configuration for a number of reasons, which include: lack of the
cluster effect, use of the full multiaxial mode power of the pump laser, ease of
mirror and cavity design, good conversion efficiency, and frequency and
output power stability. These advantages more than offset the disadvantage
of increased threshold relative to the DRO.

For the SRO in which the signal is resonant we assume $\mathscr{E}_2(z) = \mathscr{E}_{20}$ and
have for the pump and idler waves that

$$d\mathscr{E}_3^+(z)/dz = i\kappa_3 \mathscr{E}_{30} \mathscr{E}_1^+(z) \exp[-i(\Delta k \, z + \Delta\phi_+)] \tag{92a}$$

and

$$d\mathscr{E}_1^+(z)/dz = i\kappa_1 \mathscr{E}_{20} \mathscr{E}_3(z) \exp[i(\Delta k \, z + \Delta\phi_+)] \tag{92b}$$

Solving these equations with the boundary conditions $\mathscr{E}_1(0) = 0$ and $\mathscr{E}_3(0) = \mathscr{E}_{30}$ gives

$$|\mathscr{E}_1{}^+(z)|^2 = \mathscr{E}_{30}^2 \left(\frac{\omega_1 n_3}{\omega_3 n_1}\right) \frac{\sin^2 \beta z}{1 + (\Delta k/2\alpha)^2} \tag{93a}$$

and

$$|\mathscr{E}_3{}^+(z)| = \mathscr{E}_{30}^2 \left[\cos^2 \beta z + \left(\frac{\Delta k}{2\beta}\right)^2 \sin^2 \beta z \right] \tag{93b}$$

where

$$\beta^2 = \alpha^2 + (\tfrac{1}{2}\Delta k)^2 \tag{94}$$

and

$$\alpha^2 = \kappa_1 \kappa_3 |\mathscr{E}_{so}|^2 \tag{95}$$

Using Eq. (12) to relate the signal power to the idler power, Eq. (93) can be written as

$$|\mathscr{E}_2(l)|^2 - |\mathscr{E}_{20}|^2 = \kappa_1 \kappa_2 |\mathscr{E}_{30}|^2 |\mathscr{E}_{20}|^2 l^2 \operatorname{sinc}^2 \beta l$$

Finally, setting the signal gain $|\mathscr{E}_2(l)|^2/|\mathscr{E}_{20}|^2 - 1$ equal to the round trip loss $2a_2$ the signal power as a function of the incident pump power is given by the implicit relation

$$\kappa_1 \kappa_2 |\mathscr{E}_{30}|^2 l^2 \operatorname{sinc}^2 \beta l = 2a_s \qquad \text{or} \qquad \operatorname{sinc}^2 \beta l = 1/N \tag{96}$$

where N is the number of times above threshold. Note that for $\beta = 0$ this reduces to the previously derived SRO threshold condition $\Gamma^2 l^2 = 2a_s$.

At line center the SRO reaches 100% efficiency for $P_{30} = (\pi/2)^2 P_3(\text{TH})$. As the pumping level increases beyond $(\pi/2)^2$ the pump wave begins to grow at the expense of the signal. Figure 9 shows the transmitted pump power for the SRO as a function of the number of times above threshold in the low-gain plane wave approximation.

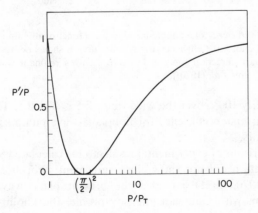

Fig. 9. Transmitted pump power $P_3{'}/P_{30}$ versus the number of times above threshold for a SRO with $\Delta k = 0$ in the plane wave limit.

The SRO can also be extended to the double-pass configuration. For the case where all three waves are reflected the threshold is reduced by two to four times in the low-gain limit depending on the phase change introduced by the reflection. If, however, only the pump and resonant wave are reflected, the threshold is reduced by up to two times without phase dependence. The threshold is independent of phase since the back-generated nonresonant wave builds from the noise in proper phase for maximum gain. Of more interest is the high conversion efficiency over a wider range of pump powers. Bjorkholm *et al.* (1970) analyzed the SRO with pump reflectivity and showed that the resonant power is determined by

$$\text{sinc}^2(\beta l) = \frac{2}{(1 + R_3{}^2 \cos \beta l)} \frac{1}{N} \tag{97}$$

where $R_3{}^2$ is the pump intensity reflectivity of the second cavity mirror.

The conversion efficiency to the nonresonant wave, proportional to $1 - [P_3{}^+(l)/P_{30}]$, and to the resonant wave, proportional to the total pump depletion is shown in Fig. 10. Of particular interest is the resonant wave

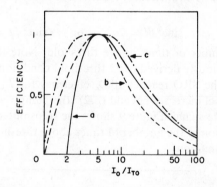

Fig. 10. Photon efficiency of a double-pass SRO as a function of number of times above threshold. Curve a shows the efficiency for $R = 0$; curve b shows nonresonant wave efficiency in the forward direction for $R = 1$; curve c shows resonant wave efficiency for $R = 1$ [after Bjorkholm *et al.* (1970)].

efficiency of nearly 100% over the wide range $2.5 \leqslant N \leqslant 8.5$. This could lead to improved conversion efficiency for nonplane wave pump sources used in actual experiments.

Early comparisons of experimental results to the calculated SRO efficiency and pump depletion did not show close agreement. Bjorkholm (1971) and later Fischer (1970, 1971) extended the present plane wave analysis to multimode beams with Gaussian intensity profile. Bjorkholm then demonstrated that many previous observations agreed very well with predictions based on the Gaussian intensity model.

The model considers a multimode plane wave pump beam with a Gaussian intensity profile

$$I(r) = I_0 \exp(-2r^2/w_0{}^2) \tag{98}$$

where w_0 is the beam radius, incident on the parametric oscillator. Each part of the beam is taken to act independently so that parts of the incident pump above threshold take part in the parametric conversion and suffer depletion while the remainder of the beam is not affected. Thus this model does *not* apply to single transverse mode parametric oscillators which are treated in Section I. The analysis consists of integrating over the Gaussian profile, treating the above threshold and below threshold regions independently.

For the DRO the transmitted intensity I_3' is given by

$$\begin{aligned} I_3' &= I_3(\text{TH}), \qquad I_{30} > I_3(\text{TH}) \\ I_3' &= I_{30}, \qquad\quad\ I_{30} < I_3(\text{TH}) \end{aligned} \tag{99}$$

The transmitted power through the DRO is therefore

$$P_3' = \int_0^\infty I'(r)\,2\pi r\,dr$$

$$2\pi \int_0^R I_3(\text{TH})\,r\,dr + 2\pi \int_R^\infty I_{30} \exp(-2r^2/w_0{}^2)\,r'\,dr' \tag{100}$$

where R is the beam radius at which $I_{30} = I_3(\text{TH})$. Carrying out the integration

$$P_3' = 2\pi I_3(\text{TH})\,R^2 + I_{30}(\tfrac{1}{2}\pi w_0{}^2) \exp(-2R^2/w_0{}^2) \tag{101}$$

The total power in a Gaussian beam is given by $P = (\tfrac{1}{2}\pi w_0{}^2)I_0$ so that the threshold power is $P_3(\text{TH}) = (\tfrac{1}{2}\pi w_0{}^2)I_3(\text{TH})$. Using these definitions Eq. (101) reduces to

$$P_3'/P_3(\text{TH}) = 1 + \ln(N) \tag{102}$$

The efficiency

$$\eta = \frac{P_1 + P_2}{P_{30}} = \frac{4}{N}\left[\sqrt{N} - 1 - \ln\sqrt{N}\right] \tag{103}$$

as a function of the number of times above threshold N is less than the plane wave efficiency by $-(4/N)\ln N$. Compared to the plane wave case, the DRO pumped with a Gaussian intensity does not show power limiting and reaches a maximum efficiency of only 41% at $N = 12.5$.

For the ring resonator the transmitted pump power is found to be

$$\frac{P_3'}{P_{30}} = 1 - \frac{4}{N}\left[\sqrt{N} - 1\right] + \frac{4}{N}\left[\ln N - (\sqrt{N} - 1)\right] \tag{104}$$

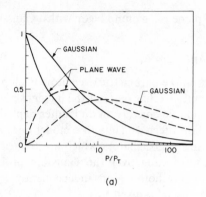

(a)

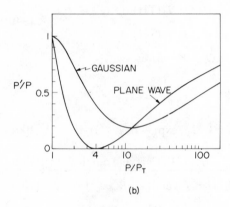

(b)

Fig. 11. (a) Transmitted pump power P_3'/P_{30} (solid line) and efficiency (dashed line) versus the number of times above threshold for a DRO; (b) Transmitted pump power P_3'/P_{30} versus the number of times above threshold for a RRO for a Gaussian intensity wave [after Bjorkholm (1971)].

and the efficiency becomes

$$\eta = 1 - (P_3'/P_{30}) \tag{105}$$

which is only 82% at $N = 4$ instead of 100% as in the plane wave analysis. Figure 11 shows the transmitted pump power and efficiency for the DRO and ring resonator with a Gaussian pump intensity. The efficiency for the plane wave analysis is shown for comparison.

The SRO can be treated in a similar manner. In this case, however, the solution remains in an integral form since the resonant signal power is not solved for explicitly. For further discussion see Bjorkholm (1971). Figure 12 shows a comparison of the transmitted pump power for the plane wave and Gaussian SRO as a function of N. For a multimode Gaussian intensity profile beam the maximum efficiency only reaches 71% at 6.5 times above threshold.

The conversion efficiencies calculated for the various optical parametric oscillator (OPO) configurations apply to operation in the low-gain essentially steady-state condition. In practice, the steady-state condition is achieved only

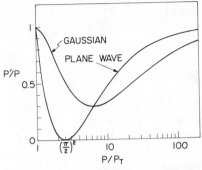

Fig. 12. Transmitted pump power P_3'/P_{30} versus the number of times above threshold for a SRO for a Gaussian intensity wave [after Bjorkholm (1971)].

after a finite build-up time. Thus the energy conversion efficiency is less than the power conversion efficiency by approximately the fraction of the pumping time that the oscillator is above threshold. Again, operation of an OPO with a short cavity length to reduce build-up time leads to improvements in the conversion efficiency.

Up to this point we have considered OPO conversion efficiency with the approximation that the resonant wave is slowly varying within the cavity. The exact solution for a three wave parametric interaction has been given by Armstrong *et al.* (1962). Bey and Tang (1972) have extended Armstrong's results to the singly resonant parametric oscillator and to an OPO with intracavity up-converter and doubler.

If we let $E_i(r, \omega) = \mathscr{E}_i e^{i\phi_i}$, the real and imaginary parts of Eq. (9) become

$$d\mathscr{E}_1/dz = -\kappa_1 \mathscr{E}_2 \mathscr{E}_3 \sin\theta, \qquad d\mathscr{E}_2/dz = -\kappa_2 \mathscr{E}_3 \mathscr{E}_1 \sin\theta$$

$$d\mathscr{E}_3/dz = \kappa_3 \mathscr{E}_1 \mathscr{E}_2 \sin\theta$$

$$\frac{d\theta}{dz} = \Delta k + \left(\frac{\kappa_3 \mathscr{E}_1 \mathscr{E}_2}{\mathscr{E}_3} - \frac{\kappa_2 \mathscr{E}_3 \mathscr{E}_1}{\mathscr{E}_2} - \frac{\kappa_1 \mathscr{E}_2 \mathscr{E}_3}{\mathscr{E}_1} \right) \cos\theta \qquad (106)$$

where

$$\theta = \Delta k\, z + \phi_3(z) - \phi_2(z) - \phi_1(z)$$

The total power flow is

$$I = \tfrac{1}{2} c\varepsilon_0 [n_1 \mathscr{E}_1{}^2 + n_2 \mathscr{E}_2{}^2 + n_3 \mathscr{E}_3{}^2] \qquad (107)$$

If we define new variables

$$u_1 = \left(\frac{n_1 \varepsilon_0 c}{2\omega_1 I} \right)^{1/2} \mathscr{E}_1, \qquad u_2 = \left(\frac{n_2 \varepsilon_0 c}{2\omega_2 I} \right)^{1/2} \mathscr{E}_2, \qquad u_3 = \left(\frac{n_3 \varepsilon_0 c}{2\omega_3 I} \right)^{1/2} \mathscr{E}_3$$

$$(108)$$

where u^2 is the photon flux and

$$\zeta = \frac{d}{c} (n_1 \mathscr{E}_1{}^2 + n_2 \mathscr{E}_2{}^2 + n_3 \mathscr{E}_3{}^2)^{1/2} \left(\frac{\omega_1 \omega_2 \omega_3}{n_1 n_2 n_3} \right)^{1/2} z$$

is a normalized length, then Eq. (106) reduces to

$$du_1/d\zeta = -u_2 u_3 \sin\theta, \qquad du_2/d\zeta = -u_3 u_1 \sin\theta, \qquad du_3/d\zeta = u_1 u_2 \sin\theta$$

$$\frac{d\theta}{d\zeta} = \Delta s \frac{d}{c^2}\cot\theta \frac{d}{d\zeta}(\ln u_1 u_2 u_3) \tag{109}$$

where $\Delta s = \Delta k\, z/\zeta$. The product $[u_p(0)\zeta]^2 = \Gamma^2 z^2$, where $u_p{}^2(0)$ is the input photon flux density. Thus ζ is a normalized nonlinear interaction length. The solution of Eq. (109) for various input and phase-matching conditions is given by Armstrong et al. (1962). Bey and Tang (1972) solved the equations for the particular case of the singly resonant parametric oscillator. The results of their analysis gives important parameters for SRO operation.

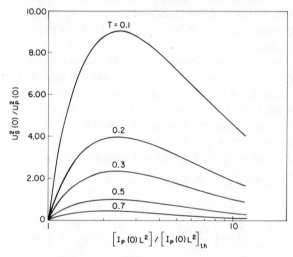

Fig. 13. Ratio of the intracavity forward-traveling signal-photon flux to the pump-photon flux versus the number of times above threshold. Here $T = 1 - R_1 R_2$, where R_1 and R_2 are the power reflectivities of the input and output mirrors [after Bey and Tang (1972)].

Figure 13 shows the signal photon flux normalized to the input pump photon flux versus the number of times above threshold. The figure shows that the circulating signal power may be many times the incident pump power especially for low-loss cavities. This is important in parametric oscillators operating near the crystal or mirror burn density limit. The circulating signal power density may be reduced by resonating the lower frequency wave since

$$I_s/I_p = (\omega_s u_s{}^2/\omega_p u_p{}^2)(A_s/A_p)$$

where ω_s/ω_p is the frequency ratio and A_s/A_p the area ratio of the signal to pump waves. Bey and Tang (1972) extended their results to include a calculation of SRO threshold versus mirror transmission and quantum efficiency versus mirror transmission. They also discussed the conditions under which a second mode of the SRO reaches threshold and begins to oscillate.

Experimental measurements of OPO conversion efficiency have been made for both cw and pulsed operation and for DRO, SRO, and PRO. Bjorkholm (1968a) and Kreuzer (1968) investigated DRO conversion efficiencies. Using a single-mode ruby pump, Kreuzer obtained 36% conversion efficiency in a LiNbO₃ parametric oscillator. Bjorkholm obtained 22% conversion for a DRO and 6% conversion for a SRO. Belyaev et al. (1969) reported similar conversion efficiencies for both a DRO and SRO. Using a ruby laser pump and a noncollinear SRO cavity arrangement, Falk and Murray (1969) obtained 70% power conversion and 50% energy conversion in a LiNbO₃ oscillator. The energy conversion was limited by the finite build-up time of the oscillator.

Early conversion efficiency results were not in very good agreement with the conversion efficiency exepcted from plane wave theory. As previously discussed, Bjorkholm (1971) investigated the spatial dependence of pump depletion for multimode Gaussian amplitude waves and showed that much better agreement was obtained if the nonuniform pumping intensity was taken into account.

Wallace (1970) described a LiNbO₃ SRO pumped by a Q-switched Nd:YAG laser. The oscillator operated at typically 50% conversion efficiency with threshold powers near 600 W. Average output power up to 70 mW has been obtained.

Using LiNbO₃ internal to a Q-switched Nd:YAG laser enabled Ammann et al. (1969) to obtain an 8% conversion efficiency at 2.13 μm with up to 17 mW of average power. Recently these results were improved to over 1.2 W of average power at 2.1 μm and 60 mW at 3.1 μm.

Efficient traveling wave oscillators have been operated in ADP and recently α-HIO₃. Yarborough and Massey (1971) reported up to 25% power conversion in an ADP oscillator pumped with the fourth harmonic of Nd:YAG laser. The output pulses were typically 2 nsec in duration at 100-kW peak power. Kovrigin and Nikles (1971) obtained 57% energy conversion at 1 to 1.10 μm in α-HIO₃ pumped with a doubled Nd:glass laser. The pulse length was 30 nsec at $P_p = 20$ MW/cm². At higher pump power densities the α-HIO₃ crystal was damaged.

In general, parametric oscillators have operated at near 50% conversion efficiency on a pulse to pulse basis. The peak power is limited by burn density of materials and the average power by thermal heating induced phase mismatch. The attainable conversion efficiency especially in the near infrared spectral region makes parametric oscillators a practical source of tunable radiation.

H. Spectral Properties

The spectral properties of a parametric oscillator include tuning range and method of tuning, gain bandwidth and the detailed spectral structure within the gain bandwidth determined by the optical cavity.

Tuning is achieved by varying the crystal birefringence while simultaneously satisfying the frequency and phase-matching conditions

$$\omega_3 = \omega_2 + \omega_1$$

and

$$\mathbf{k}_3 = \mathbf{k}_2 + \mathbf{k}_1$$

In general, temperature tuning of the index or angle tuning by changing $n^e(\theta)$ give a wide tuning range, while pressure tuning through the photoelastic effect or electric field tuning through the electrooptic effect result in small tuning ranges.

Expanding Eq. (11) in a tuning variable $\mathscr{M}$ and the signal frequency ω_2 gives the rate of tuning as a function of the tuning variable

$$\frac{d\omega_2}{d\mathscr{M}} = \frac{1}{\beta_{12}}\left[\frac{\partial k_3}{\partial \mathscr{M}} - \frac{\partial k_2}{\partial \mathscr{M}} - \frac{\partial k_1}{\partial \mathscr{M}}\right] \tag{110}$$

Here $\beta_{12} = (1/u_1) - (1/u_2)$, where u_1 and u_2 are the group velocities defined by Eq. (24a).

If we consider the pump frequency as a tuning variable, then Eq. (110) becomes

$$\frac{\partial \omega_2}{\partial \omega_3} = \frac{1}{\beta_{12}}\left[\frac{\partial k_3}{\partial \omega_3} - \frac{\partial k_1}{\partial \omega_1}\right] = \frac{\beta_{31}}{\beta_{12}} \tag{111}$$

This is the rate the signal frequency ω_2 tunes as a function of pump frequency tuning. Detailed discussion of this tuning rate in certain crystals is given by Kovrigin and Byer (1969).

As shown previously [cf. Eq. (21)] the gain bandwidth of a parametric oscillator is determined by

$$\Delta k\, l/2 = \pi$$

Thus the bandwidth is

$$|\delta\omega_2| = 2\pi/\beta_{12} l$$

where we have referred to Eq. (25) for the expansion of Δk about the phase-matching point.

If $\beta_{12} = 0$, which may occur at degeneracy ($\omega_1 = \omega_2$), then the expansion for Δk must include second order terms. For $\beta_{12} \simeq 0$ the bandwidth becomes

$$|\delta\omega_2| = |4\pi/\gamma_{12} l|^{1/2}$$

where γ_{12}, given by Eq. (25), is the group velocity dispersion. The dispersion constant β_{12} for LiNbO$_3$ has been calculated for various pump wavelengths [see Harris (1969a, Fig. 12)]. The calculation of γ_{12} shows that the second order term is not significant until tuning is within 200 cm^{-1} of degeneracy (Byer, 1968).

Comparison of Eqs. (110) and (26) show that β_{12} appears in both the tuning rate and bandwidth expressions. If β_{12} is small the tuning rate and bandwidth are both relatively large. In general a small tuning rate implies a small bandwidth. Expanding β_{12} in terms of the crystal index of refraction

$$\beta_{12} = \frac{1}{c}\left[n_2 - n_1 + \lambda_1 \frac{\partial n_1}{\partial \lambda_1} - \lambda_2 \frac{\partial n_2}{\partial \lambda_2}\right] \approx \frac{\Delta n}{c} \qquad (112)$$

and substituting into the bandwidth expression Eq. (26) we find

$$|\delta v(\text{cm}^{-1})| \approx 1/\Delta n \, l \, (\text{cm}) \qquad (113)$$

where l is the crystal length and Δn the birefringence. Thus crystals with small birefringence have larger bandwidths and tuning rates than crystals with large birefringence. Equation (113) predicts a bandwidth for 12 cm^{-1} fo a 1 cm crystal of LiNbO$_3$. The actual bandwidth is 4 cm^{-1}. Therefore the dispersion term can only be neglected to within this accuracy.

Figure 14 shows the measured pump, signal, and idler bandwidth for a 0.659 μm pumped LiNbO$_3$ parametric oscillator. The measured oscillator bandwidth of 1.55 to 1.65 cm^{-1} for a 5 cm crystal agrees very well with the calculated half power bandwidth of 1.60 cm^{-1}. This data was taken for oscillator operation approximately 10 times above threshold. For operation nearer threshold the resonant wave bandwidth decreases to a few tenths wavenumber (cf. Fig. 16).

It is interesting to take a closer look at the rate of tuning. As an example consider electric field tuning through the electrooptic effect. The phase retardation $\Delta\phi = \Delta k \, l$ is given by

$$(\pi \, \Delta V/V_\pi)(l/d) = \Delta k \, l \qquad (114)$$

where ΔV is the change in the applied field; V_π the half wave voltage, and l/d the crystal aspect ratio. Substituting into Eq. (110) and forming the ratio $d\omega_2/|\delta\omega_2|$ which is the number of bandwidths tuned we find

$$n = d\omega_2/|\delta\omega_2| = \tfrac{1}{2}(\Delta V/V_\pi)(l/d) \quad \text{(transverse field)} \qquad (115a)$$

$$= \tfrac{1}{2}(\Delta V/V_\pi) \quad \text{(longitudinal field)} \qquad (115b)$$

for a transverse and longitudinal applied electric field. For LiNbO$_3$, $V_\pi = \lambda/(n_e{}^3 r_{33} - n_o{}^3 r_{13}) \simeq 2820$ V for the transverse field. Thus for a 10 to 1 aspect

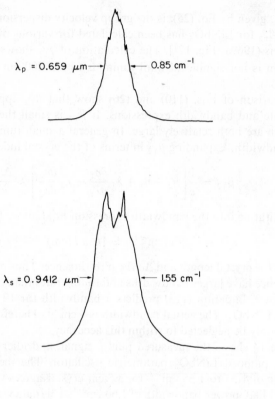

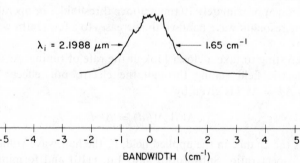

BANDWIDTH (cm⁻¹)

Fig. 14. Measured bandwidth for a 0.659-μm pumped $LiNbO_3$ SRO. The calculated linewidth for the 5-cm crystal is 1.60 cm⁻¹.

ratio crystal $n = 5\ \Delta V/V_\pi$ or approximately 5 cm⁻¹ of tuning per 2820 V for a 5 cm crystal. This tuning method may have practical importance for rapid fine tuning of an oscillator about the gain line center.

Electric field control of phase matching was first demonstrated in KDP by

Adams and Barrett (1966). Kreuzer (1967) applied the technique to a ruby pumped OPO and observed a 6.2 Å/kV cm^{-1} tuning rate at 2.0 μm which is in good agreement with the predicted rate. Krivoshchekov *et al.* (1968) extended electrooptic tuning to a KDP OPO operating near 1 μm wavelength. To achieve a significant tuning range the KDP crystal was cooled to its Curie temperature near 123°K.

Tuning curves for parametric oscillators usually are determined prior to their operation. The straightforward approach is to solve the phase matching equations [Eqs. (10) and (11)] for signal and idler frequencies at a given pump frequency as a function of the tuning variable. To carry out the calculation the indices of refraction must be given by an analytical expression. For example, KDP and ADP indices are given very accurately by Sellmeier equations which include temperature variation (Zernike, 1964). Similarly, equations have been given for LiNbO$_3$ (Hobden and Warner, 1966), CdSe (Herbst and Byer, 1971), and other crystals (Landolt Bornstein, 1969; Pressley, 1971). For accurate calculation of tuning curves the indices of refraction must be known over the

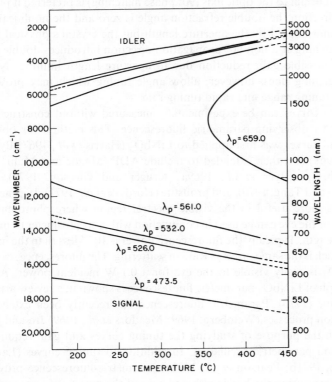

Fig. 15. LiNbO$_3$ parametric oscillator tuning curves for various doubled Nd : YAG laser pump wavelengths.

entire tuning range to an accuracy of 0.1 to 0.01%. Figure 15 shows the calculated temperature tuning curves for $LiNbO_3$ for doubled Nd:YAG laser pump wavelengths.

In calculating the phase matching curves the nonlinear polarization tensor must be taken into account and proper electric field polarizations chosen to maximize the d coefficients at the phase matched condition. Phase matching in uniaxial crystals is discussed by Boyd et al. (1965) and phase matching in biaxial crystals is treated by Hobden (1967). In general there are two types of phase matching conditions. Type I where $n_3^e \rightarrow n_2^o + n_1^o$ such as is the case for negative birefringent $LiNbO_3$, and type II where $n_3^o \rightarrow n_2^e + n_1^o$ which applies to the positive uniaxial CdSe crystal. KDP and infrared crystals of the chalcopyrite type ($42m$ point group similar to KDP) allow both types of phase matching in a non-90° propagation direction, but only type I (KDP) or type II ($CdGeAs_2$) in the 90° phase-matched direction. The effective nonlinear coefficients, which depend on the propagation direction, are listed by Boyd and Kleinman (1968) for uniaxial crystals. Phase matching with a propagation direction normal to the optic axis (90° phase matching) is preferred if possible. In that direction the double refraction angle is zero and the nonlinear interaction is not limited by the aperture length but the crystal's physical length. Phase matching at a non-90° propagation direction introduces double refraction and possible gain reduction due to aperture length limiting. Non-90° phase matching does, however, allow angle tuning which may provide an extended tuning range and rapid tuning rate.

Tuning curves can be experimentally measured without constructing an oscillator by observing parametric fluorescence. This method for obtaining tuning and curves was first applied to $LiNbO_3$ (Harris et al., 1967; Byer and Harris, 1968) and then extended to include ADP (Magde and Mahr, 1967), $Ba_2NaNb_5O_{15}$ (Byer et al., 1969a; Kruger and Gleeson, 1970), $LiIO_3$ (Campillo and Tang, 1970), and proustite (Hordvik et al., 1971). In general the method is more useful in the visible spectral region where photomultiplier detectors or the eye can be used in conjunction with a spectrometer to measure the wavelength. Typically the fluorescence power is 10^{-9} less than the incident power which is the same order as Raman scattering. The fluorescence scattering in $LiNbO_3$ is easily visible to the eye for a 0.1-W incident power. A color photograph of $LiNbO_3$ parametric fluorescence is shown in a review article by Giordmaine (1970). Parametric fluorescence has recently been extended to four-photon processes (Weinberg, 1969; Meadors et al., 1969; Ito and Inaba, 1970) with the purpose of studying the tuning curves and gain required for four-photon parametric oscillators. In addition to tuning curves (Laurence and Tittle, 1971b; Pearson et al., 1972a) parametric fluorescence provides a means of measuring the bandwidth and gain of the parametric process (Byer and Harris, 1968).

In the above examples, parametric phase matching is achieved by varying the crystal birefringence. Usually this is accomplished by crystal rotation or temperature change. These phase-matching methods are limited to crystals with birefringence adequate to overcome dispersion. Phase matching can also be achieved by using the dispersive properties of guided waves. Smith (1969) and Kuhn (1969) analyzed phase-matched processes in optical waveguides. Later Anderson et al. (1971) and Tien et al. (1970) performed phase-matched SHG in optical guides and Anderson and McMullen (1969) achieved parametric amplification in a GaAs guide. Phase-matched parametric oscillation in guides has been treated in detail by Boyd (1972). He pointed out the extreme tolerance required on the film thickness in order to maintain phase matching over reasonable lengths. Andrews (1971) has discussed the problem of guided wave phase matching in nonlinear crystals and has proposed a near symmetrical waveguide to alleviate the thickness tolerance requirement. With increased consideration being given to optical guided waves, parametric processes may become an important aspect of guided wave frequency conversion.

The spectral properties of an OPO are determined by the gain bandwidth, pump laser spectrum, and OPO cavity characteristics. The OPO has a finite acceptance bandwidth for the pump field that depends on the operating mode of the oscillator. For a DRO only a single longitudinal mode of the pump field is effective in driving the oscillator (Harris, 1966). This limitation is due to the cavity resonance conditions that have to be met simultaneously at the signal and idler fields. If the length of the OPO cavity equals that of the pumping laser, then the full multimode power of the pump laser pumps the oscillator (Harris, 1966). This condition was satisfied in the visible cw oscillator demonstrated by Byer et al. (1968). If the pump field is mode locked, then the signal and idler round trip travel time in the oscillator must equal the time between pump mode locked pulses. This case has been treated by Harris (1967).

For the SRO the maximum pump bandwidth effective in driving the oscillator depends on the resonated oscillator field. For a resonant signal field, the signal cavity fixes the signal frequency but the free idler wave frequency varies to compensate for pump frequency changes. In addition, the full multimode power of the pump effectively couples to a single frequency through corresponding idler modes to satisfy the condition $\omega_3 = \omega_2 + \omega_1$ [see Harris (1969a, Fig. 13)]. Holding the signal frequency fixed and letting the idler frequency vary with pump wavelength leads to a pump acceptance bandwidth given by

$$\Delta\omega_3 = 2\pi/\beta_{23}l \tag{116}$$

where $\beta_{23} = (\partial k_2/\partial\omega_2) - (\partial k_3/\partial\omega_3)$. For a resonant idler wave, $\beta_{23} \to \beta_{13}$. The dispersion constant β_{23} in general does not equal β_{13} so that the pump

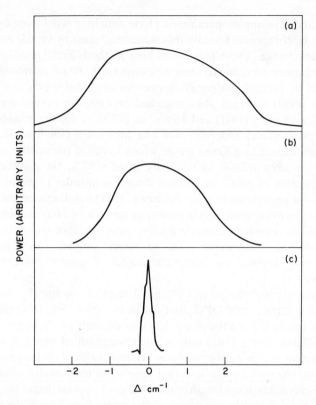

POWER (ARBITRARY UNITS)

(a)

(b)

(c)

-2   -1   0   1   2

Δ cm⁻¹

Fig. 16. Spectra of the (a) 0.473-μm pump wave, (b) 0.569-μm nonresonant wave, and (c) 2.79-μm resonant wave for a pumped LiNbO$_3$ SRO [after Young *et al.* (1971)].

bandwidth depends on the resonant wave. For example, a 3.2 cm LiNbO$_3$ crystal has $\Delta\omega_3 = 0.98$ cm^{-1} (signal resonant) and $\Delta\omega_3 = 5.8$ cm^{-1} (idler resonant) for a 0.473 μm pump wavelength. The doubled 0.946 μm Nd:YAG laser line has a 4 cm^{-1} bandwidth so that only one fourth of the pump radiation effectively pumps the oscillator if the visible wave is resonant, but all of it pumps the oscillator if the infrared idler is resonant. Young *et al.* (1971) experimentally verified the above results. In addition they showed that the resonant wave has a considerably narrowed spectra compared to either the pump or nonresonant wave. Figure 16 shows the measured spectra. This figure also demonstrates the narrow spectral width of the resonant idler field. For operation a few times above threshold, the resonant wave spectrum is typically 0.2 cm^{-1} which is considerably less than 1 cm^{-1} gain bandwidth.

The development of high power broad band laser sources led Akhmanov *et al.* (1971) to investigate the power spectral density effective in pumping Raman and parametric oscillators. The power spectral density is determined by using $\Delta\omega_3$ given above with the threshold intensity for a SRO given by Eq. (53). Since all pump modes within the line width $\Delta\omega_3$ act to pump coherently the SRO we can write

$$I_{th}(\Delta\omega_p) = I_{th}/\Delta\omega_p \qquad (117)$$

where $I_{th}(\Delta\omega_p)$ is the threshold intensity per spectral width and I_{th} is determined by $\Gamma^2 l^2 = 2a_2$. For example, if the threshold is 10 MW/cm² for a pump with bandwidth less than $\Delta\omega_p$, then for an acceptance bandwidth of 1 cm⁻¹ the power spectral density is $I_{th}(\Delta\omega_p) = 10$ MW/cm for a broad band pump source. In practice the burn density of the nonlinear material may limit the maximum pump bandwidth.

A useful variation of the broad-band pumping of parametric oscillators is pumping with a tunable source such as a dye laser. This method of tuning proposed by Herbst and Byer (1971) and recently demonstrated in LiNbO₃ by Wallace (1972) has the advantages of rapid tuning over an extended wavelength region not available to the pumping source. In addition, pump tuning remains phase matched without crystal rotation or temperature tuning since the signal and idler frequencies are not restricted. This allows the possibility of remaining at a 90° phase-match condition without adjusting the crystal temperature or angle. Figure 17 shows the theoretical and experimental tuning curve for a rhodamine 6G dye laser pumped LiNbO₃ parametric oscillater. In the experiment performed by Wallace (1972) the overall conversion efficiency was approximately 20% from the laser input through the optically pumped dye laser to the output of the parametric converter.

An extension of the parametric oscillator tuning range can be achieved by internal second harmonic generation or up conversion. The conversion efficiency has been treated theoretically by Giallorenzi and Reilly (1972) and Bey and Tang (1972). Experimentally, Wallace (1971) and Campillo and Tang (1971) demonstrated internal up conversion with up to 20% conversion efficiency. Ammann et al. (1971) discussed simultaneous parametric oscillation and second harmonic generation in LiNbO₃.

Detailed spectral properties of a parametric oscillator depend on the longitudinal mode structure of the pump and on the $c/2L$ frequency spacing of the parametric oscillator cavity modes at the signal and idler. For the DRO there is a $c/2L$ frequency spacing for both oscillator waves and for the pump. Only when $\omega_3 = \omega_2 + \omega_1$ is satisfied does the oscillator see full gain. This usually occurs in "clusters" of modes spaced a few mode intervals apart depending on crystal dispersion (Giordmaine and Miller, 1966). The DRO spectral properties were experimentally investigated by Bjorkholm (1968).

For the SRO with only one wave resonant, the nonresonant frequency is free to adjust so that $\omega_3 = \omega_1 + \omega_2$ is satisfied. Thus the SRO shows improved spectral stability compared to the DRO. This was experimentally verified by

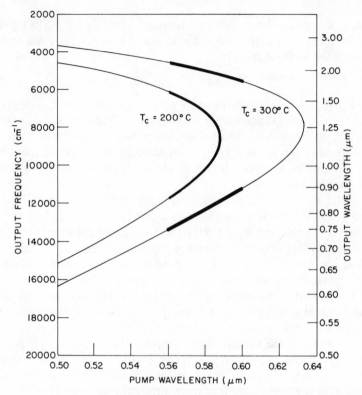

Fig. 17. Tuning curves for a 90° phase-matched LiNbO₃ parametric oscillator pumped with a tunable source. Heavy curve indicates pump range of a rhodamine 6G dye laser.

Bjorkholm (1968b). Kreuzer (1969) first narrowed the SRO spectrum to a single longitudinal mode. He used a tilted etalon in the resonant cavity of a ruby laser pumped LiNbO₃ SRO to reduce the line width from 5 cm⁻¹ to a single frequency.

Unlike other laser sources which show power loss as the bandwidth is narrowed, the OPO is expected to remain at full efficiency even for single mode operation. In this respect it behaves as a homogeneously saturated source. Recently Wallace (1971) demonstrated a long term frequency stability of better than 0.005 cm⁻¹ by use of a temperature-scanned infrared etalon internal to the oscillator cavity. In this experiment the OPO remained at 80% of its wide-band power as expected showing a slight decrease in output power due to the etalon insertion loss. Figure 18a shows a schematic of the OPO linewidth and mode structure and the etalon transmission function. Figure 18b shows the output of the oscillator operating in two- and one-cavity modes. The oscillator output spectrum was resolved with a thermally tuned external scanning interferometer.

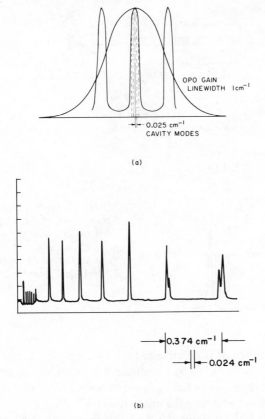

(a)

(b)

Fig. 18. (a) Schematic of the SRO gain line width and cavity mode structure. Also shown is the transmittance of an internal etalon. (b) Scanning Fabry–Perot analysis of the SRO operating with one- and two-cavity modes [after Wallace (1971a)].

The tilted etalon is useful for mode control in oscillators with relatively wide beam cross sections (Hercher, 1969). This led Pinard and Young (1972) to study an alternate interferometric mode control technique based on an interferometer proposed by Damaschini (1969) which is a duel of the better known Smith (1965) interferometer. Pinard and Young (1972) obtained a bandwidth of $0.001 \, \text{cm}^{-1}$ in the 2.5 μm spectral region tunable over a $0.015 \, \text{cm}^{-1}$ region before a second cavity mode appeared. Figure 19a shows a schematic of the parametric oscillator and the interferometer, and Fig. 19b shows the measured output spectrum. Operation of this oscillator was aided by the high gain properties of the $LiNbO_3$ oscillator which allowed oscillation off the crystal surface prior to the alignment of the interferometer.

One difficulty in obtaining single-frequency output from an OPO is the requirement that both the etalon center frequency and the cavity mirror

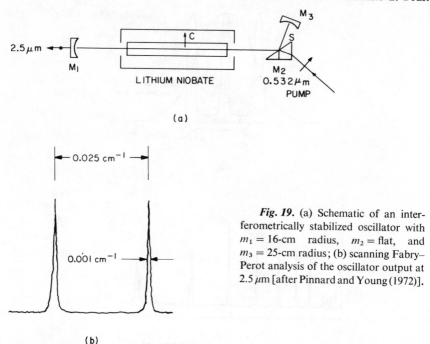

(a)

Fig. 19. (a) Schematic of an interferometrically stabilized oscillator with $m_1 = 16$-cm radius, $m_2 = $ flat, and $m_3 = 25$-cm radius; (b) scanning Fabry–Perot analysis of the oscillator output at 2.5 μm [after Pinnard and Young (1972)].

(b)

position be controlled. It is possible to tune the single-frequency output over a narrow band (~ 0.02 cm^{-1}) by simultaneously scanning the interferometer and cavity mirror. For a wider tuning range the gain bandwidth of the oscillator must be reset prior to scanning the interferometer.

For operation with an etalon it is desirable to have an OPO with a narrow gain bandwidth to avoid multiple etalon transmission bands within the gain line width. Another mode of operation, however, is to utilize a low-dispersion mirror–prism or mirror–grating combination (Bjorkholm, 1971) to frequency narrow a wide bandwidth oscillator. For this case, the signal and idler may be tuned within the wide oscillator gain line width without tuning the center of the gain band. Certain parametric oscillators show very wide bandwidths for particular pump wavelengths. In LiNbO$_3$, for example, a pump wavelength near the 0.946 μm of Nd : YAG laser generates a tuning band from 1.4 to over 2.8 μm in the infrared [see Ammann *et al.* (1971, Fig. 5)]. In CdGeAs$_2$, a pump wavelength at 4 μm generates signal and idler outputs between 6 and 15 μm with only a 1° crystal rotation. Parametric oscillators operating with such wide bandwidths have characteristics similar to wide-band dye lasers. The OPO, however, remains homogeneously saturated and must satisfy the $\omega_3 = \omega_2 + \omega_1$ condition. To date bandwidths of approximately 200 cm^{1-} have been experimentally measured for LiNbO$_3$ parametric oscillators operating near degeneracy.

I. Focusing

Up to this point the discussion of nonlinear interactions has been in the plane wave approximation. In practice, the pump beam is a mode of the pump laser cavity which must be properly focused into the nonlinear crystal. In this section we consider the important aspects of focusing, mode coupling, and mode conversion in optical parametric oscillators.

The general properties of Gaussian beams were discussed for confocal laser cavities by Boyd and Gordon (1961) and extended to general curved mirror cavities by Boyd and Kogelnik (1962). The important problem of mode coupling and conversion was treated by Kogelnik (1964). The practical aspects of Gaussian beams and laser resonators including beam transformation was later reviewed by Kogelnik and Li (1966).

Bjorkholm (1966) and Kleinman et al. (1966) have discussed focusing for second harmonic generation. Bloembergen and Pershan (1962), Kleinman (1962) and Boyd et al. (1965), extended the analysis to include double refraction. In 1965, Boyd and Ashkin (1966) considered the important problem of focusing to maximize nonlinear gain for a parametric oscillator. Their analysis, which was limited to the near-field region of a Gaussian beam, was later extended and generalized in the important paper by Boyd and Kleinman (1968) who considered the interaction of focused Gaussian beams for second harmonic generation and parametric processes. In a series of papers Asby (1969a, b, c) also considered focused beams including coupling of higher-order modes.

The following discussion considers the near-field focusing limit in detail since the near-field results apply to most experimental situations. The near-field results are then extended to the general focusing case by introducing Boyd and Kleinman's focusing parameter. Finally practical aspects of focusing are discussed.

Nonlinear interactions involve the proper focusing of diffraction-limited laser beams. The fundamental mode or Gaussian mode electric field is described by

$$\mathbf{E}(x,y,z) = \mathbf{E}_0 \frac{w_0}{w(z)} \exp[-i(kz-\phi)] \exp\left[-r^2\left(\frac{1}{w^2(z)} + \frac{ik}{2R}\right)\right] \quad (118)$$

where

$$\phi = \tan^{-1}(z/z_R)$$

is the phase factor for a fundamental (TEM_{00}) mode

$$2z_R = b = w_0^2 k \quad (119)$$

is the confocal parameter and R the beam radius of curvature. Here k is the propagation constant in the crystal and z_R half the confocal distance or Rayleigh range.

In general the beam radius and the wave front curvature are given by

$$w^2(z) = w_0{}^2[1+(z/z_R)^2] \qquad (120a)$$

and

$$R(z) = z[1+(z_R/z)^2] \qquad (120b)$$

The far-field diffraction angle of the fundamental mode is

$$\theta = \lambda/\pi w_0 \qquad (120c)$$

where w_0 is the electric field beam radius at the waist.

For $z \gtrsim z_R$ diffraction becomes important and leads to diffraction spreading of the beam. In this far-field limit the electric field shows an amplitude dependence in the z direction in addition to its radial Gaussian amplitude dependence. The z-dependent electric field leads to complications in solving the coupled nonlinear equations such that only numerical integration is possible. Solutions of the coupled equations including diffraction effects have been considered by Boyd and Kleinman (1968) and are discussed later.

In the near-field limit $z \ll z_R$, the fundamental field expression simplifies to

$$\mathbf{E}(x,y) = \mathbf{E}_0 \exp(-r^2/w_0{}^2) \qquad (121)$$

which is independent of z. In this limit $w(z) \to w_0$, $\phi \to 0$, and $R \to \infty$, so that the fundamental mode is a plane wave with a Gaussian amplitude profile. The power in the fundamental mode is

$$
\begin{aligned}
P &= \tfrac{1}{2}nc\varepsilon_0 \int |\mathbf{E}(x,y)|^2 \, dx \, dy \\
&= \tfrac{1}{2}nc\varepsilon_0 \int_0^{2\pi} \int_0^\infty |\mathbf{E}(\mathbf{r})|^2 r \, dr \, d\phi \\
&= \tfrac{1}{2}nc\varepsilon_0 |\mathbf{E}_0|^2 (\tfrac{1}{2}\pi w_0{}^2) = I_0(\tfrac{1}{2}\pi w_0{}^2) \qquad (122)
\end{aligned}
$$

where I_0 is the peak intensity and $\tfrac{1}{2}\pi w_0{}^2$ is the effective area of the Gaussian beam.

The coupled equations [Eq. (9)] can be solved in the near-field approximation. Neglecting pump depletion and loss Eqs. (9a)–(9c) become

$$\frac{dE_{10}}{dz} \exp\left(\frac{-r^2}{w_1{}^2}\right) = i\kappa_1 E_{30} \exp\left(\frac{-r^2}{w_3{}^2}\right) E_{20} \exp\left(\frac{-r^2}{w_2{}^2}\right) \exp i\,\Delta k\,z$$

$$\frac{dE_{20}}{dz} \exp\left(\frac{-r^2}{w_2{}^2}\right) = i\kappa_2 E_{30} \exp\left(\frac{-r^2}{w_3{}^2}\right) E_{10}^* \exp\left(\frac{-r^2}{w_1{}^2}\right) \exp i\,\Delta k\,z$$

$$\frac{dE_{30}}{dz} \exp\left(\frac{-r^2}{w_3{}^2}\right) = i\kappa_3 E_{20} \exp\left(\frac{-r^2}{w_2{}^2}\right) E_{10} \exp\left(\frac{-r^2}{w_1{}^2}\right) \exp(-i\,\Delta k\,z)$$

Multiplying by $\exp(-r^2/w_1{}^2)$, $\exp(-r^2/w_2{}^3)$, and $\exp(-r^2/w_3{}^2)$ and integrating over the radial coordinate we take the projection of the Gaussian fields at w_1, w_2, and w_3 into the driving polarizations which have beam waists given by

$$\frac{1}{\overline{w}_1{}^2} = \frac{1}{w_2{}^2} + \frac{1}{w_3{}^2} \tag{123a}$$

$$\frac{1}{\overline{w}_2{}^2} = \frac{1}{w_1{}^2} + \frac{1}{w_3{}^2} \tag{123b}$$

$$\frac{1}{\overline{w}_3{}^2} = \frac{1}{w_1{}^2} + \frac{1}{w_2{}^2} \tag{123c}$$

The result of the radial integration is

$$dE_{10}/dz = i\kappa_1 g_1 E_{30} E_{20}^* \exp i\,\Delta k\,z \tag{124a}$$

$$dE_{20}/dz = i\kappa_2 g_2 E_{30} E_{10}^* \exp i\,\Delta k\,z \tag{124b}$$

$$dE_{30}/dz = i\kappa_3 g_3 E_{20} E_{10} \exp(-i\,\Delta k\,z) \tag{124c}$$

where the spatial coupling factors g_1, g_2, and g_3 are

$$g_1 = 2/[1+(w_1{}^2/\overline{w}_1{}^2)], \qquad g_2 = 2/[1+(w_2{}^2/\overline{w}_2{}^2)]$$
$$g_3 = 2/[1+(w_3{}^2/\overline{w}_3{}^2)] \tag{125}$$

Only in the limit of very large beam radii do the coupling factors approach unity. For the typical focusing case, the driving polarization beam radii are less than the driven field radii so that coupling is never complete. Equations (124a)–(124c) are identical to Eqs. (9a)–(9c) except for the coupling factors. Therefore for Gaussian beams in the near-field approximation the parametric gain coefficient [Eq. (13)] becomes

$$\Gamma^2 = \frac{\omega_1 \omega_2 |d|^2 |E_3|^2}{n_1 n_2 c^2} \cdot g_1 g_2 \tag{126}$$

where

$$g_1 g_2 = 4w_3{}^2 [w_1 w_2 w_3/(w_1{}^2 w_2{}^2 + w_2{}^2 w_3{}^2 + w_3{}^2 w_1{}^2)] \tag{127}$$

from Eqs. (123) and (125). Writing $g_1 g_2 = 4w_3{}^2 M^2$ the parametric gain in terms of the pump power becomes

$$\Gamma^2 = \frac{16}{\pi} \cdot \frac{\omega_1 \omega_2 |d|^2}{n_1 n_2 n_3 \varepsilon_0 c^3} \cdot P_{p0} M^2 \tag{128}$$

where P_{p0} is the peak pump power given by Eq. (122).

To maximize M for a fixed w_1 and w_2 the pump beam radius should be

$$1/w_3{}^2 = (1/w_1{}^2) + (1/w_2{}^2) \tag{129}$$

so that

$$M_{max}^2 = \frac{1}{4} \frac{1}{w_1^2 + w_2^2} \tag{130}$$

According to Eq. (130), M^2 is maximum for the smallest beam radii w_1 and w_2. The near-field approximation, however, limits w_1 and w_2 to the confocal spot size given by Eq. (119). Boyd and Ashkin (1966) discussed this solution and pointed out that the maximum gain should occur near the confocal focusing condition. In addition, if the two resonant parametric waves are confocally focused, then the pump beam is also confocally focused according to Eq. (129). The exact analysis by Boyd and Kleinman (1968) showed that the parametric gain is 20% less at $l/b = 1$ (the confocal condition) than at the $l/b = 2.86$ where the peak gain occurs. Here l is the crystal length and b the confocal parameter. Experimentally l/b is usually less than unity so that the near-field analysis for parametric gain is a good approximation.

Substituting M_{max} into Eq. (128) the parametric gain $G = \Gamma^2 l^2$ becomes

$$G = \frac{4\omega_1 \omega_2 |d|^2 l^2}{n_1 n_2 n_3 \varepsilon_0 c^3} \cdot \frac{P_{p0}}{\pi(w_1^2 + w_2^2)} \tag{131}$$

At degeneracy where $\omega_1 = \omega_2 = \omega_0$ and $w_1 = w_2 = w_0$ the gain simplifies to

$$G_{deg} = 2\omega_0^2 |d|^2 l^2 P_{p0} / n_0^2 n_3 \varepsilon_0 c^3 \pi w_0^2 \tag{132}$$

where $w_0^2 = l\lambda_0 / 2\pi n_0$.

The gain increases as l^2 until the crystal length equals the confocal parameter. At longer crystal lengths the gain increases linearly with crystal length since the spot size must also increase to keep within the near field approximation. For a 1-cm $LiNbO_3$ crystal with a degeneracy wavelength of 1 μm and confocal focusing, the gain is $G_{deg} = 0.005 P_{p0}$, where P_{p0} is the peak pump power in watts.

The analysis thus far has assumed that both the signal and idler fields have a well-defined radial variation. This is the case for a DRO. For the SRO only one of the waves is resonant. The nonresonant wave propagates freely and assumes a radial variation determined by its driving polarization. If we assume that the field E_{10} is not resonated, then $g_1 = 1$ in Eq. (124a) and the spot size becomes

$$1/\overline{w}_1^2 = 1/w_1^2 = (1/w_2^2) + (1/w_3^2)$$

where w_2 is the spot size determined by the cavity and w_3 the focused pump radius. The equations describing the SRO thus become

$$dE_1/dz = i\kappa_1 E_2^* E_3 \tag{133a}$$

and

$$dE_2/dz = i\kappa_2 g_{SRO} E_1^* E_3 \tag{133b}$$

where

$$g_{SRO} = (1 + w_2{}^2/w_3{}^2)^{-1}$$

Solving for the single-pass gain we find

$$G_{SRO} = \frac{4\omega_1\omega_2|d|^2l^2}{n_1 n_2 n_3 \varepsilon_0 c^3} \frac{P_{p0}}{\pi(w_3{}^2 + w_2{}^2)} \tag{134}$$

which is similar to gain for the DRO case given by Eq. (131) except for the replacement $w_1{}^2 \to w_3{}^2$ in the area factor. This result has been derived previously (Byer, 1968), Asby (1969a) and has been discussed by Herbst (1972).

For the DRO case the gain can be maximized with respect to the pump spot size. For the SRO this is not the case. In addition, the gain for the DRO is symmetric in the signal and idler waves. For the SRO the gain is higher if the resonated wave is the higher frequency with the smaller spot size. Resonating the higher frequency wave, however, also increases the circulating oscillator power and thus the intensity at the nonlinear crystal which may lead to crystal damage problems. In practice the gain of SRO may be increased by approximately a factor of two by resonating the high-frequency wave if the oscillator operates far from degeneracy.

We now proceed to consider the exact analysis for focused nonlinear interactions. The analysis is discussed in detail by Boyd and Kleinman (1968) and we refer to their results. Boyd and Kleinman define a general second harmonic generation efficiency reduction factor $h(B, \xi)$ such that the second harmonic power is given by

$$P_{SH} = (2\omega_F{}^2 d^2/\pi n_F{}^2 n_{SH} \varepsilon_0 c^3) P_F{}^2 l k_F h(B, \xi) \tag{135}$$

where P_F is the fundamental power, l the crystal length, and k_F the fundamental wavevector in the material. The second harmonic generation efficiency P_{SH}/P_F equals the parametric gain coefficient for small gains so that we can write

$$G = \Gamma^2 l^2 = (2\omega_0{}^2 d^2/\pi n_0{}^2 n_3 \varepsilon_0 c^3) P_{30} l k_0 (1 - \delta^2)^2 \bar{h}(B, \xi) \tag{136}$$

where $\bar{h}(B, \xi)$ is the gain reduction factor. The bar expresses coupling only to the Gaussian signal and idler mode in the gain expression. Here $(1 - \delta^2)^2$ is the degeneracy factor defined by

$$\omega_1 = \omega_0(1 - \delta), \quad \omega_2 = \omega_0(1 + \delta), \quad \omega_1\omega_2 = \omega_0{}^2(1 - \delta^2) \tag{137}$$

where ω_0 is the degeneracy frequency.

We now proceed to discuss the gain reduction factor and its limiting forms. For second harmonic generation, Eq. (135) shows that the total second harmonic efficiency is reduced by $h(B, \xi)$. In parametric processes we are usually interested in the parametric gain coupled into a Gaussian signal and idler mode. For this case $h(B, \xi)$ is reduced further to $\bar{h}(B, \xi)$ used in Eq. (136). Figure 20 shows the values of $h(B, \xi)$ and $\bar{h}(B, \xi)$ as a function of the double refraction parameter B where

$$B = \tfrac{1}{2}\rho (lk_0)^{1/2} \tag{138}$$

and

$$\tan \rho = \frac{n_0{}^{\circ}}{2}\left[\frac{1}{[n_3{}^{e}(\theta)]^2} - \frac{1}{(n_0{}^{\circ})^2}\right]\sin 2\theta \tag{139}$$

is the double refraction angle. Here we have assumed a negative uniaxial crystal with ordinary and extraordinary indices of refraction n° and $n^{e}(\theta)$ where θ is the propagation direction with respect to the optic axis. Then B can be written in terms of the double refraction aperture length

$$l_a = w_0 \sqrt{\pi}/\rho \tag{140}$$

as

$$B = \tfrac{1}{2}\sqrt{\pi}\,(l/l_a)\xi^{-1/2} \tag{141}$$

where the focusing parameter $\xi = l/b$ is the ratio of the crystal length to confocal focusing parameter.

Figure 20 shows that for small B, or equivalently for small walk-off angles or for crystal lengths less than the aperture length, $h(B, \xi) \to \bar{h}(B, \xi)$. For second harmonic generation this corresponds to the total second harmonic output becoming less multimode due to Poynting vector walk-off and more fundamental mode.

For parametric gain calculations $\bar{h}(B, \xi)$ can usually be approximated by one of its limiting forms. In the near-field approximation with negligible

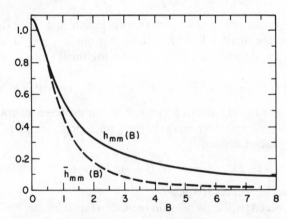

Fig. 20. Conversion efficiency reduction factor for total SHG given by $h_{mm}(B)$ and gain reduction factor for parametric generation $\bar{h}_{mm}(B)$. The dashed curve is closely approximated by Eq. (143) [after Boyd and Kleinman (1968)].

double refraction

$$\bar{h}(B, \xi) \to \xi, \qquad \xi < 0.4, \quad \xi < 1/6B^2 \qquad (142)$$

Substituting this limit into Eq. (136) reduces it to the previous near-field result given by Eq. (132). In the limit of negligible double refraction, except for confocal focusing where $\xi = 1$, the gain reduction factor becomes

$$\bar{h}(0, 1) \approx 1$$

which corresponds to the previous near-field result first derived by Boyd and Ashkin (1966). In fact, the maximum value of $\bar{h}(B, \xi) = \bar{h}_{mm}(0, 2.84) = 1.068$ occurs at $\xi = 2.84$ instead of $\xi = 1$. Practical considerations, however, usually limit $\xi < 1$ so that the confocal approximation is an excellent one.

In the tight-focusing limit, neglecting double refraction, the gain reduction factor becomes

$$\bar{h}(0, \xi) \to 1.187\pi^2/\xi \qquad 80 < \xi < \pi^2/16B$$

which shows that the gain reduction factor $\bar{h}(0, \xi)$ is proportional to b/l in this limit so that the gain reduces to b^2.

When double refraction is important the gain reduction factor can be represented by the empirical approximation

$$\bar{h}_{mm}(B) \approx \bar{h}_{mm}(0)/[1 + (4B^2/\pi)\bar{h}_{mm}(0)] \qquad (143)$$

to better than 10% accuracy over the entire range of B. Here $h_{mm}(0) = 1.068$ is the maximum value of $\bar{h}(B, \xi)$ at optimum focusing where $\xi = 2.84$. The condition

$$(4B^2/\pi)\bar{h}_{mm}(0) \approx 1$$

designates the point at which the gain begins to limit due to double refraction effects. We use this condition to define an effective useful crystal length

$$l_{eff}(\rho) = \lambda_0/2n_0\rho^2\bar{h}_{mm}(0) \approx \lambda_0/2n_0\rho^2 \qquad (144)$$

For a crystal longer than this length the spot size must be increased to keep $l_a \geqslant l$ so that the parametric gain per input power remains constant.

In terms of l_{eff} the gain reduction factor becomes

$$\bar{h}_{mm}(B) \approx \bar{h}_{mm}(0)/[1 + l/l_{eff}]$$

In the limit of strong double refraction such that $l_{eff} \ll l$ or $(4B^2/\pi)\bar{h}_{mm}(0) \gg 1$ the gain reduction factor becomes

$$\bar{h}_m(B, \xi) \to \pi/4B^2, \qquad B^2/4 > \xi > 2/B^2 \qquad (145)$$

Referring to the gain expression given by Eq. (136), we can see that the strong double refraction limit has the effect of replacing the crystal length l by l_{eff}.

In addition, the large double refraction limit maintains $\bar{h}_m(B, \xi)$ at a nearly constant value over a wide focusing range. Thus to minimize crystal damage problems ξ can be chosen such that $\xi \gtrsim 2/B^2$ which is approximately the condition given by Eq. (144). This condition corresponds to a focal spot size

$$w_0 \lesssim (1/2\sqrt{2})\rho l \tag{146a}$$

and effective area

$$\pi w_0^2/2 \lesssim (\pi/16)\rho^2 l^2 \tag{146b}$$

For larger spot sizes the gain is reduced and for smaller spot sizes the gain is practically constant but the intensity increases. Focusing for minimum intensity has been discussed by Harris (1969c) and by Rutt and Smith (1972).

Figure 21 shows the gain reduction factor as a function of the focusing parameter ξ for various values of the double refraction parameter B. The figure shows the significant reduction in gain for nonzero values of B. In addition the near field limit $\bar{h}(0, \xi) \to \xi$ is evident for small ξ.

The gain reduction factor and the effective crystal length are seriously reduced for noncollinear phase matching. Referring to Eqs. (132) and (136) we can write a crystal figure of merit for parametric interactions in the limits of collinear and noncollinear phse matching as

$$M_{\rho=0} = d^2/n_0^2 n_3 \tag{147a}$$

and

$$M_{\rho\neq0} = (d^2/n_0^2 n_3)(l_{\text{eff}}/l) \tag{147b}$$

For large double refraction the gain is reduced by $\pi/4B^2 \approx l/l_{\text{eff}}$. For LiNbO$_3$ at room temperature with phase matching achieved by angle tuning to $\theta = 43°$ for a 1.06 μm pump source, $\rho = 0.037$ rad and $B = 4.7l^{1/2}$. The gain is reduced by 28 times for a 1-cm crystal length and 140 times for a 5-cm

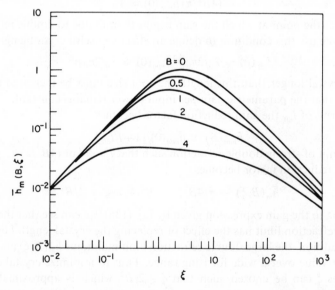

Fig. 21. Gain reduction factor $\bar{h}_m(B, \xi)$ versus ξ [from Boyd and Kleinman (1968)].

crystal length compared to the 90° phase-matched case. For this double refraction angle the maximum useful crystal length is only $l_{\text{eff}} \approx 0.36$ mm. For a 1-cm crystal the focusing parameter ξ can vary between $5.5 > \xi > 0.09$ without affecting the gain. This corresponds to a confocal parameter variation between 11 and 0.2 cm. For experimental ease and reduction of the intensity at the crystal surface the larger value of b would be utilized. For a 1-mm crystal length the gain remains the same but the confocal parameter varies between $1.1 < b < 1.8$ mm, and thus the intensity is considerably increased. The pumping intensity decreases with crystal length as l^2 in agreement with Eq. (146).

In choosing a cavity for a parametric oscillator it becomes immediately obvious that one has to choose between maximizing the OPO gain and ease of cavity construction. For example, the gain is maximum for $\xi = 2.84$ or in the near-field approximation for $\xi = 1$. To achieve confocal focusing with a crystal of index n within the optical cavity requires mirrors on the crystal surface or external mirrors spaced toward a spherical resonator configuration. The spherical resonator, however, requires critical mirror alignment approaching that of a plane parallel cavity of a few seconds of arc (Bloom, 1963).

If maximizing gain is essential, such as for cw OPO operation, the spherical cavity must be used. If, however, gain is high for normal input pump powers, such as is the case for Q-switched pump sources, then a cavity can be chosen which is more tolerant to misalignment. In practice the confocal cavity is usually chosen. The confocal cavity has the highest tolerance for mirror misalignment θ' given by

$$\theta' = (4\lambda/\pi R)^{1/2} \tag{148}$$

where R is the mirror radius of curvature and $R = L'$ where L' is the mirror separation including the crystal optical distance. Thus for a 10-cm confocal cavity at 1-μm wavelength $\theta' \simeq 3$ mrad or 12 min. This corresponds to one full turn of a micrometer for a typical mirror mount. The ease of alignment gained by use of a confocal cavity results in a gain reduction proportional to $(l/n)/R$. For a 5-cm LiNbO$_3$ crystal within a 10-cm confocal cavity the gain reduction is 1/4.4. This reduction in gain is tolerable for most Q-switched pumped parametric oscillators.

In some experimental situations it is desirable to use a hemispherical cavity for the parameteric oscillator. The hemispherical cavity alignment tolerance is similar to that for the confocal cavity. The gain, however, is reduced by a factor of two in the absence of double refraction (Ammann and Montgomery. 1970; Boyd and Nash, 1971). For $B = 1$, $\bar{h}_{mm}(B = 1)$ has the values 0.48 and 0.31 for the confocal and hemispherical cases, respectively. For $B = 4$, the gain reduction factor $\bar{h}_{mm}(B = 4)$ is 0.049 and 0.046, respectively. Thus as B increases, the thresholds for a confocal (symmetric) and hemispherical (half symmetric) cavities become nearly equal.

Up to this point we have limited the discussion to circular symmetric modes. In uniaxial crystals, walk-off or double refraction occurs in a plane determined by the direction of propagation and optic axis. By elliptical focusing the parametric gain can be considerably increased. The problem of elliptical focusing was first treated by Volosov (1970) for second harmonic generation and recently by Kuizenga (1972) for parametric generation.

In treating elliptical focusing, Kuizenga defined the effective signal to idler confocal parameter $b_{12} = (b_x \cdot b_y)^{1/2}$ and the focusing parameter $\xi = l/b_{12}$. In addition, the ratio of the signal to idler confocal parameter $\alpha_{12} = (b_y/b_x)^{1/2}$ is important as is the ratio of the signal–idler confocal parameter in the x and y transverse dimensions to the pump confocal parameter, $\alpha_x = b_x/b_{\xi y}$. Carrying out the maximization of $\bar{h}_m$, Kuizenga shows that the signal–idler and pump confocal parameters are the same in the non-walk-off direction. For no double refraction ($B = 0$) the optimum conditions are $\alpha_x = \alpha_y = 1$ and $\alpha_{12} = 1$ in agreement with Boyd and Kleinman's analysis.

For focusing with double refraction elliptical focusing increases the gain. Figure 22 shows a comparison of $\bar{h}_{mm}(B)/\bar{h}_{mm}(0)$ versus B for various focusing conditions. Elliptical focusing of the pump improves the parametric gain considerably for large B. Figure 22 shows that for $B > 4$, $\bar{h}_m \propto 1/B$ for elliptical

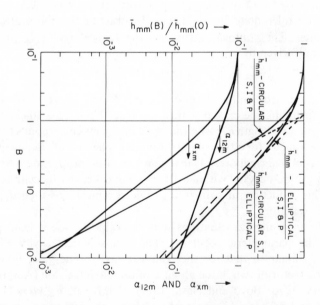

Fig. 22. Comparison of optimum threshold parameter $\bar{h}_{mm}(B)$ as a function of B for elliptical focusing. α_{12m} and $\alpha_{\lambda m}$ are the ellipticity ratios of the signal–idler and pump waves. Here the signal–idler are assumed ordinary waves and the pump is an extraordinary wave [after Kuizenga (1972)].

focusing and $\bar{h}_m \propto 1/B^2$ for circular focusing. Thus elliptical focusing allows the use of longer crystals to lower the threshold pump power. For example, with circular focusing in a 50° cut LiNbO$_3$ crystal the gain is constant for crystal lengths longer than 5 mm assuming a 1.06-μm pump source. With elliptical focusing and longer crystals (~ 2 cm) the gain is increased by 5.8.

Figure 22 also shows the elliptical focusing parameters for optimum gain. For reasonable values of B, α_{12m} remains near unity so that the signal and idler can be circularly focused. This simplifies the practical problem of elliptical focusing considerably since only the pump beam need be elliptically focused to a value given by α_{xm}.

The ratio of the crystal length to pump beam aperture length in the walk off direction is

$$l/l_{\mathrm{ax}} = 2B(\xi\alpha_{12}\alpha_x/\pi)^{1/2} \tag{149}$$

For $B > 4$ and for optimum focusing this ratio is within 10% of unity. Since α_{12} is close to unity we can restrict it to unity and find the gain reduction factor. This is shown by the dotted line in Fig. 22. In this approximation $\alpha_x \simeq \pi/4B^2$. The actual optimum value of $\bar{h}$ occurs at $\xi = 1.55$ and gives $\alpha_{xm} = 0.44/B^2$ for $B > 4$.

Since in the low gain limit the parametric gain and second harmonic conversion efficiency are equal, the use of elliptical focusing can be extended to improve the SHG efficiency for double refraction-limited crystals.

The operation of a parametric oscillator involves the important consideration of optimum coupling or mode matching the pump laser source into the OPO cavity. The procedure for mode matching is straightforward and is treated by Kolgelnik and Li (1966). An example, however, helps to illustrate

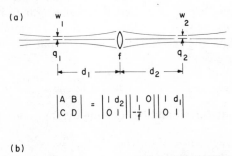

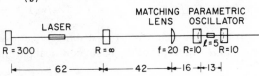

Fig. 23. (a) Gaussian beam transformation by a thin lens. (b) Schematic of the laser, mode-matching lens, and parametric oscillator.

the steps involved. Figure 23a illustrates mode matching by a lens and Fig. 23b shows a schematic of a laser source, mode-matching lens, and parametric oscillator cavity. The problem is to focus the pump beam with parameters determined by the laser cavity into the parametric oscillator cavity at the optimum location and with proper spot size.

Gaussian beams transform according to the $ABCD$ law (Kolgelnik, 1965)

$$q_2 = (Aq_1 + B)/(Cq_1 + D) \tag{150}$$

where $ABCD$ are elements of the ray transfer matrix and the beam parameter q is given by

$$\frac{1}{q} = \frac{1}{R} - i\frac{\lambda}{\pi w^2} \tag{151}$$

Equation (150) relates the input Gaussian with beam parameter q_1 to the resultant output Gaussian beam with parameter q_2 after passing through an optical system characterized by the $ABCD$ matrix. [For a list of $ABCD$ matrices see Kogelnik and Li (1966, Table I)]. For a multiple element optical system the resultant $ABCD$ matrix is the product of the $ABCD$ matrix for each optical component and distance between components. For example, the transformation of a Gaussian beam from a distance d_1 prior to a thin lens of focal length f to a distance d_2 following the lens is given by

$$q_2 = \frac{(1 - d_2/f)q_1 + (d_1 + d_2 - d_1 d_2/f)}{-(q_1/f) + (1 - d_1/f)} \tag{152}$$

To mode match the pump beam into the parametric oscillator a lens of focal length larger than a characteristic focal length f_0 is chosen and the distances d_1 and d_2 are set to results given below. The minimum focal length f_0 results from equating the real parts of the input and transformed beam parameters and is given by

$$f_0 = \pi w_1 w_2 / \lambda \tag{153}$$

The distances d_1 and d_2, found by equating the imaginary parts of the beam parameters, are

$$d_1 = f \pm (w_1/w_2)(f^2 - f_0^2)^{1/2}, \qquad d_2 = f \pm (w_2/w_1)(f^2 - f_0^2)^{1/2} \tag{154}$$

where w_1 refers to the input beam waist at distance d_1 and w_2 refers to the focused beam waist at d_2. In terms of the beam confocal parameters b_1 and b_2 we have

$$f_0^2 = \tfrac{1}{4} b_1 b_2 \tag{155}$$

and

$$d_1 = f \pm \tfrac{1}{2}b_1 [(f^2/f_0^2) - 1]^{1/2}, \qquad d_2 = f \pm \tfrac{1}{2}b_2 [(f^2/f_0^2) - 1]^{1/2} \quad (156)$$

In this form the wavelength does not appear explicitly.

As an example we consider at 10-cm confocal cavity parametric oscillator operating with a 90° phase-matched 5-cm-long LiNbO$_3$ crystal. We assume a 0.6589-μm pump source (the second harmonic of a Nd:YAG 1.32-μm laser line) incident on the parametric oscillator with a crystal temperature of 376°C. At this temperature the signal and idler wavelengths are 1.11 and 1.62 μm, respectively. An assumed laser cavity 62 cm in length with a 300-cm flat mirror gives a calculated spot size at the output flat mirror of 504 μm. The 5-cm LiNbO$_3$ crystal increases the physical cavity length of the oscillator by $l[1-(1/n)]$, where n is the index of refraction ($n \approx 2.2$) so that for 10 cm radius of curvature mirrors the cavity is confocal at a physical mirror spacing of 12.7 cm. The calculated parametric oscillator signal and idler spot sizes lead to a pump spot size of 102 μm. At this pump spot size the parametric gain is $\Gamma^2 l^2 = 2.2/\text{kW}$ of pump power. The maximum gain limited by the crystal burn density of 80 MW/cm^2 is $\Gamma^2 l^2 = 28.5$ and occurs at 13 kW input pump power. In practice this oscillator will oscillate from the Fresnel reflections of the LiNbO$_3$ crystal surfaces.

To carry out the matching calculations we can first find the apparent parametric oscillator mode size and location as viewed through the negative lens formed by the 10 cm radius of curvature oscillator mirror. Using the resulting apparent waist size and the laser spot size we find $f_0 = 19.67$ cm. Any lens of focal longer than f_0 effects the matching. In practice the finite length of the optical bench limits the choice of a matching lens focal length as slightly ($\sim 20\%$) greater than f_0. For this example we choose $f = 20$ cm and calculate a laser output mirror to lens distance of 42 cm and a lens to oscillator input mirror distance of 15.37 cm. Due to the negative lens formed by the oscillator input mirror and the relatively small oscillator spot size, the lens to oscillator mirror separation is the most critical distance.

Experimentally the laser pump source provides a convenient way to align the parametric oscillator if it is visible or can be imaged. The procedure is to center the oscillator mirrors on the pump beam and then insert the matching lens in its proper location and on center. The front oscillator mirror is aligned by monitoring its back reflected spot. To do this either a "dirty" glass microscope slide or a card with a hole in it can be used as an imaging plane between the laser and oscillator cavity. Next the crystal is centered and aligned and finally the second cavity mirror is aligned. The reflected spots from the cavity mirrors indicate the success of the mode matching. If the oscillator is properly mode matched, the reflected spots are the same radius as the incident pump laser spot. Alignment of parametric oscillators by this method is usually good to a degree well within the misalignment tolerance of the confocal cavity.

III. PARAMETRIC OSCILLATOR DEVICES

A. LiNbO₃ Parametric Oscillators

In 1965 Giordmaine and Miller observed the first optical frequency parametric oscillation. The nonlinear crystal used in that experiment was $LiNbO_3$. Since then $LiNbO_3$ has played an important part in the development of parametric oscillators, and is now used in a commercial singly resonant parametric oscillator which is pumped by a Q-switched Nd:YAG laser source and is tunable over a spectral range from 0.56 to 3.5 μm.

$LiNbO_3$ is a ferroelectric material (Matthias and Remeika, 1949) with a Curie temperature approximately 40°C below its melting point of 1253°C. Since the recognition of the unique electrooptical (Peterson et al., 1964) and nonlinear optical (Boyd et al., 1964) properties of $LiNbO_3$ in 1964, it has been extensively studied. The growth and physical properties of $LiNbO_3$ have been discussed in a series of papers by K. Nassau et al. (1966) and Abrahams et al. (1966). The electrooptic coefficients for $LiNbO_3$ have been measured by a number of workers (Lenzo et al., 1966; Turner, 1966; Zook et al., 1967; Hulme et al., 1969) as have the indices of refraction (Hobden and Warner, 1966; Miller and Savage, 1966; Boyd et al., 1967). A particularly useful form for the refractive indices, including temperature dependence, is given by Hobden and Warner (1966).

Early work with $LiNbO_3$ showed two potentially troublesome optical properties: optically induced inhomogenities in the refractive index (Ashkin et al., 1966; Chen, 1969; Johnston, 1970) and growth-dependent birefringent variations (Bergman et al., 1968; Fay et al., 1968; Midwinter, 1968; Miller et al., 1971). The optically induced index inhomogenities were found to be self-anealling for crystal temperatures above approximately 180°C for visible radiation and 100°C for near infrared radiation. Attempts to eliminate the induced index inhomogenities have not been successful so that $LiNbO_3$ parametric oscillators usually operate above 180°C.

The growth-dependent birefringent variations were more difficult to eliminate. The problem was solved by growth of $LiNbO_3$ from its congruent melting composition near a lithium-to-niobium ratio of 0.48 mole % (Lerner et al., 1968; Carruthers et al., 1971). The growth of $LiNbO_3$ crystals from a congruent melt plus improved optical quality tests (Midwinter, 1968a; Byer et al., 1970) led to uniform high-quality single crystals of over 5 cm in length.

Ferroelectric $LiNbO_3$ has a large variation of birefringence with temperature. This allows SHG at 90° phase matching for fundamental wavelengths between 1 and 3.8 μm at temperatures between 0 and 550°C. Figure 24 shows SHG curves versus crystal temperature for $LiNbO_3$ at various phase-matching angles. Conversely, $LiNbO_3$ 90° phase matches for parametric oscillation for

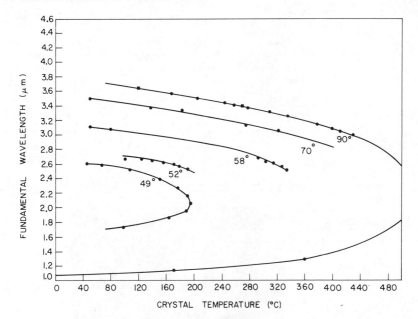

Fig. 24. Second harmonic generation curve versus LiNbO₃ crystal temperature for 90° and off-angle phase matching.

a number of pump wavelengths and can be temperature tuned over a broad spectral region. Figure 15 shows the parametric oscillator tuning curves for various pump wavelengths of an internally doubled Q-switched Nd:YAG laser. These LiNbO₃ phase-matching curves were calculated by using the Hobden and Warner (1966) index of refraction expressions in the energy and momentum equations [Eqs. (10) and (11)]. Additional LiNbO₃ tuning curves, including angular dependence, are given by Harris (1969a,c) and Ammann et al. (1971).

The internally doubled Q-switched Nd:YAG pumped LiNbO₃ parametric oscillator is the best developed parametric oscillator at this time. This system is described by Wallace (1970). The Nd:YAG laser operates with an internal LiIO₃ doubling crystal, acoustooptic Q-switch, and two Brewster angle prisms for wavelength selection. The combination of four doubled Nd:YAG pump wavelengths and temperature tuning allows the OPO to cover the 0.54 to 3.65 μm spectral region. Table V gives the OPO spectral output wavelengths, laser pump wavelengths, and output power levels. the values used in this table are useful for comparative purposes since they reflect the characteristics of the particular Q-switched Nd:YAG pump laser and are not LiNbO₃ oscillator limits.

The Q-switched Nd:YAG laser is a multiaxial mode pump source. Thus

Table V

LiNbO₃ SRO *Operating Parameters*

Tuning range (μm)	Peak power (W)	Average power (mW)	Pulse length (nsec)	Laser pump wavelength (μm)
3.50–2.50	80–150	5–10	70	0.532
0.975–2.50	250–350	30	80	0.659
0.725–0.975	250	20	200	0.562
0.623–0.760	250	50	150	0.532
0.546–0.593	300	3–5	200	0.473

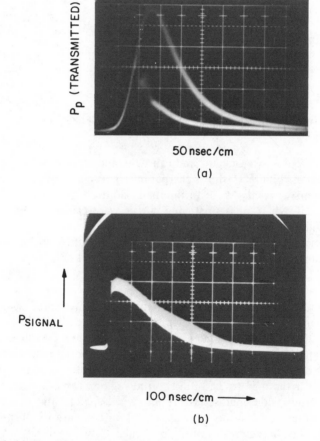

Fig. 25. (a) The 67% energy conversion of a LiNbO₃ SRO pumped with 0.695 μm. (b) Oscilloscope photograph of more than 100 oscillator output pulses at $\lambda_s = 0.72$ μm for a 0.561-μm pump [after Wallace (1971a)].

previously calculated conversion efficiencies for the single-mode SRO do not apply directly. For the singly resonant oscillator, however, all modes of the pump laser are effective in pumping the oscillator (Harris, 1966) as long as the pump bandwidth does not exceed the oscillator acceptance bandwidth (Young *et al.*, 1971). For all the lines of a Nd : YAG laser except the second harmonic of 0.946 μm at 0.473 μm (2–4 cm^{-1} bandwidth) the laser bandwidth is less than the oscillator acceptance bandwidth. If the 0.473-μm pumped SRO is operated with the infrared wave resonant, the pump acceptance bandwidth of 3.7 cm^{-1} for a 5-cm crystal is adequate. It it is operated with the signal or visible wave resonant, the oscillator operates poorly due to the much narrower 0.63 cm^{-1} pump acceptance bandwidth.

Threshold for the LiNbO$_3$ SRO with a 5-cm-long 90° phase-matched crystal is typically between 300 and 600 W. The peak power conversion efficiency is near 50% for an oscillator operating a few times above the threshold. Due to the finite pump pulse width and oscillator build-up time, the energy conversion efficiency is approximately 30%. Much higher conversion efficiencies, however, have been reported. Figure 25a shows 67% pump depletion for a LiNbO$_3$ SRO pumped at 0.659 μm. Figure 25b shows the peak-to-peak stability at $\lambda_s = 0.72$ μm for a 0.561-μm pumped LiNbO$_3$ oscillator. The oscilloscope photo is a multiple exposure of over 100 consecutive pulses. The long-term average power stability is shown in Fig. 26. Here an Eppley thermopile monitors the 30-mW average power output of a 0.532-μm pumped LiNbO$_3$ oscillator operating at 70 pps. The average power stays within 3% of its set value for most of the 1½-hr duration.

LiNbO$_3$ exhibits a nonlinear absorption for pump wavelengths in the visible. For power densities normally used to pump a LiNbO$_3$ parametric oscillator near 10 MW/cm^2, the nonlinear absorption approaches 30% in a 5-cm crystal (Wallace and Dere, 1972). More measurements need to be made,

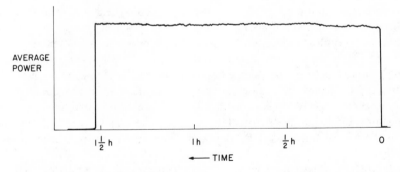

AVERAGE POWER

1½ h   1 h   ½ h   0

◄—— TIME

Fig. 26. Eppley thermopile recording of the output of a LiNbO$_3$ SRO operating with a 3.8-kW peak, 30-mW average power, and 0.532-μm pump source [after Wallace and Dere (1972)].

but preliminary work shows that the nonlinear absorption does not occur for wavelengths at 0.66 μm or longer. The effect reduces the oscillator's conversion efficiency and in some crystals is large enough to prevent the oscillator from reaching threshold.

The calculated gain line width [cf. Eq. (26)] for a $LiNbO_3$ parametric oscillator is near 1 cm^{-1} for a 5-cm crystal. In practice the resonant wave of the oscillator oscillates over a small fraction of the total bandwidth. Measured resonant wave bandwidths are typically 0.1–0.2 cm^{-1} for an oscillator operating two to six times above threshold. The nonresonant wave bandwidth approaches the full gain bandwidth and also reflects the pump bandwidth and axial mode characteristics. Thus parametric oscillator line narrowing is usually done at the resonant wave frequency. For an oscillator operating far above threshold, both the resonant and nonresonant wave bandwidth increases to the full gain bandwidth of the oscillator (cf. Fig. 14).

Narrowing of the resonant wave bandwidth has been successfully done using Fabry–Perot etalons in the oscillator cavity. The small beam waist precludes tilting the etalon as is normally done for tuning, so that the etalon spacing is thermally scanned. In this way bandwidths to 30 MHz have been achieved for a period of over eight hours. The frequency can be tuned over 1500 MHz or 0.05 cm^{-1} by scanning the oscillator cavity length. The etalon and cavity length must both be scanned to increase the tuning range further.

The parametric oscillator can be continuously tuned at between 2 and 10 cm^{-1} min^{-1} by scanning the oven temperature at 1°C/min. This tuning rate may be slow for some applications, but at 0.2 cm^{-1} bandwidth for the resonant wave it is quite rapid tuning for a pulsed tunable source. Still, scanning the crystal temperature may not satisfy some tuning requirements. In this case, the use of a tunable pump source such as a rhodamine 6G dye laser provides an alternative. Figure 17 shows the experimental tuning range achieved at fixed $LiNbO_3$ temperatures.

An extension of the $LiNbO_3$ tuning range can also be achieved by internal up-conversion first reported by Wallace (1971c). Using $LiIO_3$ as the up-conversion crystal Wallace achieved 15% peak power and 10% energy conversion of laser input to up-converted output. Figure 27 shows an experimental tuning curve of a 0.659-μm pumped up-converted $LiNbO_3$ oscillator. A single $LiIO_3$ crystal at near 30° phase-matching angle was able to phase match the two processes shown. Up-conversion has also been used to extend the tuning range of a $LiIO_3$ parametric oscillator by Campillo and Tang (1971) and has been theoretically discussed by Bey and Tang (1972) and Giallorenzi and Reilly (1972).

$LiNbO_3$ has been used in internal optical parametric oscillator (IOPO) experiments. Oshman and Harris (1968) first analyzed the IOPO. Their analysis included laser saturation but did not include the temporal variation

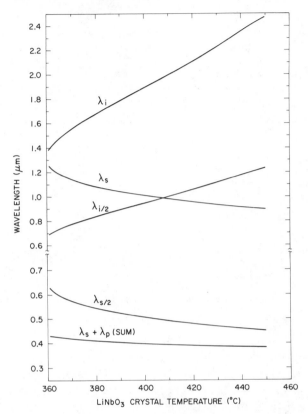

Fig. 27. Experimental tuning curves for a 0.659-μm pumped LiNbO$_3$ parametric oscillator. Also shown are the up-converted wavelengths obtained with an internal LiIO$_3$ crystal.

of the laser population difference. Smith and Parker (1970) experimentally observed many of the solutions predicted by Oshman and Harris in a cw Ba$_2$NaNb$_5$O$_{15}$ parametric oscillator. Ammann *et al.* (1970) studied a LiNbO$_3$ internal oscillator pumped with a repetitively Q-switched Nd:YAG laser. The analysis of internal parametric oscillators was extended by Falk *et al* (1971) to include the temporal fluctuations of the pumping laser. Experimental studies using a Q-switched Nd:YAG laser as a pump source for an internal LiNbO$_3$ oscillator verified the theory.

Figure 28 shows the internal LiNbO$_3$ parametric oscillator experimental arrangement. Figure 29a and (b) show the computed pump power, oscillator signal power, population inversion, and phase of the laser and oscillator for initial normalized inversion of 1.2 and 1.4. The dynamics occur due to the close coupling of the oscillator and laser gain medium. The laser field builds

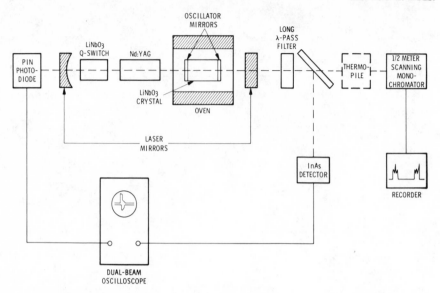

Fig. 28. Internal LiNbO₃ parametric oscillator experimental arrangement [after Ammann *et al.* (1970)].

up following the opening of the Q switch and begins to pump the oscillator. The parametric oscillator reaches threshold and in a short time builds up its fields. The oscillator fields deplete the pump field and thus temporarily drive it below laser threshold. If the laser pumping is strong enough, there still is enough population inversion remaining to start the process over again. Figures 30a and (b) show two and four strong oscillator spikes at normalized inversions of 1.2 and 1.4, respectively. The dynamics of the internal parametric oscillator are observed experimentally as shown in Fig. 30. In each case, the pump is harder than the preceding case. Significantly, experimental observation of both the internal DRO and internal SRO show that the spiking behavior remains the same.

In addition to interesting dynamics, the internal parametric oscillator provides high conversion efficiency to the signal and idler waves. In the above experiment, the oscillator is more than 70% efficient. The high efficiency has led to high average powers. To date Yarborough (1972) has reported up to 1.8 W average power from a 1.06 μm Nd:YAG pumped internal parametric oscillator.

Efforts are continuing to improve the LiNbO₃ SRO. In particular, methods to narrow the bandwidth and improve the ease of fine tuning are being studied. The rhodamine 6G dye laser is an almost ideal pump source for the LiNbO₃ oscillator so that further development along this direction can be expected. The LiNbO₃ tuning range has been extended toward the visible and near

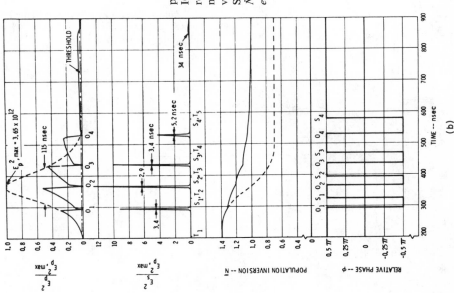

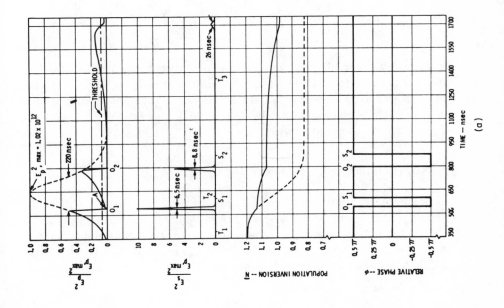

Fig. 29. (a) Computer solutions for IOPO for 2% oscillator round trip loss. Initial normalized pump inversion $\bar{N}^e - 1.2$. (b) Same as for (a) except $\bar{N}^e = 1.4$ [after Falk *et al.* (1971)].

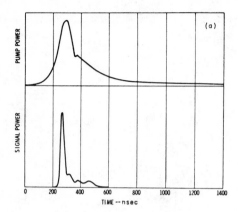

Fig. 30. Experimentally observed pump and signal pulse shapes for successively harder pumping [after Falk *et al.* (1971)].

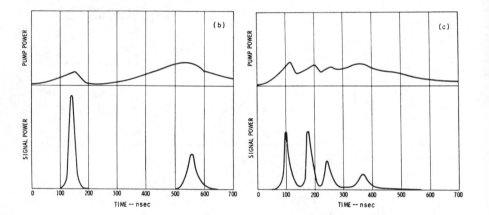

ultraviolet by internal up conversion. We may expect that extension into the infrared by mixing will receive more attention. Finally, LiNbO$_3$ devices pumped with picosecond pulses have operated in a superfluorescence mode with reasonable efficiency. The LiNbO$_3$ crystal thus provides a potential widely tunable picosecond pulse source.

B. KDP, ADP, and LiIO$_3$ Parametric Devices

KDP, ADP, and their isomorphs are ferroelectric crystals (Jona and Shirane, 1962) belonging to the 42m tetragonal point group above their Curie temperatures. KDP crystals are transparent from 0.21 to 1.4 μm for non-deuterated and 0.21 to 1.7 μm for the deuterated material (Milek and Wells, 1970). ADP has a similar transparency range. KDP and its isomorphs are negative birefringent and are phase matchable over most of their transparency range. Over small regions of wavelength the crystals 90° phase match with temperature as the variable. For a complete reference to crystal indices of refraction see the work of Milek and Wells (1970), Landolt and Bornstein (1969) and Pressley (1971).

Because of the availability of large, high optical quality crystals, KDP and ADP were the subject of early nonlinear optical experiments by Miller *et al.* (1963) and Miller (1964). Later Francois (1966) and Bjorkholm and Siegman (1967) made accurate cw measurements of ADPs nonlinearity. In addition, Bjorkholm (1968) has made comparative nonlinear coefficient measurements of other crystals relative to KDP and ADP.

Parametric amplification was first reported by Wang and Racette (1965) in ADP. Their report was closely followed by reports of parametric gain in KDP and ADP by Akhmanov *et al.* (1965c). In 1966 shortly after Giordmain and Miller (1965) reported parametric oscillation in LiNbO$_3$, Akhmanov *et al.* (1966) achieved parametric oscillation in KDP pumped by 0.53 μm from a KDP doubled Nd : glass laser. That work was extended to the demonstration of parametric superfluorescence in a multipass traveling wave oscillator using ADP and KDP by Akhmanov *et al.* (1968c, 1969).

The large electrooptic effect in KDP led to studies of electrooptic controlled SHG phase matching by Adams and Barrett (1966) and electric field tuned parametric oscillators by Krivoshchekov *et al.* (1968). In 1967 Magde and Mahr observed parametric fluorescence in ADP pumped with a doubled ruby laser. Later Dowley (1969) measured visible tuning curves for ADP and KDP pumped by 2573 Å, the second harmonic of 5145 Å from an argon–ion laser. Observation of three frequency parametric fluorescence was extended to four frequency processes by Ito and Inaba (1970).

KDP and ADP have played a significant role as efficient second harmonic generators for both cw and pulsed sources. For cw doubling ADP and KDP can be temperature tuned to the 90° phase matching condition over a limited range of fundamental wavelengths between 0.54 and 0.49 μm at temperatures between $+60$ and -80°C (Dowley and Hodges, 1968; Huth and Kiang, 1969). In particular ADP and KDP 90° phase match for doubling 5145 Å at -9.2 and -11.0°C. Other isomorphs of KDP 90° phase match over different wavelength regions. For example, rubidium dihydrogen arsenate (RDA) 90° phase matches for doubling the 0.694 μm ruby laser and cesium dihydrogen arsenate (CDA) 90° phase matches for doubling 1.06 μm. Figure 31 shows temperature phase matching regions for various isomorphs of KDP (Wallace, 1973).

Using 90° phase matched ADP, Dowley and Hodges obtained up to 100 mW of 2573 in 1 msec pulses and 30–50 mW of cw power. The doubling was performed internally to an argon–ion laser cavity to take advantage of the high

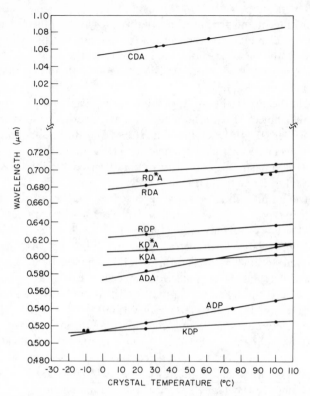

Fig. 31. 90° phase-matching tuning curves for KDP isomorph crystals [after Wallace (1973)].

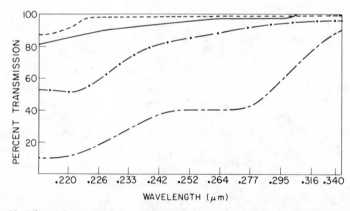

Fig. 32. Ultraviolet transmission of ADP (dashed line) and KDP crystals. Each crystal was 5 cm long. The KDP crystals used were Isomet (—), Clevite (–··–), and Harshaw (———). ADP crystals from all suppliers had similar low loss [after Dowley and Hodges (1968)].

circulating fields. The SHG efficiency was strongly dependent on crystal losses. Figure 32 shows measured ultraviolet transmission of KDP and ADP crystals used by Dowley and Hodges (1968).

The ultraviolet transparency and phase matching characteristics of KDP isomorphic crystals make them useful for ultraviolet generation by sum and second harmonic generation. Huth *et al.* (1968) were the first to demonstrate this capability by externally doubling a dye laser source. Similarly Yeung and Moore (1971) and Sato (1972) generated tunable ultraviolet between 3044 to 3272 Å by suming a ruby pumped due laser and a ruby laser.

Wallace (1971c) reported an intracavity doubled dye laser that tunes between 2610 and 3150 Å. This source uses a Q switched internally doubled Nd : YAG laser as a pump source for the rhodamine 6G and sodium fluorescein dye laser. The ADP intracavity doubled dye laser produced 32 mW of average power at 2900 Å in a 2–3/cm bandwidth. The conversion efficiency from input doubled Nd : YAG to ultraviolet power was 4.3%. At the 65° phase matching angle used for doubling the rhodamine 6G dye laser, the ADP double refraction angle is $\rho = 0.025$ rad which results in a conversion efficiency to the second harmonic of 1% per 100 W of input power.

Recently a KDP internally doubled dye laser and dye laser plus 1.06 μm wavelength sum generation source has been commercially developed which tunes over a 0.26–0.70 μm spectral region.[†] The device demonstrates the wide tuning range available using angle phase matched KDP as the sum and second harmonic generator.

ADP 90° phase matches for doubling the 0.532 μm second harmonic of Nd : YAG at 49°C. Wallace (1971a) reported peak power conversion efficiencies of 50 to 55% and energy conversion efficiencies of 38% for doubling 0.5320 μm. The output at 0.266 μm reached 1.5 kW peak and 80 mW average power. The power was limited by induced nonlinear absorption in the ADP crystal. The 0.266 μm power was high enough to allow Wallace to use it as a pump source for an ADP parametric oscillator (Wallace, 1971a).

Yarborough and Massey (1971) used the fourth harmonic of a Nd : YAG laser to generate high gain parametric oscillation in ADP. Figures 33a and (b) show the generated tuning curves and Fig. 34 shows a photograph of the oscillator. This high gain oscillator generated 100-kW peak power and 10-mW average power output across the visible spectrum. Overall average power 1.06 Nd : YAG source to the ultraviolet was 5.3% yielding an average power of 30 mW at 30 pps at 0.2662 μm. The peak ultraviolet power was 200 kW and the power density at the ADP generator crystal was 750 MW/cm^2.

Recently Yarborough (1972) obtained considerably higher pulse energies and average powers. For the 1.06 μm source Yarborough uses a flashlamp

[†] Chromatix, Inc.

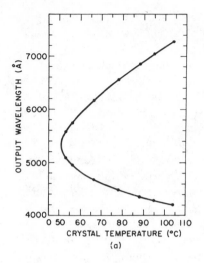

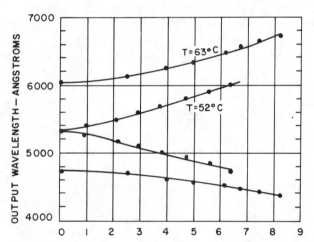

Fig. 33. (a) Temperature-tuning curve for an ADP OPO for a pump wavelength at 2662 Å; (b) Angle-tuning curve for ADP at two fixed crystal temperatures. [After Yarborough and Massey (1971)].

pumped $\frac{1}{4}$-in. diameter rod Nd:YAG oscillator followed by $\frac{1}{4}$-in. diameter and $\frac{3}{8}$-in. diameter rod amplifiers which generated $1J$ pulses at 10 pps. This source is doubled in a 90° phase matched CD*A crystal at 45% efficiency yielding 450 mJ, 0.532 μm pulses. The CD*A phase matches at 103°C and the incident 1.06-μm power is unfocused into the 20 mm CD*A crystal. The green 0.532-μm source is doubled into the ultraviolet in a 2-cm-long 90°

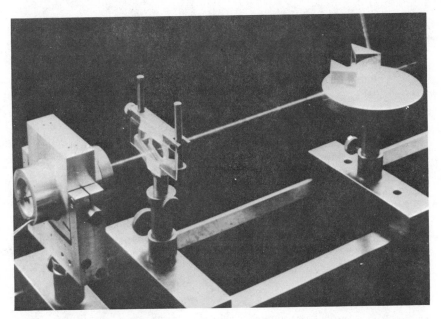

Fig. 34. Photograph of an ADP oscillator operating in the visible and pumped with the fourth harmonic of a 1.06-μm Nd : YAG laser (courtesy of J. M. Yarborough, Sylvania Inc., Mountain View, California).

phase matched ADP crystal. The conversion efficiency for this step is 22% yielding 100-mJ pulses at 0.266 μm. Again the doubling is done without focusing in the ADP crystal. The generated pulses are 10–15 nsec with peak powers of 5 to 10 MW. The harmonic generation in both CD*A and ADP takes place without damage to the crystals. The power levels reported are limited by the laser source becoming unpolarized at high pumping levels. Simultaneously, thermal heating of the doubling crystals is beginning to limit the second harmonic generation efficiency. The ultraviolet source has been used as a pump for an ADP parametric oscillator. In addition, the green output at 0.532 μm has been used to pump dye lasers with up to 60% conversion efficiency.

The high energy, high average power source reported by Yarborough illustrates the optical quality of KDP and its isomorphs. In addition, the use of 90° phase matching for efficient second harmonic generation demonstrates its advantage in these experiments. Although KDP-type crystals have not been utilized extensively in parametric oscillator studies, they play an important role in generating tunable ultraviolet radiation by second harmonic and sum generation of tunable visible sources.

In 1968, Kurtz and Perry applied a technique based on the measurement of SHG of powders to the search for nonlinear materials. That search led to the evaluation of α-iodic acid (α-HIO$_3$) for phase matched second harmonic generation Kurtz et al. (1968). Measurement of the nonlinearity of α-HIO$_3$ showed that it had a nonlinear coefficient approximately equal to that of LiNbO$_3$. The favorable nonlinear properties of α-HIO$_3$ led to consideration of other AIO$_3$ crystals (Bergman et al., 1969). One of the first crystals considered was lithium iodate LiIO$_3$. Its optical and nonlinear optical properties were studied by Nath and Haussuhl (1969a) and Nash et al. (1969). The more favorable optical quality of LiIO$_3$ compared to α-HIO$_3$ has led to its use in a number of nonlinear applications spanning its entire transparency range from 0.35 to 5.5 μm.

LiIO$_3$ (point group 6) has a measured nonlinear coefficient slightly greater than that of LiNbO$_3$ (Campillo and Tang, 1970; Bjorkholm, 1968c; Jerphanon, 1970; Pearson et al., 1972c). Its indices of refraction and large birefringence ($\Delta n = 0.15$) have been measured as having phase-matching angles for various pump wavelengths. LiIO$_3$ phase matches at 52° for doubling 0.6943 μm of a ruby laser (Pearson et al., 1972; Nath and Haussuhl, 1969) and at 30° for doubling a 1.06 μm Nd : YAG laser.

LiIO$_3$ has proved particularly useful as a high optical quality nonlinear crystal for internal second harmonic generation of a Nd : YAG laser. For this application the laser mirrors are highly reflecting in the infrared but are transparent at the second harmonic. To obtain efficient doubling the laser operates Q switched. In this mode, the LiIO$_3$ acts as a nonlinear output coupler and efficiently doubles the Nd : YAG. By operating with a prism in the laser cavity any one of 15 Nd : YAG laser lines can be selected and efficiently doubled by rotating the LiIO$_3$ to the phase-matching angle. In this way wavelengths at 0.473, 0.532, 0.579, and 0.659 μm can be generated. For example, peak powers of over 10 kW and average powers of greater than 1 W have been obtained from the internally doubled Nd : YAG laser source.

For the low gain Nd : YAG transitions the LiIO$_3$ substantially overcouples the laser output to the second harmonic. This overcoupling leads to stretching of the Q switch pulse as discussed by Murray and Harris (1970). Young et al. (1971) experimentally confirmed pulse stretching for the 0.946 μm line of Nd : YAG (Wallace and Harris, 1969). In their experiment the unfocused Nd : YAG cavity spot size was 0.6 mm and the second harmonic was generated in a 1.4-cm-long LiIO$_3$ crystal. The pulse stretched doubled Nd : YAG laser is a useful pump source for dye lasers and for parametric oscillators.

In 1970, Campillo and Tang studied spontaneous parametric scattering in LiIO$_3$ and Dobrzhanskii et al. (1970) carried out similar measurements in

α-HIO$_3$. Shortly afterward, Goldberg (1970) constructed a LiIO$_3$ SRO pumped with a ruby laser and Izrailenko *et al.* (1970) demonstrated SRO using LiIO$_3$ and α-HIO$_3$ pumped by a doubled Nd: Glass laser.

Goldberg (1970) reported a threshold of 60 MW/cm^2 for a 0.8 cm LiIO$_3$ crystal oriented at 21°. The doubled refraction angle is $\rho = 3°$. The calculated cw threshold assuming SRO operation with 5% loss is 5 MW/cm^2. The observed oscillator rise time of 10 nsec corresponds well with the cavity length and round trip gain of 1.8 dB. Pumping 1.3 times above threshold, pump depletion of up to 15% was observed. The oscillator was limited to operation

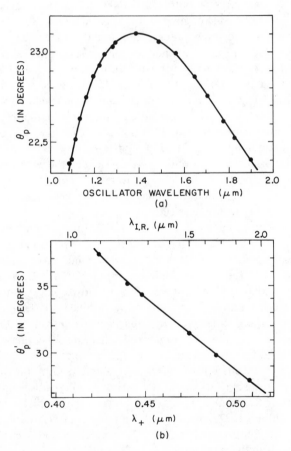

Fig. 35. (a) Measured tuning curve for a doubly resonant LiIO$_3$ oscillator with $\lambda_p =$ 6943 Å. (b) Measured up-converted wavelength and phase-matching angle with $\omega_p + \omega_{IR} \rightarrow \omega_+$ [after Campillo and Tang (1970)].

near threshold by the 125 MW/cm² measured LIIO₃ surface damage intensity. Tuning the oscillator by crystal rotation covered the range 0.84 to 0.96 μm and 4.0 to 2.5 μm.

Izrailenko *et al.* (1970) reported a 10 MW/cm² threshold for a 1.6 cm LiIO₃ crystal SRO pumped by a doubled Nd : glass laser. At 45 MW/cm² the measured conversion efficiency was 8%. In addition, the measured bandwidth varied from 40 Å near degenerate to 2 Å at $\lambda_s = 0.68$ μm.

The LiIO₃ oscillator can be compared with an off angle phase matched LiNbO₃ oscillator. For equal loss and crystal lengths and $\theta = 50°$ for LiNbO₃, the ratio of the oscillator gains is $\Gamma^2 l^2 (\text{LiNbO}_3)/\Gamma^2 l^2 (\text{LiIO}_3) \simeq 3.6$. The LiIO₃ oscillator compares favorably to LiNbO₃ due to its higher damage intensity of 125 MW/cm² compared to approximately 80 MW/cm² for LiNbO₃. LiIO₃ does, however, suffer from internal damage due to inclusions that may occur at lower intensities than for the surface damage.

The threshold for a doubly resonant ruby pumped LiIO₃ parametric oscillator reported by Campillo and Tang (1971) is reduced to 5 MW/cm² compared to the SRO threshold of 60 MW/cm². At a level of 25 MW/cm², 50% pump depletion was observed. Figure 35 shows the oscillator and LiIO₃ up-converted tuning curve for this DRO. The oscillator was externally up-converted resulting in up to 10 kW of up converted power. Further studies of the DRO LiIO₃ oscillator by Campillo (1972) showed the expected "cluster" effect in its output spectrum. The clusters were 1 Å wide, separated by 40 Å at 1.15 μm, 80 Å at 1.25 μm, and 120 Å at 1.3 μm. The measured spectral width of the oscillator at degenerate was 600 Å.

Campillo also externally doubled the DRO LiIO₃ oscillator using an 8 mm LiIO₃ crystal cut at 21.4°. Figure 40 shows the phase-matching angles obtained in that experiment over a range of fundamental wavelengths between 1.1 and 1.8 μm. The second harmonic output at 100-W peak power tuned between 0.560 and 0.915 μm. Figure 36 also shows phase matching data obtained by Herbst (1973) in an internally doubled LiNbO₃ parametric oscillator. The tuning curve for that 0.659 μm pumped SRO LiNbO₃ oscillator is shown in Fig. 27. Figure 36 illustrates the phase matching properties of LiIO₃ for up-conversion over a broad spectral region.

The infrared transmission of LiIO₃ allows interactions out to 5.5 μm. Parametric oscillation is possible but not useful for idler wavelengths this far in the infrared due to low gain. LiIO₃ does, however, phase match for mixing. Meltzer and Goldberg (1972) demonstrated mixing in LiIO₃ internal to a ruby pumped dye laser. Output powers of 100 W were generated over the 4.1–5.2 μm region by mixing the ruby source with the wavelengths from a DTTC dye laser. The spectral width of the dye laser was 6 Å in its 0.802–0.835 μm region. Although LiIO₃ is transparent and phase matchable in the near infrared spectral region, its low effective nonlinearity due to the 19° phase-matching

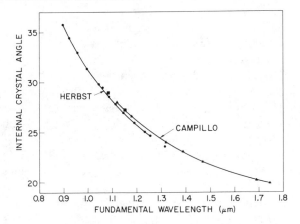

Fig. 36. Measured LiIO₃ phase-matching angles versus fundamental wavelength. Data between 1.1 and 1.8 μm from Campillo (1972). Data between 0.9 and 1.3 μm from Herbst (1973).

angle reduces its conversion efficiency to the point where other materials may prove more useful for infrared generation by mixing.

Recently Goldberg (1972) has investigated an 1.06-μm Nd:YAG pumped internal LiIO₃ parametric oscillator. He obtained up to 20 mW average output power in 13-KW, 14-nsec pulses over a 1.95–2.34 μm spectral region. The oscillator operated both doubly and singly resonant. In the singly resonant configuration output was obtained near 1.45 μm and from 3.8 to 4.2 μm. At higher pump powers the dynamic effects previously observed in internal LiNbO₃ parametric oscillators were evident.

C. Infrared Parametric Oscillators

LiNbO₃ and LiIO₃ parametric oscillators have been tuned to 3.7 and 4.5 μm in the infrared. Infrared oscillation is limited to these wavelengths due to the onset of crystal absorption. Infrared oscillators, as discussed in this section, means the capability of extended infrared tuning to the two phonon absorption limit beyond 10 μm. All materials that satisfy this extended infrared transparency range are semiconductors with bandgaps between 2 and 0.5 eV.

Proustite (Ag_3AsS_3) is an extensively studied infrared nonlinear crystal (Hulme *et al.*, 1967). Its large nonlinear coefficient (approximately twice that of LiNbO₃) and birefringence allow phase matched nonlinear interactions over its entire 0.6–13 μm transparency range. Proustite has been used to double a 10.6 μm CO_2 laser (Boggett and Gibson, 1968) and to up convert 10.6 μm into the visible using a ruby laser pump source (Warner, 1968). Its indices of refraction have been measured (Hobden, 1969). Proustite's large birefringence ($\Delta n = 0.22$) allows off-angle phase-matching near $\theta = 20°$. Propagation at such a phase-matching angle results in a reduced effective nonlinear coefficient $d_{\text{eff}} = d_{31}\sin\theta + d_{22}\cos\theta$ that is approximately twice d_{31} for LiNbO$_3$. In addition, the large birefringence results in a large double refraction angle ($\rho = 0.08$) which gives a short effective interaction length [cf. Eq. (144)]. In spite of these drawbacks and the low crystal burn density near 25 MW/cm^2, proustite has operated as a doubly resonant parametric oscillator near 2.1 μm pumped by a 1.06-μm Q-switched source.

Ammann and Yarborough (1970) first reported oscillation in proustite. They used a 2-kHz repetitively Q-switched Nd : YAG laser for a pump source. The 3.8 mm proustite crystal parametric oscillator operated within a hemispherical cavity at power densities up to 450 KW/cm^2. At this intensity surface damage occurred on the antireflection coated crystal surface.

Using higher-quality material, Hanna *et al.* (1972a) obtained doubly resonant oscillator operation without crystal damage for incident intensities up to 8.5 MW/cm^2 at 26-nsec pulse lengths. To obtain stable DRO operation, the pump laser was operated in a single frequency. With a 5% output coupling,

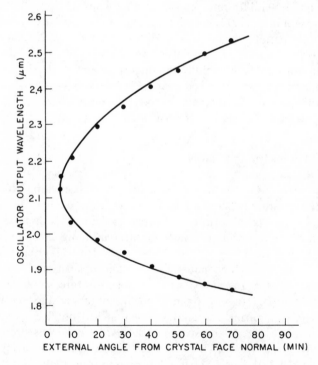

Fig. 37. Measured proustite DRO tuning curve pumped with a single-frequency 1.06-μm source [after Hanna *et al.* (1972a).

oscillator efficiencies up to 1% at peak powers of 1 kW were obtained. Figure 37 shows the measured tuning curve for the 1.06 pumped proustite oscillator. Calculated tuning curves show that phase-matching is possible over a 1.2–9.5 μm region. Operation of a DRO, however, is very difficult over this extended region due to mirror coating requirements and the increase in threshold for off degeneracy operation. These difficulties can be bypassed by mixing in proustite using a near infrared tunable source. Hanna *et al.* (1971) have generated 0.1-W peak output power in the 10.1–12.7 μm region using a ruby pumped dye laser mixed against the ruby laser as a source. The low burn density and off-angle phase matching limited the conversion efficiency.

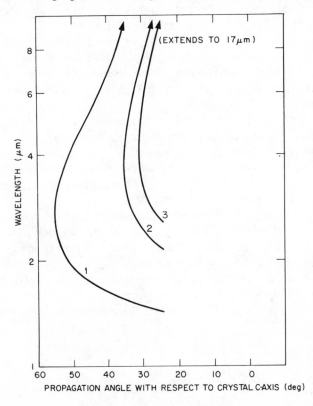

Fig. 38. Calculated Tl_3AsSe_3 parametric oscillator tuning curves [after Feichtner and Roland (1972)]. Curve 1 is for $\lambda_p = 1.32$ μm and $\rho = 0.0535$; curve 2 is for $\lambda_p = 1.833$ μm and $\rho = 0.0554$; curve 3 is for $\lambda_p = 2.06$ μm and $\rho = 0.0520$.

In addition to proustite, two closely related crystals have been investigated for infrared nonlinear optical applications. These materials are pyrargyrite (Ag_3SbS_3) reported by Gandrud *et al.* (1970) and McFee *et al.* (1970), and

Tl$_3$AsSe$_3$ described by Feichtner and Roland (1972). Pyrargyrite has properties very similar to proustite. It belongs to the same point group (3*m*), is
transparent between 0.6 and 13 μm, and has a large enough birefringence to
phase match over its transparency region. The measured phase-matching
angle for doubling 10.6 μm is 29°. Its nonlinear cofficient is very close to that
of proustite.

The material Tl$_3$AsSe$_3$ appears useful as an infrared nonlinear material.
Its transparency range extends from 1.26 to 17 μm and the crystal has been
grown in sizes up to 1.2 cm in diameter and 3 cm in length. The crystal has
large birefringence similar to proustite and phase matches for doubling 10.6 μm
at 22°. The measured nonlinear coefficient is 3.3 times that of proustite and
pyrargyrite. In addition, Feichtner and Roland measured the crystal burn
density at 10.6 μm to be 32 MW/cm^2 compared to 20 MW/cm^2 for proustite
and 14 MW/cm^2 for pyrargyrite. Figure 38 shows the calculated phase-
matching curves for Tl$_3$AsSe$_3$ parametric oscillators pumped at 1.32, 1.833,
and 2.06 μm. Experiments to construct a Tl$_3$AsSe$_3$ parametric oscillator are
in progress (Feichtner, 1972).

Unlike proustite and its related crystals, cadmium selenide (CdSe) 90°
phase matches over an extended infrared region. Wurtzite (6 mm) CdSe does
not have adequate birefringence for second harmonic generation, however,
it does phase match for mixing and off-degenerate parametric oscillator
operation (Harris, 1969a). In addition to its high optical quality, CdSe has a
nonlinear coefficient three times that of LiNbO$_3$.

Herbst and Byer (1971) performed the first phase matched interaction in
CdSe by mixing 10.6 μm against the 1.833-μm Nd : YAG line (Wallace, 1971b)
to generate 2.2 μm. The measured angle for the allowed type II phase matching
was 77°. The observed mixing efficiency was 35% in a 0.6 cm crystal at 48
MW/cm^2 which is less than the measured crystal burn density of 60 MW/cm^2.

The extended 0.75–25 μm transparency range of CdSe makes it useful as a
tunable mixing source between 8 and 25 μm. Figure 39 shows the transmission

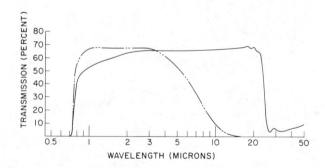

Fig. 39. CdSe transmittance; dotted curve is for as-grown crystals, solid curve is for
selenium-compensated crystals [after Herbst and Byer (1971)].

of as-grown uncompensated CdSe and selenium compensated CdSe. The calorimetrically measured loss at 10.6 μm is 0.03/cm which compares favorably with the 0.02/cm GaAs loss and 1/cm proustite loss. In a recent experiment using a LiNbO$_3$ parametric oscillator source, Herbst and Byer (1973) generated tunable radiation between 10 and 13.5 μm by mixing in CdSe. The measured peak output power of 2 W agreed with the expected conversion efficiency. Figure 40 shows the calculated phase matching curve for mixing in CdSe and the experimentally measured points. The mixed output was not measured beyond 13 μm due to the detector response cutoff.

In an extension of the mixing experiment in CdSe, Herbst and Byer (1972) demonstrated the first singly resonant infrared parametric oscillator in CdSe pumped by the 1.833-μm line of Nd : YAG. Davydov *et al.* (1972) also reported achieving singly resonant oscillation in CdSe. The observed threshold for the Herbst–Byer oscillator was 550 W and pump depletions of up to 40% were obtained. Figure 41a shows a schematic of the CdSe parametric oscillator, and Fig. 41b shows the calculated and measured tuning curves. Figure 42 is a photograph of the CdSe SRO showing the 2-cm CdSe crystal within the 5.7-cm confocal cavity.

The CdSe oscillator operated with pump intensities that were below the 60 MW/cm^2 burn density. The crystal did, however, show surface damage on both the front and back surfaces due to the high circulating signal fields at 2.2 μm. An estimate of the circulating signal power is 7–8 times the incident pump power [cf. Fig. 13 and the discussion following Eq. (109)]. By resonating the idler wave, the circulating power is reduced by the frequency ratio ω_s/ω_i and by the area ratio. For the CdSe oscillator this factor is $(4.5)^2$. Herbst and Byer (1972) succeeded in resonating the idler wave. High optical losses due to poorer quality mirrors and antireflection coatings resulted, however, in a higher threshold. To compensate, a round trip SRO was constructed in which the pump, signal, and idler were reflected back through the crystal. This lowered the oscillator threshold as expected [cf. Eq. (97)]. The increased pump intensity required to reach the higher threshold exceeded the 60 MW/cm^2 crystal damage threshold, however.

At this time CdSe crystals up to 3 cm in length are commercially available (Herbst and Byer, 1971), with longer (74-mm) higher-quality crystals having been reported (Abagyan *et al.*, 1972). CdSe singly resonant oscillators should operate without crystal burning for 4-cm-long crystals. Crystals of this length have one-third the gain of 4-cm LiNbO$_3$ crystals and should operate as reliably in the SRO configuration.

Recently considerable effort has gone into growing new nonlinear materials especially materials that are useful in the infrared. One group of materials with the chalcopyrite (CuFeS$_2$, point group 42m) structure has particularly useful nonlinear properties. Since oscillation has not been achieved in any of these materials, a discussion of their properties is considered in the next section.

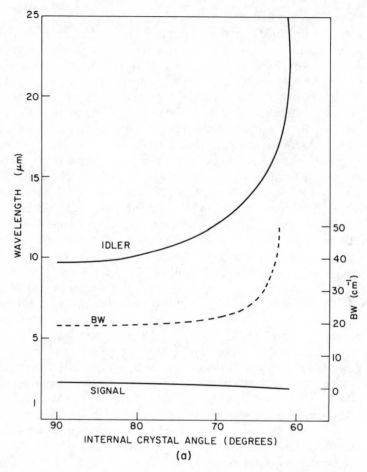

Fig. 40a. Calculated mixing curve for CdSe for a fixed 1.833-μm source using index data from Herbst and Byer (1971).

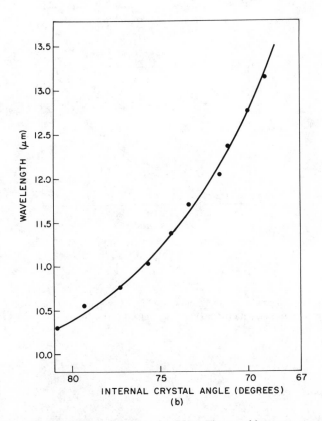

Fig. 40b. Measured difference output in CdSe. The tunable pump source was two LiNbO$_3$ SRO operating within a single cavity pumped by 0.659 μm of a doubled Nd:YAG laser. One oscillator was tuned while the second was held at 1.833 μm.

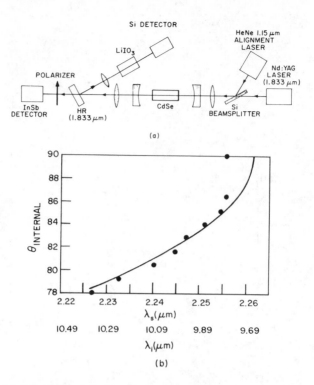

(a)

(b)

Fig. 41. (a) Schematic of the singly resonant CdSe infrared parametric oscillator. (b) Experimental tuning data (points) and theoretical curve (line) with $\lambda_p = 1.833$ μm [after Herbst and Byer (1972)].

Fig. 42. Photograph of a CdSe SRO showing the 2-cm-long CdSe crystal within the 5.7-cm confocal cavity.

IV. APPLICATIONS AND FUTURE DEVICE DEVELOPMENT

A. Nonlinear Materials

The development of parametric oscillators and the extension of parametric oscillation to new frequency ranges has been and still is severely materials limited. At this time only a few materials and their closely related isomorphs are of high enough optical quality for use in parametric oscillators. These crystals are KDP, $LiNbO_3$, $LiIO_3$, CdSe, and perhaps Ag_3AsS_3.

The optical quality requirements for nonlinear materials include low absorption loss, birefringence and phase matchability, crystal uniformity, high nonlinearity, and high crystal burn density. All of these properties must be met if the crystal is to be useful. Aside from acentricity, the largest majority of crystals lack sufficient birefringence to phase match within their transparency range (Byer, 1970). Of those crystals that phase match, only a few have the necessary high optical quality or are amenable to crystal growth in useful sizes and quality. The problem is further compounded by the requirement to withstand high optical intensities of typically greater than $10 \, \text{MW/cm}^2$ without damage or large induced absorption. Fortunately, parametric devices have wide tuning ranges so that ultimately only a handful of well developed crystals may be required to cover the entire ultraviolet to far infrared frequency range.

In evaluating new nonlinear crystals, the first optical properties usually determined are the transparency range and indices of refraction. The crystal transparency immediately gives the useful frequency range. For low powers, the transparency range is limited at the short wavelength by the band edge and at the long wavelength by two phonon absorption at twice the restrahl frequency. In practice, induced nonlinear absorption may limit the pump field to frequencies approximately one-half the bandgap. Induced nonlinear absorption has been noted in $LiNbO_3$ (Wallace and Dere, 1972), KDP and ADP (Dowley and Hodges, 1968), and semiconductor materials (Lee and Fan, 1972).

The crystal indices of refraction must be measured over an extended frequency range in order to calculate phase-matching angles for SHG or parametric tuning. The indices of refraction are usually measured by the minimum deviation prism method. The apparatus and procedure for this measurement have been described by Bond (1971). The crystal point group together with the sign of the crystal birefringence determine the type of phase matching that is allowed and the effective nonlinear coefficients (Boyd and Kleinman, 1968). This information allows the design of an experiment to measure the nonlinear coefficients of the crystal and to verify the predicted phase matching. In addition to standard linear optical quality measurements. the phase-matched nonlinear process can also be used to evaluate crystal quality and uniformity (Byer *et al.* ,1970). Finally, by use of the appropriate laser source, the crystal damage threshold can be measured.

The above information is adequate to characterize the potential of a new material for nonlinear optical application. Recently, crystal evaluation along these lines has been applied to the chalcopyrite group of crystals (Parthe, 1964). These tetragonal symmetry ($\bar{4}2m$) ternary semiconductor crystals are potentially useful infrared nonlinear crystals.

The chalcopyrite crystals form two groups which are ternary analogs of the II–VI and III–V binary semiconductors. The first group is the I–III–VI$_2$ compounds of which AgGaS$_2$ and AgGaSe$_2$ are examples. The second group is the II–IV–V$_2$ compounds which include CdGeAs$_2$ and ZnGeP$_2$. Goryunova *et al.* (1965) were the first to investigate the semiconducting and nonlinear optical properties of chalcopyrite crystals. In 1971, Chemla *et al.* reported on AgGaS$_2$ and later that year Boyd *et al.* (1971a) measured the properties of ZnGeP$_2$ and carried out a phase matched up-conversion experiment at 10.6 μm (Boyd *et al.*, 1971b). Byer *et al.* (1971) measured the properties of CdGeAs$_2$. The properties of these chalcopyrites and other potentially useful crystals of the same class were extensively studied by (Boyd *et al.*, 1971c, 1972a, b) in a series of papers. As a result of these and other measurements (Kildal, 1972; Bahr and Smith, 1972) four crystals belonging to the chalcopyrite semiconductors have been identified as phase matchable and potentially useful for infrared nonlinear optics. These crystals are CdGeAs$_2$, ZnGeP$_2$, AgGaSe$_2$, and AgGaS$_2$.

Table VI lists these chalcopyrite crystals along with other nonlinear materials useful in parametric processes. The first column of Table VI lists the presently accepted values for the crystal nonlinear coefficients. These values are based upon Levine and Bethea's (1972) best value for GaAs. The parametric gain is proportional to the material parameters $d^2/n_0{}^2 n_3$, where d is the effective nonlinear coefficient at the angle of phase matching and n_0 and n_3 are the indices of refraction for the degenerate and pump waves. Column 8 of Table VI lists the material figure of merit. For better visual reference, Fig. 43 shows the material figure of merit and the crystal transparency range.

Of more interest for comparison of nonlinear materials is the parametric gain. In 90° phase-matched crystals the gain is proportional to crystal length for confocal focusing [cf. Eq. (136)]. For non 90° phase matching double refraction may limit the effective interaction length to l_{eff} given by Eq. (144). The parametric gain is then proportional to l_{eff} and not the crystal length. Table VI shows the maximum interaction length for various crystals. In addition to length, parametric gain varies as $\omega_0{}^3$ for 90° phase matched confocally focused crystals and as $\omega_0{}^2$ for double refraction-limited crystals. The pump wavelengths used to calculate parametric gain are shown in Table VI along with the calculated gain per watt of pump power. For parametric

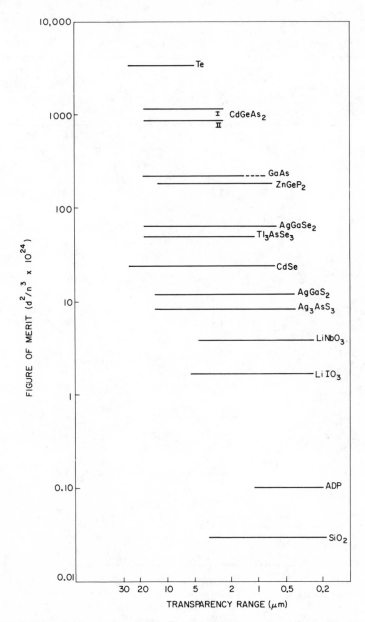

Fig. 43. Material figure of merit $d^2/n_0^2 n_3$ and transparency range for nonlinear crystals.

Table VI

Nonlinear Coefficient, Figure of Merit, and Gain for Nonlinear Crystals

Material (point group) pump wavelength)	$d \times 10^{12}$ (m/V)	n_o n_e	$n_e - n_o$	θ_m (deg)	p	$d_{eff} \times 10^{12}$
Te (32) $\lambda_p = 5.3~\mu m$	$d_{11} = 649^a$	6.25 4.80	-1.45	14	0.10	638 ($d \cos^2 \theta_m$)
CdGaAs$_2$ ($\bar{4}2m$) $\lambda_p = 5.3~\mu m$	$d_{36} = 236^{b,c}$	3.51 3.59	$+0.086$	II 55 I 35	0.021 0.021	193 ($d \sin \theta$) 212 ($d \sin 2\theta$)
GaAs ($\bar{4}3m$)	$d_{36} = 90.1^d$	3.30	0	—	—	—
ZnGeP$_2$ ($\bar{4}2m$) $\lambda_p = 1.83~\mu m$	$d_{36} = 75^e$	3.11 3.15	$+0.038$	II 90 I 62	0.0 0.01	d_{36} 62.2 ($s \sin 2\theta$)
Tl$_3$AsSe$_3$ (3m) $\lambda_p = 1.83~\mu m$	$d_+ = 40^f$	3.34 3.15	-0.182	36	0.055	d_+
AgGaSe$_2$ ($\bar{4}2m$) $\lambda_p = 1.83~\mu m$	$d_{36} = 33^g$	2.62 2.58	-0.32	I 55 I 90	0.01 0.0	27 ($d \sin \theta$) d_{36}
CdSe (6mm) $\lambda_p = 1.83~\mu m$	$d_{31} = 19^{e,h}$	2.45 2.47	$+0.019$	90	0.0	d_{31}
AgGaS$_2$ ($\bar{4}2m$) $\lambda_p = 0.946~\mu m$	$d_{36} = 12^i$	2.42 2.36	-0.054	I 64 I 90	0.17 0.0	10.8 ($d \sin \theta$) d_{36}
Ag$_3$SbS$_3$ (3m) $\lambda_p = 1.06~\mu m$	$d_+ = 12^a$	2.86 2.67	-0.19	—	—	d_+
Ag$_3$AsS$_3$ (3m) $\lambda_p = 1.06~\mu m$	$d_+ = 11.6^f$	2.76 2.54	-0.223	30	0.078	d_+
LiIO$_3$ (6) $\lambda_p = 0.694~\mu m$	$d_{31} = 7.5^j$	1.85 1.72	-0.135	23	0.071	3.04 ($d \sin \theta$)
LiNbO$_3$ (3m) $\lambda_p = 0.532$	$d_{31} = 6.25^k$	2.24 2.16	-0.081	90	0.0	d_{31}
ADP ($\bar{4}2m$) $\lambda_p = 0.266$	$d_{36} = 0.57^{l,m}$	1.53 1.48	-0.0458	90	0.0	d_{36}
KDP ($\bar{4}2m$) $\lambda_p = 0.266$	$d_{36} = 0.50^h$	1.51 1.47	-0.0417	90	0.0	d_{36}
SiO$_2$ (32)	$d_{11} = 0.33^n$	1.55 1.56	$+0.0095$	--	—	d_{11}

[a] McFee *et al.* (1970)   [b] Byer *et al.* (1971)   [c] Boyd *et al.* (1972a)
[d] Levine and Bethea (1972).   [e] Boyd *et al.* (1971a)   [f] Feichtner and Roland (1972).
[g] Boyd *et al.* (1972b).   [h] Herbst and Byer (1971).   [i] Boyd *et al.* (1971c).
[j] Campillo and Tang (1970).   [k] Byer and Harris (1968)   [l] Francois (1966).
[m] Bjorkholm and Siegman (1967).   [n] Jerphanon and Kurtz (1970).

$d_{\text{eff}}^5/n_0{}^2 n_3$ $\times 10^{24}$	$l(\rho)_{\text{eff}}$ (cm)	$\Gamma^2 l^2$ (1 W)	$\Gamma^2 l^2$ (1 MW/cm^2)	I_{burn} (MW/cm^2)	Transmission range (μm)
3634	0.011	0.95×10^{-4}	0.18	40–60	4–25
861 1039	0.34	3.8×10^{-4} 4.6×10^{-4}	0.033 0.040	20–40	2.4–17
226	$l_{\text{coh}} = 104\ \mu$m	—	—	60	$\left.\begin{array}{c}0.9\\1.4\end{array}\right\}$–17
187 127	$l = 1$ cm 0.59	7.1×10^{-3} 1.8×10^{-3}	0.21 0.05	>4	0.7–12
51	0.019	3.7×10^{-5}	0.022	32	1.2–18
42 63.4	0.71 $l = 1$ cm	5.52×10^{-4} 1.98×10^{-3}	0.02 0.07	>10	0.73–17
24	$l = 2$ cm	1.3×10^{-3}	0.09	60	0.75–25
9.2 11	0.14 $l = 1$ cm	2.3×10^{-4} 2.3×10^{-3}	0.013 0.045	12–25	0.60–13
7.6	—	8.2×10^{-6}	0.01	14–50	0.60–14
8.2	0.007	9.0×10^{-6}	0.011	12–40	0.60–13
1.88	0.008	5.6×10^{-6}	5.5×10^{-3}	125	0.31–5.5
3.88	$l = 5$ cm	2.1×10^{-2}	1.28	50–140	0.35–4.5
0.100	$l = 5$ cm	2.9×10^{-3}	0.131	>1000	0.20–1.1
0.079	$l = 5$ cm	2.30×10^{-3}	0.103	>1000	0.22–1.1
0.029	$l_{\text{coh}} = 14\ \mu$m	—	—	>1000	0.18–3.5

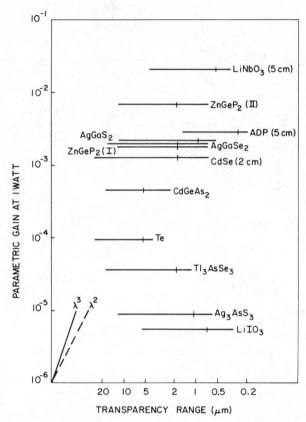

Fig. 44. Parametric gain at 1-W pumping power and transparency range of nonlinear crystals for optimum focusing. The crystal lengths are 1 cm unless specified. For non-90° phase matching the interaction length is limited to l_{eff} [cf. Eq. (144)].

processes in which the pump power is limited, the gain per watt is the important material parameter. Figure 44 shows the gain per watt and crystal transparency for the nonlinear materials listed in Table VI. Also shown by the vertical tick mark is the pump wavelength. For 90° phase matched crystals the crystal length is 1 cm unless shown otherwise. For non-90° phase-matched crystals the effective interaction length l_{eff} is the limiting length. Figure 44 shows the very favorable parametric gains or, equivalently, second harmonic generation efficiency achieved by the four selected chalcopyrite crystals, and by $LiNbO_3$ and ADP.

In many applications the pump laser has more than adequate power. In this case the quantity of interest is the parametric gain per input intensity and the crystal burn intensity. The parametric gain per pump intensity is crystal length dependent. For 90° phase matching the gain improves as l and the area increases as l so that for confocal focusing the improvement is as l^2. For non-90° phase matching the gain improves up to the length l_{eff} [cf. Eq. (144)]. The optimum focusing for minimum intensity varies as l^2 according to Eq. (146b). Therefore the gain per intensity varies as crystal length squared. The calculated gain given in Table VI assumes a 1-cm-length crystal unless stated otherwise. Figure 45 shows the parametric gain at 1 MW/cm² pump intensity for the crystals listed in Table VI. Again the four chalcopyrite crystals and LiNbO₃ and ADP show high gains. Using these gain results and the listed crystal burn densities, the maximum parametric gain limited by crystal damage can be calculated. In this comparison, ADP and LiNbO₃ show the largest gains.

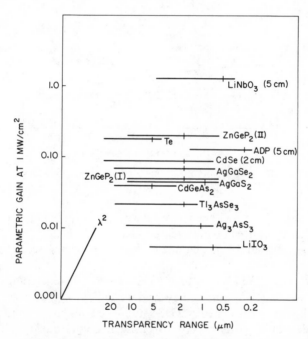

Fig. 45. Parametric gain at 1 MW/cm² input intensity and transparency range of nonlinear crystals listed in Table VI. The pump wavelengths are indicated by the vertical tick. The crystal legnths are taken to be 1 cm unless specified.

AgGaS₂ is transparent between 0.6 and 13 μm. It is negative birefringent and phase matchable over a wide range of the infrared (Chemla et al., 1971; Boyd et al. (1971c, 1972a). Figure 46 shows the calculated parametric tuning curves for various pump wavelengths based on the index data of Boyd et al. (1971c). The calculated gain for 90° phase matching is adequate to obtain singly resonant oscillation assuming a 1-cm-length quality crystal is available.

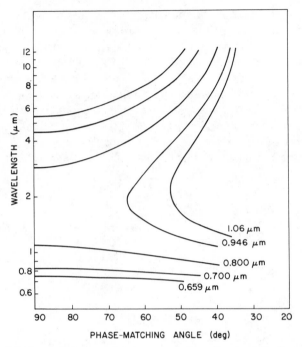

Fig. 46. Calculated type I parametric tuning curves for AgGaS$_2$ for various pump wavelengths.

At this time crystals up to 1 cm in size have been grown, but optical quality is not yet sufficient for use in parametric oscillators. The material may be useful for infrared generation by mixing or for up-conversion and second harmonic generation.

AgGaSe$_2$ is a close analogy to AgGaS$_2$. Its transparency range (0.73–17 μm) and phase-matching curves, however, are shifted to longer wavelengths. Figure 47 shows the calculated AgGaSe$_2$ tuning curves based on the index of refraction data of Boyd *et al.* (1972b). Present crystal quality is again not adequate for use in parametric oscillators. Parametric mixing may, however, be possible in presently available crystals for a widely tunable infrared source.

Unlike the negative birefringent AgGaS$_2$ and AgGaSe$_2$, ZnGeP$_2$ and CdGeAs$_2$ are positive birefringent. Thus the type II phase-matching $[n_p{}^o\omega_p = n_s{}^e(\theta)\omega_s + n_i{}^o\omega_i]$ is allowed at 90° where type I phase matching has a zero effective nonlinear coefficient. The effective nonlinearity, however, is maximum for type I phase matching $[n_p{}^o\omega_p = n_s{}^e(\theta)\omega_s + n_i{}^e(\theta)\omega_i]$ at $\theta = 45°$. Type II phase matching increases the birefringent needed to achieve phase matching by almost a factor of two since its averages the birefringence at the longer wavelengths. ZnGeP$_2$ does not have enough birefringence to type II

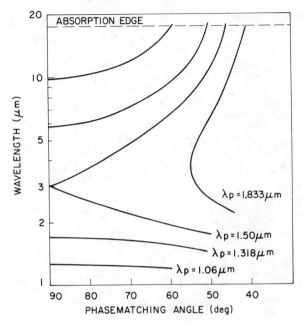

Fig. 47. Calculated type I parametric tuning curves for $AgGaSe_2$ for various pump wavelengths.

phasematch for SHG. It does, however, phase match for off-degenerate parametric oscillation as shown by Fig. 48a. Figure 48b shows the calculated type I phase-matching curves based on the index data of Boyd *et al.* (1971a). The behavior of the 2.5-μm pump phase matching curve is due to phase matching near the minimum dispersion region of the crystal transparency range. Complete type I SHG phase-matching curves for $ZnGeP_2$ are given by Boyd *et al.* (1972b). $ZnGeP_2$ has been used for infrared up conversion (Boyd *et al.*, 1971b).

CdGeAs$_2$ has the highest parametric figure of merit of any known crystal except tellurium. In addition it is particularly useful for SHG of a 10.6 μm CO$_2$ laser or for parametric oscillation with a 5.3-μm pump source (Byer *et al.*, 1971). Recent work (Kildal *et al.*, 1972) has demonstrated the potential of CdGeAs$_2$ by SHG in which an atmospheric pressure CO$_2$ laser was doubled with 2% efficiency in a 2-mm-long crystal. That same work demonstrated that phase-matched third harmonic generation is also possible.

Figures 49a and (b) show the calculated type II and type I phase-matching curves for CdGeAs$_2$. The crystal has a large enough birefringence to phase match through the degeneracy point for type II phase matching. This tuning characteristic is unique to CdGeAs$_2$ at this time. It leads to the possibility of maintaining a narrow-band parametric gain even though degenerant since,

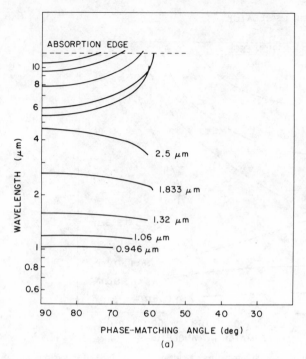

Fig. 48a. Calculated type II parametric phase-matching curves for ZnGeP₂.

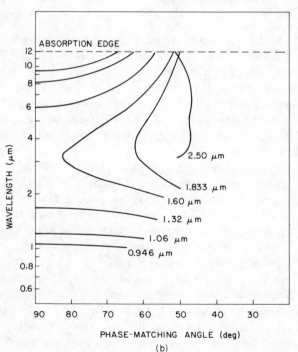

Fig. 48b. Calculated type I parametric phase-matching curves for ZnGeP₂.

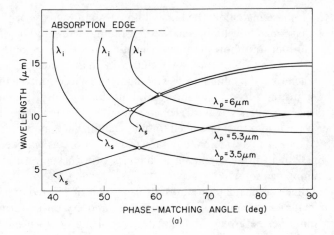

(a)

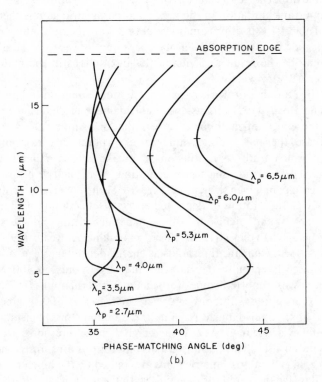

(b)

Fig. 49. (a) Calculated type II parametric phase-matching curves for CdGeAs$_2$. (b) Calculated type I parametric phase-matching curves for CdGeAs$_2$ [after Kildal (1972)].

unlike type I phase matching, the bandwidth does not increase at that point. The type I phase-matching tuning curves show tuning increasing and decreasing with angle for long and short pump wavelengths. At the 4.0 μm pump region, the tuning curve is very wide for a small change in phase matching angle. That tuning characteristic is similar to a wide-band dye laser source. Instead of tuning by adjusting phase matching conditions, the oscillator frequency can be controlled by dispersive cavity elements at either the signal or idler wave. The other wave would then be forced to tune such that energy conservation is satisfied. To date such a wide band parametric oscillator has not been demonstrated, but it may provide tuning advantages over the more usual phase-matched tuning methods.

At this time the chalcopyrite crystals satisfy all of the requirements for nonlinear materials except low absorption loss and crystal uniformity. The crystals cannot yet be grown in adequate sizes of high optical quality. The very useful nonlinear properties, however, especially in the infrared, lend considerable importance to chalcopyrite crystal growth efforts.

An important limitation to the maximum parametric gain is the material damage. Table VI lists approximate values for damage in nonlinear crystals based on reported damage intensity measurements. Since laser-induced damage may be the result of a number of interactions in the material, the reported values vary widely.

Recently a number of workers have given more attention to laser damage mechanisms. Ready (1971) discusses possible damage mechanisms in a recent book on the effects of high power laser radiation. Among the mechanisms discussed are thermal heating, induced absorption due to multiphoton absorption which leads to heating or to breakdown, stimulated Brillouin scattering, self focusing, surface preparation, and dielectric breakdown. Thermal heating, in which the temperature rise of the material is proportional to the input pulse duration or deposited energy, is a common damage mechanism in metals and highly absorbing materials. Semiconductors behave more like metals than dielectrics for this form of damage. Ready (1971, p. 305) lists semiconductor damage thresholds as energy dependent near 5–10 J/cm^2. For a 100-nsec laser pulse the damage intensity is thus 50–100 MW/cm^2 which agrees with reported values for silicon and germanium (Hanna et al., 1972). For 1-nsec pulses the power density increases by 100 to greater than 1 GW/cm^2 if thermal damage remains the dominant mechanism. There is experimental evidence in GaAs that such high damage thresholds are reached for pulses near 1 nsec. This has important implications for infrared nonlinear optics since, in general, parametric gains are limited by the much lower 20–60 MW/cm^2 damage threshold for long pulses. In the limit of cw pumping, semiconductor materials have damage thresholds as low as 1 kW/cm^2 due to surface heating and melting.

Damage thresholds for dielectric materials are generally much higher than for semiconductors. For example, sapphire damages near 25 GW/cm^2 for Q-switch pulse irradiation. Glass and Guenther (1972) reviewed damage studies in dielectrics. They pointed out that nonlinear materials show a marked decrease in damage threshold for phase-matched second harmonic generation. For example, $LiIO_3$ shows surface damage at 400 MW/cm^2 for 10 nsec pulses. When phase matched it damages at 30 MW/cm^2 for the fundamental and 15 MW/cm^2 for the second harmonic. Similar results have been noted for $LiNbO_3$ and $Ba_2NaNb_5O_{15}$.

In a series of very careful studies, Ammann (1972) measured $LiNbO_3$ surface damage thresholds with various quarter-wave coatings. He found that uncoated $LiNbO_3$ damages at 40 MW/cm^2 at 1.08 μm for a 0.5-W, 4-kHz, 140-nsec repetitively Q-switch laser source. The damage threshold was increased to 150 MW/cm^2 for a $\lambda/4$ layer of Al_2O_3-coated $LiNbO_3$ crystal. Additional measurements on thin ZnS films showed the importance of over-coating and the lowering of the damage threshold with time on exposure to air (Ammann and Wintemute, 1973). Similar lowering of damage threshold for ThF_2 overcoated CdSe crystals used in an infrared parametric oscillator has been noted (Herbst, 1972). In that case the damage threshold decreased from 60 to 10 MW/cm^2 over a period of a few days. In addition, damage occurred much more readily along surface scratches indicating that surface preparation is important.

Bass and Barrett (1972) proposed a probabilistic model for laser induced damage based on an avalanche breakdown model. For this damage, the laser field acts as an ac analog to dc dielectric breakdown. A laser power density of 10 GW/cm^2 corresponds to 4×10^6 V/cm which is close to the measured dc dielectric breakdown fields near 30×10^6 V/cm. Bass and Barrett (1972) presented the laser damage threshold in a probabilistic way such that the probability to induce surface damage is proportional to $\exp(-K/E)$, where K is a constant and E is the rms optical field strength. Measured dielectric breakdown intensities lie near 25 GW/cm^2 for glasses and fused silica and between 2 to 4 GW/cm^2 for nonlinear crystals. It appears that if other damage mechanisms do not limit the laser intensity at lower levels, then laser induced dielectric breakdown determines the maximum incident intensity. Efforts to understand the mechanisms of laser induced damage have been increasing. Hopefully these studies will lead to a better understanding of the material's damage intensity limits.

For high average power applications thermally induced focusing and change in birefringence become limiting factors. The problem of temperature rise in a laser heated slab has been treated by Brugger (1972). These solutions can be applied to find the local temperature rise of a nonlinear crystal and its local change in index of refraction. The change in index may lead to thermal focusing or defocusing of the laser beam (Jasperse and Gianino, 1972) and therefore affect the nonlinear gain through the change in focal spot size and interaction length.

A more serious limitation is the breaking of phase matching due to the thermal change in crystal birefringence. This problem has been treated by Okada and Ieiri (1971) for second harmonic generation. They show that the optimum phase matching temperature shifts with increasing average laser power. The break in phase matching is proportional to $\delta(\Delta n)/\delta T$, where Δn is the crystal birefringence. For temperature phase-matched crystals such as $LiNbO_3$ this value is quite large being approximately $10^{-4}/°C$. In these crystals thermal breaking of phase matching is a serious consideration at power levels near 1 W. On the other hand, crystals such as $LiIO_3$ have a very small $\delta(\Delta n)/\delta T$ and are capable of much higher average input powers.

The properties of a large number of nonlinear materials have been measured. Only a few of these crystals, however, have adequate quality for use in parametric devices. At this time potentially useful infrared nonlinear crystals have been found and characterized and are now being studied to improve their quality. $LiNbO_3$ is perhaps the best developed nonlinear material useful in the visible and near infrared for parametric oscillator devices. $LiIO_3$ is widely used for internal second harmonic generation of Nd:YAG lasers and for up-conversion. Finally, ADP and related crystals are being used for ultraviolet sources by sum and harmonic generation. Although the nonlinear crystals listed in Table VI allow nonlinear interactions over the spectral range from 0.25 to 25 μm, there remain large classes of materials which have not been investigated. Among these are organic materials (Jerphanon, 1972) and crystals potentially useful in the ultraviolet at wavelengths less than 0.2 μm.

B. Applications

The principle use of tunable radiation generated by parametric oscillators or parametric frequency conversion processes is for spectroscopic studies. The spectroscopic applications cover chemical, biological, and physical systems. Chemical spectroscopy includes fluorescence decay, photodetachment, energy transfer rate studies, isotope separation, and photoionization. Biological applications mentioned in a recent survey of parametric oscillator users include in situ fluorescence studies, holographic illumination at a particular wavelength, and microscopic illuminated fluorescence studies. Physical spectroscopy includes studies of gas, liquid, and solid-state systems. In addition, the relatively high power available from a parametric oscillator in the near infrared allows consideration of tunable laser radar applications for air pollution measurements and upper atmospheric studies. Examples of these applications are discussed briefly below.

In a recent experiment, Smyth and Brauman (1972) used a parametric oscillator to measure the photodetachment energy of an electron from the phosphide ion. The parametric oscillator tuned across a 0.87-μm range with a bandwidth of 0.1 to 0.2 nm as compared to the conventional xenon arc lamp source bandwidth of 23.8 nm. In presenting their results, Smyth and Brauman compared the tunable laser source with the xenon arc lamp.

In a series of experiments, Moore has used tunable laser sources for energy transfer studies. In one experiment, a doubled ruby laser pumped dye laser-excited formaldehyde for measurements of lifetimes, energy transfer, and quenching rates for single vibronic states (Yeung and Moore, 1973). In another experiment, a frequency-narrowed $LiNbO_3$ parametric oscillator was used as a source for measurements of vibrational–vibrational energy transfer and fluorescence decay time measurements of HCl (Leone and Moore, 1972). Another application discussed by Yeung and Moore (1973) is isotopic separation by photopredissociation using a tunable momochromic source for selective excitation of isotopic lines.

An interesting biological application reported by Donaho et al. (1972) is holography of the microcirculation. In this application a long coherence length Q switched pulsed source is needed to "stop" the blood cell motion through the field of view of a microscope. In their experiment Donaho et al. (1972) observed 5–15 μm blood cells with magnifications between 500 to 1000. The holographic image was reconstructed using a He–Ne laser source.

The use of tunable lasers in solid-state spectroscopy includes measurements of the dispersion of nonlinear susceptibilities, two photon absorption studies, excitation decay studies, photoluminescence, and excited state spectroscopy. Mercier (1973) utilized a parametric oscillator in the spectral region from 0.55 to 0.58 μm to make photoluminescence studies of CaSe and to study the excitonic levels. In addition, Mercier plans to study the second-order susceptibility dispersion of GaSe in the 1.08–1.28 μm region.

Solid—state spectroscopy has been extended to a more applied form by McMahon et al. (1972). They utilized a parametric oscillator to probe a high power Nd : glass amplifier system for excited state absorption that may limit the gain of the system. The experiment demonstrated that superfluorescence and not excited state absorption was the dominant mechanism in limiting the amplifier gain.

An interesting spectroscopic application is the measurement of molecular absorption spectra which are of interest in planetary atmospheric studies. In this application long path lengths of several kilometers are required to observe the very weak overtone absorption bands. For example, a methane absorption band at 0.68 has been observed in the spectra of Jupiter, Saturn, Uranus, and Neptune. The study of this band is valuable for the determination of temperature, pressure, and composition in the planetary upper atmosphere (Fox, 1973).

In atmospheric studies on earth, the parametric oscillator has been

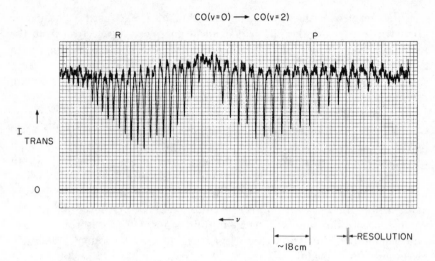

Fig. 50. Overtone spectrum of 3 atm CO at 2.35 μm with a 4-cm path length taken with LiNbO$_3$ SRO. (Courtesy of P. Sackett, Air Force Cambridge Research Labs.)

proposed as a source of tunable infrared for remote pollutant measurements. The sensitivity and tunable laser requirements have been reviewed by Kildal and Byer (1971), Measures and Pilon (1972), and recently by Byer and Garbuny (1973). Figure 50 illustrates the use of an infrared parametric oscillator for molecular spectral analysis. This figure shows the first overtone spectra of CO taken with a LiNbO$_3$ parametric oscillator (Sacket, 1972). The use of parametric oscillator sources for molecular spectroscopy should increase rapidly as the frequency range is extended further into the infrared and the bandwidth is reduced.

The applications reviewed here are examples of the wide range of potential use for parametric devices. In reviewing the applications, the need for extended tuning range, especially in the infrared, careful control of bandwidth, and higher powers were mentioned as desired improvements. With further development, parametric sources should find use in an even greater variety of applications.

V. CONCLUSION

In this article I have reviewed the developments in optical parametric oscillators through 1972, with particular attention to theoretical and experimental advances since the review article by Harris (1969a). These include the development of a practical, widely tunable LiNbO$_3$, singly resonant parametric oscillator, superradiant parametric conversion, and short pulse propagation, a better theoretical understanding of threshold, rise time, conversion efficiency, and spectral properties, new parametric oscillator devices, and the discovery of new nonlinear materials.

The optical parametric oscillator is treated theoretically in Section II, which includes a discussion of parametric amplification and short pulse propagation. A space–time analogy is introduced which clarifies the relation between spatial focusing and picosecond pulse propagation. Experimental studies in this area are just beginning and should prove very interesting as they progress. Theoretical consideration is also given to parametric oscillator operations including threshold, rise time, conversion efficiency, and spectral properties. When possible the theoretical treatment is illustrated by experimental results.

The spectral properties of parametric oscillators are important in device application. The tuning and bandwidth of parametric oscillators are discussed and, in addition, recent experimental results are reviewed which illustrate the important single-frequency operation of a parametric oscillator. Operation of a parametric oscillator with a tunable pump source is also discussed since it offers the potential of a rapidly tunable, extended spectral range source.

Focusing is discussed in the final part of Section II. The treatment includes optimum focusing for doubly and singly resonant oscillator operation in the confocal limit. The aspects of Gaussian beam focusing as treated by Boyd and Kleinman (1968) are discussed including focusing with and without the presence of double refraction and focusing for maximum gain per incident pump power. The treatment of focusing concludes with a practical example of mode matching a pump beam into a parametric oscillator cavity.

In Section III, I review recent optical parametric oscillator device development. In particular the $LiNbO_3$ singly resonant parametric oscillator is discussed in detail. The discussion includes a review of internal $LiNbO_3$ parametric oscillator operation and internal upconversion of a $LiNbO_3$ parametric oscillator using a $LiIO_3$ crystal. Progress in ADP, $LiIO_3$, and infrared parametric oscillators using proustite and CdSe is also reviewed including the use of these materials for mixing and sum generation.

The final section reviews recent research in nonlinear materials and their potential application to parametric oscillators. The important properties of nonlinear materials are discussed and a detailed comparison of important nonlinear materials is made. The comparison includes a determination of parametric gain per watt and gain per incident intensity for the selected materials. Particular attention is paid to the recently characterized chalcopyrite class of infrared nonlinear materials which show very favorable phase-matching properties with high nonlinearity. The important aspects of crystal damage are also considered. Finally, applications, which were among those mentioned during a survey of over thirty parametric oscillator users, are briefly discussed.

At this time the basic theory of optical parametric oscillators is well known and the operation of a number of parametric oscillators has experimentally defined their characteristics. Among the goals of present parametric oscillator work is the extension of the available tuning range especially in the infrared region, the development to a higher degree the known nonlinear materials, and the search for new materials. In addition, there is interest in obtaining higher peak and average output powers and better frequency and bandwidth control. Narrow bandwidth operation is important for many spectroscopic applications.

Parametric oscillators and frequency converters are useful across the widest frequency range (from the ultraviolet to the far infrared) of any tunable device. In addition their efficiency and high power capability assure future applications as tunable coherent sources.

REFERENCES

Abagyan, S. A., Ivanov, G. A., Kartushina, A. A., and Koroleva, G. A. (1972). *Sov.Phys. Semicond.* **5**, 1425.
Abrahams, S. C., Ready, J. M., and Bernstein, J. L. (1966a). *J. Phys. Chem. Solids* **27**, 997.
Abrahams, S. C., Hamilton, W. C., and Ready, J. M. (1966b). *J. Phys. Chem. Solids* **27**, 1013.
Abrahams, S. C., Levinstein, H. J., and Ready, J. M. (1966c). *J. Phys. Chem. Solids* **27**, 1019.
Adams, III, N. I., and Barrett, J. J. (1966). *IEEE J. Quantum. Electron.* **QE-2**, 430.
Akhmanov, S. A., and Khoklov, R. V. (1962). *Zh. Eksp. Teor. Fiz.* **43**, 351 [*English Transl;* (1963). *Sov. Phys.-JETP* **16**, 252].
Akhmanov, S. A., and Khoklov, R. V. (1964). "Problem y Nyelineynov Optiki." *Akad. Nauk SSSR, Moscow.*
Akhmanov, S. A., Kovrigin, A. I., Khoklov, R. V., and Chunaev, O. N. (1963). *Zh. Eksp. Teor. Fiz.* **45**, 1336 [*English Transl.;* (1964). *Sov. Phys.-JETP* **18**, 919].
Akhmanov, S. A., Dmitriev, V. G. and Modenov V. P. (1965a). *Radiotekh. Electron.* **10** 649 [*English Transl.;* (1965). *Radio Eng. Electron. Phys.* **10** 552].
Akhmanov, S. A., and Khoklov, R. V. (1968). *Usp. Fiz. Nauk* **95**, 231 [*English Transl.;* (1968). *Sov. Phys.-Usp.* **11**, 394].
Akhmanov, S. A., Kovrigin, A. I., Piskarskas, A. S., Fadeev, V. V., and Khokhlov, R. V. (1965b). *Zh, EKsp. Teor. Fiz. Pis. Red.* **2**, 300 [*English Transl.:* (1965). *JETP Lett.* **2**, 191.]
Akhmanov, S. A., Ershov, A. G., Fadeev, V. V., Khoklov, R. V. Chunaev, O. N., and Shvom, E. M. (1965c). *Zh, EKsp. Teor. Fiz. Red.* **2**, 485 [*English Transl.:* (1965). *JETP Lett.* **2**, 285]. **2**, 285].
Akhmanov S. A. Kovrigin A. I. Kolosov V. A. Piskarskas A. S. Fadeev V. V. and Khokhlov R. V. (1966). *JETP Lett.* **3** 241.
Akhmanov, S. A., Chunaev, O. N., Fadeev, V. V., Khokhlov, R. V., Klyshko, D. N., Kovrigin, A. I., and Piskarskas, A. S. (1967a). Parametric Generators of Light, presented at the *1967 Symp. Mod. Opt.* Polytechnic Inst. of Brooklyn, New York, March.

Akhmanov, S. A., Sukhorukov, A. P., and Khokhlov, R. V. (1967b). *Usp. Fiz. Nauk* **93**, 19 [*English Transl.;* (1968). *Sov. Phys.-Usp.* **10**, 609].

Akhmanov, S. A., Kovrigin, A. I., Sukhorukov, A. P., Khokhlov, R. V., and Chirkin, A. S. (1968a). *JETP Lett.* **7**, 182.

Akhmanov, S. A., Chirkin, A. S., Drabovich, K. N., Kovrigin, A. I., Khokhlov, R. V., and Sukhorukov, A. P. (1968b). *IEEE J. Quantum Electron.* **QE-4**, 598.

Akhmanov, A. G., Akhmanov, S. A., Khokhlov, R. V., Kovrigin, A. I., Piskarskas, A. S., and Sukhorukov, A. P. (1968c). *IEEE J. Quantum Electron.* **QE-4**, 828.

Akhmanov, S. A., Sukhorukov, A. P., and Chirkin, A. S. (1969). *Sov. Phys.-JETP* **28**, 748.

Akhmanov, S. A., D'Yakov, Yu. E., and Chirkin, P. S. (1971). *JETP Lett.* **13**, 514.

Amman, E. O. (1972). *Tech. Rep. AFAL-TR-72-177, Air Force Avionics Lab., Wright Patterson AFB, Ohio.*

Ammann, E. O., and Montgomery, P. C. (1970). *J. Appl. Phys.* **41**, 5270.

Ammann, E. O., and Wintemute, J. D. (1973), "Damage to ZnS thin Films from 1.08μ Laser Radiation" (to be published).

Ammann, E. O., and Yarborough, J. M. (1970). *Appl. Phys. Lett.* **17**, 233.

Ammann, E. O., Foster, J. D., Oshman, M. K., and Yarborough, J. M. (1969). *Appl. Phys Lett.* **15**, 131.

Ammann, E. O., Yarborough, J. M., Oshman, M. K., and Montgomery, P. C. (1970) *Appl. Phys. Lett.* **16**, 309.

Ammann, E. O., Yarborough, J. M., and Falk, J. (1971). *J. Appl. Phys.* **42**, 5618.

Anderson, D. B., and Boyd, J. T. (1971). *App . Phys. Lett.* **19**, 266.

Anderson, D. B., and McMullen, J. D. (1969). *IEEE J. Quantum Electron.* **QE-5**, 354.

Anderson, D. B., Boyd, J. T., and McMullen, J. D. (1971). *Proc. Symp. Submillimeter Waves* MRI Ser. Vol. XX, p. 191. Polytechnic Inst. of Brooklyn Press, New York.

Andrews, R. A. (1971a). *IEEE J. Quantum Electron.* **QE-6**, 68.

Andrews, R. A. (1971b). *IEEE J. Quantum. Electron.* **QE-7**, 523.

Armstrong, J. A., Bloembergen, N., Ducuing, J., and Pershan, P. S. (1962). *Phys. Rev.* **127**. 1918.

Armstrong, J. A., Jha, S. S., and Shiren, N. S. (1970). *IEEE J. Quantum Electron.* **QE-6**, 123.

Asby, R. (1969a). *Opto-Electron.* **1**, 165.

Asby, R. (1969b). *Phys. Rev.* **187**, 1062.

Asby, R. (1969c). *Phys. Rev.* **187**, 1070.

Ashkin, A., Boyd, G. D., and Dziedzic, J. M. (1966a). *IEEE J. Quantum Electron.* **QE-2**, 109.

Ashkin, A., Boyd, G. D., Dziedic, J. M., Smith, R. G., Ballman, A. A., Levinstein, J. J., and Nassau, K. (1966b). *App . Phys. Lett.* **9**, 72.

Bahr, G. C., and Smith, R. C. (1972). *Phys. Status Solidi* **13**, 157.

Bass, M., and Barrett, H. H. (1972). *IEEE J. Quantum Electron.* **QE-8**, 338.

Belyaev, Y. N., Kiselev, A. M., and Freidman, G. I. (1969). *JETP Lett.* **9**, 263.

Bergman, J. G., Ashkin, A., Ballman, A. A., Dziedzic, J. M., Levinstein, H. J., and Smith, R. G. (1968). *Appl. Phys. Lett.* **12**, 92.

Bergman, J. G., Jr., Boyd, G. D., Ashkin, A., and Kurtz, S. K. (1969). *J. Appl. Phys.* **40** 2860.

Bey, P. P., and Tang, C. L. (1972). *IEEE J. Quantum Electron.* **QE-8**, 361.

Bjorkholm, J. E. (1966). *Phys. Rev.* **142**, 126.

Bjorkholm, J. E. (1968a). *Appl. Phys. Lett.* **13**, 53.

Bjorkholm, J. E. (1968b). *Appl. Phys. Lett.* **13**, 399.

Bjorkholm, J. E. (1968c). *IEEE J. Quantum Electron.* **QE-4**, 970, and

Bjorkholm, J. E. (1969a). *IEEE J. Quantum Electron.* **QE-5**, 260 (correction to above).

Bjorkholm, J. E. (1969b). *IEEE Quantum Electron.* **QE-5**, 293.

Bjorkholm, J. E. (1971). *IEEE J. Quantum Electron.* **QE-7**, 109.
Bjorkholm. J. E., and Danielmayer, H. G. (1969). *Appl. Phys. Lett.* **15**, 300.
Bjorkholm, J. E. and Siegman, A. E. (1967). *Phys. Rev.* **154**, 851.
Bjorkholm, J. E., Ashkin, A., and Smith, R. G. (1970). *IEEE J. Quantum Electron.* **QE-6**, 797.
Bjorkholm, J. E., Damen, T. C., and Shah, J. (1971). *Opt. Commun.* **4**, 283.
Bloembergen, N. (1965). "Nonlinear Optics." Benjamin, New York.
Bloembergen, N. (1967). *Amer. J. Phys.* **35**, 989.
Bloembergen, N., and Pershan, P. S. (1962). *Phys. Rev.* **128**, 606.
Bloom, A. L. (1963). Spectra Phys. Tech. Bull. Number 2.
Bonch-Bruevich, A. M., and Khodovoi, V. A. (1965). *Usp. Fiz. Nauk* **85**, 3 [*English Transl.;* (1965). *Sov. Phys.-Usp.* **8**, 1].
Bond, W. L. (1971). *J. Appl. Phys.* **36**, 1674.
Boggett, D. M., and Gibson, A. F. (1968). *Phys. Lett.* **28A**, 33.
Boyd, G. D., and Ashkin, A. (1966). *Phys. Rev.* **146**, 187.
Boyd, G. D., and Gordon, J. P. (1961). *Bell Syst. Tech.* **40**, 489.
Boyd, G. D., and Kleinman, D. A. (1968). *J. Appl. Phys*, **39**, 3597.
Boyd, G. D., and Kogelnik, H. (1962). *Bell Syst. Tech. J.* **41**, 1347.
Boyd, G. D., and Nash, F. R. (1971). *J. Appl. Phys.* **42**, 2815.
Boyd, G. D., Miller, R. C., Nassau, K., Bond, W. L., and Savage, A. (1964). *Appl. Phys. Lett.* **5**, 234.
Boyd, G. D., Ashkin, A., Dziedzic, J. M., and Kleinman, D. A. (1965). *Phys. Rev.* **137**, A1305.
Boyd, G. D., Bond, W. L., and Carter, H. L. (1967). *J. Appl. Phys* **48**, 1941.
Boyd, G. D., Buehler, E., and Stortz, F. G. (1971a). *Appl. Phys. Lett.* **18**, 301.
Boyd, G. D., Grandrud, W. B., and Buehler, E. (1971b). *Appl. Phys. Lett.* **18**, 446.
Boyd, G. D., Kaspar, H., and McFee, J. H. (1971c). *IEEE J. Quantum Electron.* **QE-7**, 563.
Boyd, G. D., Buehler, E., Storzt, F. G., and Wernick, J. H. (1972a). *IEEE J. Quantum Electron.* **QE-8**, 419.
Boyd, G. D., Kaspar, H. M., McFee, J. H., and Stortz, F. G. (1972b). *IEEE J. Quantum Electron.* **QE-8**, 900.
Boyd, J. T. (1972). *IEEE J. Quantum Electron.* **QE-8**, 788.
Brugger, K. (1972). *J. Appl. Phys.* **43**, 577.
Burneika, K., Ignatavicius, M., Kabelka, V., Piskarskas, A., and Stabinis, A. (1972). *IEEE J. Quantum Electron.* **QE-8**, 511.
Butcher, P. N. (1965). Nonlinear Optical Phenomena. Ohio State Univ., Columbus, Ohio.
Byer, R. L. (1968). Parametic Fluorescence and Optical Parametic Oscillation, Ph.D. dissertation, Stanford Univ. Stanford, California.
Byer, R. L. (1970). *Opt. Spectra* September, 42.
Byer, R. L., and Garbuny, M, (1973). Pollutant Detection by Absorption Using Mie Scattering and Topographical Targets as Retroreflectors. To be published in *Appl. Opt.*
Byer, R. L., and Harris, S. E. (1968). *Phys. Rev.* **168**, 1064.
Byer, R. L., Oshman, M. K., Young, J. F., and Harris, S. E. (1968). *Appl. Phys. Lett.* **13**, 109.
Byer, R. L., Harris, S. E., Kuizenga, D. J., and Young, J. F. (1969a). *J. Appl. Phys.* **40**, 444.
Byer, R. L., Kovrigin, A., and Young, J. F. (1969b). *Appl. Phys. Lett.* **15**, 136.
Byer, R. L., Young, J. F., and Feigelson, R. S. (1970). *J. Appl. Phys.* **41**, 2320.
Byer, R. L., Kildal, H., and Feigelson, R. S. (1971). *Appl. Phys. Lett.* **19**, 237.
Campillo, A. J. (1972). *IEEE J. Quantum Electron.* **QE-8**, 809.
Campillo, A. J., and Tang, C. L. (1970). *Appl. Phys. Lett.* **16**, 242. See also: (1968). *ibid.* **12**, 376.

Campillo, A. J., and Tang, C. L. (1971). *Appl. Phys. Lett.* **19**, 36.

Carruthers, J. R., Peterson, G. E., Grasso, M., and Bridenbaugh, P. M. (1971). *J. App. Phys.* **42**, 1846.

Chemla, D. S., Kupecek, P. J., Robertson, D. S., and Smith, R. C. (1971). *Opt. Commun.* **1**, 29.

Chen, F. S. (1969). *J. Appl. Phys.* **40**, 3389.

Comly, J., and Garmire, E. (1968). *Appl. Phys. Lett.* **12**, 7.

Damaschini, R. (1969). *C. R. Acad. Sci. Paris* **268**.

Davidov, A. A., Kulevskii, L. A., Prokorov, A. M., Savel'ev, A. D., and Smirnov, V. V (1972). *Zh. Eksp. Teor. Fiz. Pis. Red.* **15**, 725 [*English Transl.;* (1972). *JETP Lett.* **15**, 513].

Dewey, C. F., Jr., and Hocker, L. O. (1971). *Appl. Phys. Lett.* **15**, 58.

Dobrzhanskii, G. F., Kitaeva, V. F., Kuleskii, L. A., Polivanov, Yu. N., Poluektov, S. N. Prokhorov, A. M., and Sobolev, N. N. (1970). *JETP Lett.* **12**, 353.

Donoho, P. L., Bond, T. P., and Guest, M. M. (1972). Holography of the Microcirculation *Eur. Conf. Microcircul. 7th* Aberdeen, Scotland, August 26–September 1.

Dowley, M. W. (1969). *Opto.-Electron.* **1**, 179.

Dowley, M. W., and Hodges, E. B. (1968). *IEEE J. Quantum Electron.* **QE-4**, 552.

Falk, J. (1971). *IEEE J. Quantum. Electron.* **QE-7**, 230.

Falk, J., and Murray, J. E. (1969). *Appl. Phys. Lett.* **14**, 245.

Falk, J., Yarborough, J. M., and Ammann, E. O. (1971). *IEEE J. Quantum Electron.* **QE-7**, 359.

Faries, D. W., Gerhring, K. A., Richards, P. L., and Shen, Y. R. (1960). *Phys. Rev.* **180**, 363.

Fay, H., Alford, W. L., and Dess, H. M. (1968). *Appl. Phys. Lett.* **12**, 89.

Feichtner, J. D. (1972). Private communication.

Feichtner, J. D., and Roland, G. W. (1972). *Appl Opt.* **11**, 993.

Fischer, R. (1970). Zun Wirkungsgrad Parametrisch Abstimmbaier Laser, *In "Laser und ihre Anwendungen."* Int. Tagung Dresden 10.6–17.6 *Tiel 13*, 773–856.

Fischer, R. (1971). *Exp. Tech Phys.* **XIX**, 193.

Fox, K. (1973). Univ. of Tennessee, Private communication.

Francois, G. E. (1966). *Phys. Rev.* **143**, 597.

Franken, P. A., and Ward, J. F. (1963). *Rev. Mod. Phys.* **35**, 23.

Franken, P. A., Hill, A. E., Peters, C. W., and Weinreich, G. (1961). *Phys. Rev. Lett.* **7**, 118

Gandrud, W. R., Boyd, G. D., McFee, J. H., and Wehmeier, F. H. (1970). *Appl. Phys. Lett.* **16**, 59.

Giallorenzi, T. G., and Reilly, M. H. (1972). *IEEE J. Quantum Electron.* **QE-8**, 302.

Giallorenzi, T. G., and Tang, C. L. (1968). *Phys. Rev.* **166**, 225.

Giordmaine, J. A. (1962). *Phys. Rev. Lett.* **8**, 19.

Giordmaine, J. A. (1969). *Phys. Today* **22**, 38.

Giordmaine, J. A., and Miller, R. C. (1965). *Phys. Rev. Lett.* **14**, 973.

Giordmaine, J. A., and Miller, R. C. (1966a). Optical Parametric Oscillation in $LiNbO_3$ *In "Physics of Quantum Electronics"* (P. L. Kelley, B. Lax, and P. E. Tannenwald, eds.) pp. 31–42. McGraw-Hill, New York. Also available (1965). *Proc. Phys. Quant. Electron. Conf.* San Juan, Puerto Rico, June 28–30.

Giordmaine, J. A., and Miller, R. C. (1966b). *Appl. Phys. Lett.* **9**, 298.

Glass, A. J., and Guenther, A. H. (1972). *Appl. Opt.* **11**, 832.

Glenn, W. H. (1967). *Appl. Phys. Lett.* **11**, 333.

Glenn, W. H. (1969). *IEEE J. Quantum Electron.* **QE-5**, 284.

Goldberg, L. S. (1970). *Appl. Phys. Lett.* **17**, 489.

Goldberg, L. S. (1972). A Repetitively-Pulsed $LiIO_3$ Internal Optical Parametric Oscillator, presented at the *Int. Quant. Electron. Conf. 7th* Montreal Canada.

Goryunova, N. A., Ryvkin, S. M., Fishman, I. M., Shpen'kov, G. P., and Yaroshetskii, I. D. (1965). *Sov. Phys.-Semicond.* **2**, 1272 See also, Goryunova, N. A. (1965). "The Chemistry of Diamond Like Semiconductors". M.I.T. Press, Cambridge, Massachusetts.

Hagen, W. F., and Magnante, P. C. (1969). *J. Appl. Phys.* **40**, 219.

Hanna, D. C., Smith, R. C., and Stanley, C. R. (1971). *Opt. Commun.* **4**, 300.

Hanna, D. C., Luther-Davies, B., Rutt, H. N., and Smith, R. C. (1972a). *Appl. Phys. Lett.* **20**, 34.

Hanna, D. C., Luther-Davies, B., Rutt, H. N., Smith, R. C., and Stanley, C. R. (1972b). *IEEE J. Quantum Electron.* **QE-8**, 317.

Harris, S. E. (1966). *IEEE J. Quantum Electron.* **QE-2**, 701.

Harris, S. E. (1967). *IEEE J. Quantum. Electron.* **QE-3**, 205.

Harris, S. E. (1969a). *Proc. IEEE* **57**, 2096.

Harris, S. E. (1969b). *Appl. Phys. Lett.* **14**, 335.

Harris, S. E. (1969c). Harmonic Generation. Chromatix Tech. Lett. No. 1.

Harris, S. E., Oshman, M. K., and Byer, R. L. (1967). *Phys. Rev. Lett.* **18**, 732.

Harris, S. E., Carson, D., Young, J. F. (1972). Stanford Univ. Private communication.

Herbst, R. L. (1972). Cadmium Selenide Infrared Parametric Oscillator, Ph.D. dissertation, Stanford Univ. Stanford, California; (available as Microwave Lab. Rep. No. 2125).

Herbst, R. L. (1973). Private communication.

Herbst, R. L., and Byer, R. L. (1971). *Appl. Phys. Lett.* **19**, 527.

Herbst, R. L., and Byer, R. L. (1972). *Appl. Phys. Lett.* **21**, 189.

Herbst, R. L., and Byer, R. L. (1973). to be published.

Hercher, M. (1969). *Appl. Opt.* **8**, 1103.

Hobden, M. V. (1967). *J. Appl. Phys.* **38**, 4365.

Hobden, M. V. (1969). *Opto-Electron.* **1**, 159.

Hobden, M. V., and Warner, J. (1966). *Phys. Lett.* **22**, 243.

Hordvik, A., Schlossberg, H. R., and Stickley, C. M. (1971). *Appl. Phys. Lett.* **18**, 448.

Hulme, K. F., O'Jones, Davies, P. H., and Hobden, M. V. (1967). *Appl. Phys. Lett.* **10**, 133.

Hulme, K. F., Davies, P. H., and Cound, V. M. (1969). *J. Phys. Chem. (Solid State Phys.)* **2**, 855.

Huth, B. G., and Kiang, Y. C. (1969). *J. Appl. Phys.* **40**, 4976.

Huth, B. G., Farmer, G. I., Taylor, L. M., and Kagan, M. R. (1968). *Spec. Lett.* **1**, 425.

Ito, H., and Inaba, H. (1970). *Opto-Electron.* **2**, 81.

Izrailenko, A. I., Kovrigin, A. I., and Nikles, P. V. (1970). *JETP Lett.* **12**, 331.

Jasperse, J. R., and Gianino, P. D. (1972). *J. Appl. Phys.* **43**, 1686.

Jerphanon, J. (1970). *Appl. Pyhs. Lett.* **16**, 298.

Jerphanon, J. (1972). Private communication.

Jerphanon, J., and Kurtz, S. K. (1970). *Phys. Rev.* **1B**, 1738.

Johnston, W. D., Jr., (1970) *J. Appl. Phys.* **41**, 3279.

Jona, F., and Shirane, G. (1962). "Ferroelectric Crystals." McMillan, New York.

Keilich, S. (1970). *Opto-Electron.* **2**, 125.

Kildal, H. (1972). CdGeAs$_2$ and CdGeP$_2$ Chalcopyrite Crystals for Infrared Nonlinear Optics. Ph.D. dissertation, Stanford Univ., Stanford, California; (available as Microwave Lab. Rep. No. 2118).

Kildal, H., and Byer, R. L. (1971). *Proc. IEEE* **59**, 1644.

Kildal, H., Begley, R. F., Choy, M. M., and Byer, R. L. (1972). *J. Opt. Soc. Amer.* **62**, 1398.

Kingston, R. H. (1962). *Proc. IRE* **50**. 472.

Kleinman, D. A. (1962a). *Phys. Rev.* **126**, 1977.

Kleinman, D. A. (1962b). *Phys. Rev.* **128**, 1761.

Kleinman, D. A. (1968). *Phys. Rev.* **174**, 1027.

Kleinman, D. A., and Boyd, G. D. (1969). *J. Appl. Phys.* **40**, 546.

Kleinman, D. A., Ashkin, A., and Boyd, G. D. (1966). *Phys. Rev.* **145**, 338.

Kogelnik, H. (1964). Coupling Coefficients and Conversion Coefficients for Optical Modes, *Sym. Quasi-Opt.* p. 333. Polytechnic Inst. of Brooklyn, June 8–10.

Kogelnik, H. (1965). *Bell Syst. Tech. J.* **44**, 455.

Kogelnik, H., and Li, T. (1966). *Appl. Opt.* **5**, 1550.

Kovrigin, A. I., and Byer, R. L. (1969). *IEEE J. Quantum Electron.* **QE-5**, 384.

Kovrigin, A. I., and Nikles, P. V. (1971). *JETP Lett.* **13**, 313.

Kreuzer, L. B. (1967). *Appl. Phys. Lett.* **10**, 336.

Kreuzer, L. B. (1968). *Appl. Phys. Lett.* **13**, 57.

Kreuzer, L. B. (1969a). Single and Multimode Oscillation of the Singly Resonant Optical Parametric Oscillator, *Proc. Joint Conf. Lasers Opto. Electron.* p. 53. Univ. of Southampton, Southampton, England.

Kreuzer, L. B. (1969b). *Appl. Phys. Lett.* **15**, 263.

Krivoshchekov, G. V., Kruglov, S. V., Marennikov, S. I., and Polivanov, Yu. N. (1968). *JETP Lett.* **7**, 63. See also, *Zh. Eksp. Teor. Fiz.* **55**, 802.

Kroll, N. M. (1962). *Phys. Rev.* **127**, 1207.

Kruger, J. S., and Gleason, T. J. (1970). *J. Appl. Phys.* **41**, 3903.

Kuhn, L. (1969). *IEEE J. Quantum Electron.* **QE-5**, 383.

Kuizenga, D. J. (1972). *Appl. Phys. Lett.* **21**, 570.

Kurtz, S. K., and Perry, T. T. (1968). *J. Appl. Phys.* **39**, 3798.

Kurtz, S. K., Perry, T. T., and Bergman, J. G. Jr., (1968). *Appl. Phys. Lett.* **12**, 186.

Laurence, C., and Tittel, F. (1971a). *J. Appl. Phys.* **42**, 2137.

Laurence, C., and Tittel, F. (1971b). *Opto-Electron.* **3**, 1.

Landolt-Börnstein (1969). Numerical Data and Functional Relationships in Science and Technology, "New Series, Vol. 2, Group III Elastic, Piezoelectric, Piezooptica, Electrooptic Constants, and Nonlinear Dielectric Susceptibilities of Crystals" (K. H. Hellwege and A. M. Hellwege, eds.) Springer-Verlag, Berlin and New York.

Lee, C. C., and Fan, H. Y. (1972). *Appl. Phys. Lett.* **20**, 18.

Lenzo, P. V., Spencer, E. G., and Nassau, K. (1966). *J. Opt. Soc. Amer.* **56**, 633.

Leone, S. R., and Moore, C. B. (1972). *J. Opt. Soc. Amer.* **62**, 1358; see also (1973). V → V Energy Transfer in HCl with Tunable Optical Parametric Oscillator Excitation, (to be published).

Lerner, P., Legras, C., and Dumas, J. P. (1968). Stoechiometrie des Monocristaux de Metaniobate de Lithium, published in "Crystal Growth," *1968 Proc. Int. Conf. Crystal Growth, 2nd Birmingham, U. K., July* (F. C. Frank, J. B. Mullin, and H. S. Peiser, eds.), p. 231. North-Holland Publ., Amsterdam.

Levine, B. F., and Bethea, C. G. (1972). *Appl. Phys. Lett.* **20**, 272.

Louisell, W. H. (1960). "Coupled Mode and Parametric Electronics." Wiley, New York.

McFee, J. H., Boyd, G. D., and Schmidt, P. H. (1970). *Appl. Phys. Lett.* **17**, 57.

McMahon, J. M., Emmett, J. L., Holzrichter, J., and Trenholm. J. B. (1972). Disc Laser Test Results. Naval Res. Lab. Rep., March.

Magde, D., and Mahr, H. (1967). *Phys. Rev. Lett.* **18**, 905.

Maker, P. D., Terhune, R. W., Nisenoff, N., and Savage, C. M. (1962). *Phys. Rev. Lett.* **8**, 21.

Martin, M. D., and Thomas, E. L. (1966). *IEEE J. Quantum Electrom.* **QE-2**, 196.

Mathias, B. T., and Reimeika, J. P. (1949). *Phys. Rev.* **76**, 1886.

Meadors, J. G., Savage, W. T., and Damon, E. K. (1969). *Appl. Phys. Lett.* **14**, 360.

Measures, R. M., and Pilon, G. (1972). *Opto-Electron.* **4**, 141.

Meltzer, D. W., and Goldberg, L. S. (1972). *Opt. Commun.* **5**, 209.

Mercier, A. (1972). Ecole Polytechnique Federale de Lausanne, Private communication.

Midwinter, J. E. (1968a). *Appl. Phys. Lett.* **11**, 128.

Midwinter, J. E. (1968b). *J. Appl. Phys.* **39**, 3033.

Midwinter, J. E., and Warner, J. (1965). *Brit. J. Appl. Phys.* **16**, 1135.

Milek, J. T., and Welles, S. J. (1970). Linear Electrooptic Modulator Materials. Electronic Properties Information Center, Hughes Aircraft Co.

Miles, R. (1972). Private communication.

Miller, R. C. (1964). *Appl. Phys. Lett.* **5**, 17.

Miller, R. C. (1968). *Phys. Lett.* **26A**, 177.

Miller, R. C., and Nordland, W. A. (1967). *Appl. Phys. Lett.* **10**, 53.

Miller, R. C., and Savage, A. (1966). *Appl. Phys. Lett.* **9**, 169.

Miller, R. C., Kleinman, D. A., and Savage, A. (1963). *Phys. Rev. Lett.* **11**, 146.

Miller, R. C., Nordland, W. A., and Bridenbaugh, P. M. (1971). *J. Appl. Phys.* **42**, 4145.

Minck, R. W., Terhune, R. W., and Wang, C. C. (1966). *Appl. Opt.* **5**, 1595.

Murray, J. E., and Harris, S. E. (1970). *J. Appl. Phys.* **41**, 609.

Nash, F. R., Bergman, J. G., Jr., Boyd, G. D., and Turner, E. H. (1969). *J. Appl. Phys.* **40**, 5201.

Nassau, K., Levinstein, H. J., and Loiacono, G. M. (1966). *J. Phys. Chem. Solids* **27**, 983, 989.

Nath, G., and Haussühl, S. (1969a). *Appl. Phys. Lett.* **14**, 154.

Nath, G., and Haussühl, S. (1969b). *Phys. Lett.* **29A**, 91.

Nye, J. F. (1960). "Physical Properties of Crystals," Chapter VII. Oxford Univ. Press, London and New York.

Okada, M., and Ieiri, S. (1971). *IEEE J. Quantum Electron.* **QE-7**, 560.

Oshman, M. K., and Harris, S. E. (1968). *IEEE J. Quantum Electron.* **QE-4**, 491.

Ovander, L. N. (1965). *Usp. Fiz. Nauk* **86**, 3 [*English Transl.;* (1965). *Sov. Phys.-Usp.* **8**, 337].

Parthe, E. (1964). "Crystal Chemistry of Tetrahedral Structures." Gordon and Breach, New York.

Pearson, J. E., Ganiel, U., and Yariv, A. (1972a). *IEEE J. Quantum Electron.* **QE-8**, 383.

Pearson, J. E., Ganiel, U., and Yariv, A. (1972b). *IEEE J. Quantum Electron.* **QE-8**, 433.

Pearson, J. E., Evans, G. A., and Yariv, A. (1972c). *Opt. Commun.* **4**, 366.

Pershan, P. S. (1963). *Phys. Rev.* **130**, 919.

Pershan, P. S. (1966). Nonlinear Optics," *Prog. Opt. V*, 85–114.

Peterson, G. E., Ballman, A. A., Lenzo, P. V., and Bridenbaugh, P. M. (1964). *Appl. Phys. Lett.* **5**, 62.

Pinard, J., and Young, J. F. (1972). *Opt. Commun.* **4**, 425.

Pressley, R. J. (ed.) (1971). "Handbook of Lasers with Selected Data on Optical Technology." Chemical Rubber Co., Ohio.

Rabson, T. A., Ruiz, H. J., Shah, P. L., and Tittle, F. K. (1972). *Appl. Phys. Lett.* **21**, 129.

Rank, D. H. (1970). *J. Opt. Soc. Amer.* **60**, 443.

Ready, J. F. (1971). "Effects of High-Power Laser Radiation." Academic Press, New York.

Rutt, H. N., and Smith, R. C. (1972). Intensity Thresholds of Optical Parametric Oscillators" (to be published).

Sackett, P. (1972). Air Force Cambridge Res. Lab. Private communication.

Sato, T. (1972). *J. Appl. Phys.* **43**, 1837.

Shapiro, S. L. (1968). *Appl. Phys. Lett.* **13**, 19.

Seigman, A. E. (1962). *Appl. Opt.* **1**, 739.

Smith, G. E. (1969). *IEEE J. Quantum Electron.* **QE-5**, 383.

Smith, P. W. (1965). *IEEE J. Quantum Electron.* **QE-1**, 343.

Smith, R. G. (1970). *J. Appl. Phys.* **41**, 4121.

Smith, R. G. (1973). Optical Parametric Oscillators. (to be published).

Smith, R. G., and Parker, J. V. (1970). *J. Appl. Phys.* **41**, 3401.

Smith, R. G., Nassau, K., and Galvin, M. F. (1965). *Appl. Phys. Lett.* **7**, 256.

Smith, R. G., Geusic, J. E., Levinstein, H. J., Singh, S., and Van Uitert, L. G. (1968a). *J. Appl. Phys.* **39**, 4030.

Smith, R. G., Geusic, J. E., Levinstein, H. J., Rubin, J. J., Singh, S., and Van Uitert, L. G. (1968b). *Appl. Phys. Lett.* **12**, 308.

Smyth, K. C., and Brauman, J. I. (1972). *J. Chem. Phys.* **56**, 1132.

Starunov, V. S., and Fabelinskii, I. L. (1969). *Usp. Fiz. Nauk* **98**, 441 [*English Transl.;* (1970). *Sov. Phys.-Usp.* **12**, 463].

Sukhorunkov, A. P., and Shchednova, A. K. (1971). *Sov. Phys.-JETP* **33**, 677.

Sushchik, M. M., Fortus, V. M., and Freidman, G. I. (1970). *Radiophysica* **13**, 631.

Terhune, R. W., and Maker, P. D. (1968). Nonlinear Optics in Lasers, *In* "Advances in Lasers" (A. K. Levine, ed.), Vol. II, pp. 295–370.Dekker, New York.

Tien, P. K., Ulrich, R., and Martin, R. J. (1970). *Appl. Phys. Lett.* **17**, 447. See also, Tien, P. K. (1971). *Appl. Opt.* **10**, 2395.

Turner, E. H. (1966). *Appl. Phys. Lett.* **8**, 303.

Volosov, V. D. (1970). *Sov. Phys.-Tech. Phys.* **14**, 1652.

Wagner, W. G., and Hellwarth, R. W. (1964). *Phys. Rev.* **133**, A915.

Wallace, R. W. (1970). *Appl. Phys. Lett.* **17**, 497. See also Wallace, R. W., and Harris, S. E. (1970). Laser Focus, November, p. 42.

Wallace, R. W. (1971a). *IEEE Conf. Laser Appl.* Washington, D. C.

Wallace, R. W. (1971b). *IEEE J. Quantum Electron.* **QE-7**, 203.

Wallace, R. W. (1971c). *Opt. Commun.* **4**, 316.

Wallace, R. W. (1972). *IEEE J. Quantum Electron.* **QE-8**, 819.

Wallace, R. W. (1973). Second Harmonic 90° Phasematching of KDP Isomorphs (to be published).

Wallace, R. W., and Dere, D. (1972). Chromatix, Inc. Private communication.

Wallace, R. W., and Harris, S. E. (1969). Extending the Tunability Spectrum, Laser Focus (1970). p. **000**; see also, Wallace R. W. (1969). *Appl. Phys. Lett.* **15**, 111.

Wang, C. C., and Racette, G. W. (1965). *Appl. Phys. Lett.* **6**, 169.

Warner, J. (1968). *Appl. Phys. Lett.* **12**, 222.

Warner, J. (1971). *Opto-electron.* **3**, 37.

Weinberg, D. L. (1969). *Appl. Phys. Lett.* **14**, 32.

Weller, J. F. and Andrews, R. A. (1972). CW Parametric Oscillator Studies with $Ba_2NaNb_5O_{15}$Pumped by 0.5145 μ Radiation (to be published).

Wright, J. K. (1963). *Proc. IEEE* **51**, 1663.

Yajima, T., and Inave, K. (1968). *Phys. Lett.* **26A**, 281.

Yarborough, J. M. (1972). Private communication.

Yarborough, J. M., and Massey, G. A. (1971). *Appl. Phys. Lett.* **18**, 438.

Yarborough, J. M., Sussman, S. S., Puthoff, H. E., Pantell, R. H., and Johnson, B. C. (1969). *Appl. Phys. Lett.* **15**, 102.

Yariv, A., and Pearson, J. E. (1969). Parametric Processes, *In* "Progress in Quantum Electronics" (J. H. Sanders and K. W. H. Stevens, eds.), Vol. I, pp. 1–49. Pergamon, Oxford.

Yeung, E. S., and Moore, C. B. (1971). *J. Amer. Chem. Soc.* **93**, 2059.

Yeung, E. S., and Moore, C. B. (1973a). Photochemistry of Single Vibronic States: An Application of Nonlinear Optics (to be published).

Yeung, E. S., and Moore, C. B. (1973b). Isotonic Separation by Photo-Predissociation (to be published).

Young, J. F., Murray, J. E., Miles, R. B., and Harris, S. E. (1971a). *Appl. Phys. Lett.* **18**, 129.

Young, J. F., Miles, R. B., and Harris, S. E. (1971b). *J. Appl. Phys.* **42**, 497.

Zernike, F. (1964). *J. Opt. Soc. Amer.* **54**, 1215.

Zernike, F., and Berman, P. R. (1965). *Phys. Rev. Lett.* **15**, 999.

Zook, J. D., Chen, D., and Otto, G. N. (1967). *Appl. Phys. Lett.* **11**, 159.

10

Difference Frequency Generation and Up-Conversion

JOHN WARNER

Royal Radar Establishment
Malvern, Worcestershire
United Kingdom

I. INTRODUCTION

In this chapter we consider how nonlinear optics may be used to generate and detect infrared radiation. In both cases the nonlinear nature of the optical susceptibility of acentric crystals provides a coupling mechanism between electromagnetic waves passing simultaneously through a crystal allowing the generation of sum or difference frequencies. The coupling is essentially the same as that used for second harmonic generation and parametric oscillation; these processes are considered in detail in Chapters 8 and 9.

The magnitude of the nonlinear susceptibility is rather small and consequently very intense optical beams, such as one might find using lasers, are required to enable significant mixing at optical and infrared frequencies. An infrared wave may be generated by beating two visible or near infrared laser beams together to generate a third beam at the difference frequency. If the frequency of one of the two laser beams is tunable (e.g., a dye laser or parametric oscillator), then the generated difference frequency would also be tunable. The possibilities of generating infrared radiation by frequency mixing are discussed more fully in Section III.

Infrared radiation may be detected with the aid of nonlinear optics by using the nonlinear susceptibility to facilitate changing the frequency of the infrared beam to the visible spectrum so that an efficient visible detector (e.g., photomultiplier) may be used. The infrared beam is mixed with an intense visible or near infrared laser beam to generate light at the sum frequency (occasionally at the difference frequency). Such a frequency-changing process is known as up-conversion; the up-converted light has an intensity proportional to that of the incident infrared so the output from a visible detector of the sum frequency light gives a measure of the intensity of the incident infrared beam.

The characteristics of up-conversion are discussed more fully in Section II. We can say here in summary that, when compared to other methods of infrared detection, it is a potentially better detector for spectrometer-type applications where a small bandwidth and small signal-to-noise ratio are required. It is a room-temperature detection system whereas most other infrared detectors must be cooled, but it must be remembered that the price to pay is the provision of a laser beam to "pump" the nonlinear crystal.

Spatial or directional information carried by an infrared beam as it passes through the mixing crystal is transferred to the sum frequency in an up-conversion process. Infrared objects may therefore be viewed either directly with the eye or with the aid of an image intensifier or low light level TV camera if an up-converter is interposed between the object and the viewer. The infrared radiation from the object may be provided by a source of illumination, such as a carbon dioxide laser, or it may be the thermal radiation radiated from the object itself. The expected performance and experimental achievements made so far are also reviewed in Section III.

We have not attempted to consider the subject in great depth in this chapter. The most important features and results have been summarized. The reader is referred to the appropriate references for further study.

II. UP-CONVERSION

A. Theory

The formal theory of sum-frequency generation was first given by Armstrong et al. (1962) for three interacting plane waves. A more recent study by Boyd and Kleinman (1968) includes sections on sum-frequency and difference-frequency mixing of two Gaussian beams in which they indicate how to optimize the geometry for maximum conversion efficiency. We will base our calculations on a plane wave rather than a Gaussian beam theory first because infrared beams from real objects can be expressed in terms of a series of plane waves and second because the plane wave theory shows clearly the importance, in terms of achieving efficient mixing, of properly arranging the phases of the three interacting waves.

1. CONVERSION EFFICIENCY FOR UP-CONVERSION

"Up-conversion" is the name given to the special case of sum-frequency generation ($\omega_s = \omega_p + \omega_{ir}$), where one of the waves (at ω_p) contributing to the sum frequency is very much more intense than the other (at ω_{ir}) and consequencly does not appreciably change in amplitude as it passes through the nonlinear dielectric medium (i.e., the mixing crystal). The amount of sum-frequency radiation at the start of the crystal is assumed to be small. Given these boundary conditions one may solve the coupled differential equations relating the amplitudes of the three waves to give the following expression for the sum-frequency field amplitude within the crystal [see, for example, Armstrong $et\ al.$ (1962)]:

$$\mathscr{E}_s(esu) = \left(\frac{\omega_s^2 k_{ir}}{\omega_{ir} k_s}\right)^{1/2} \mathscr{E}_{ir} \sin\left\{\frac{2\pi L}{c^2}\left(\frac{\omega_{ir}^2 \omega_s^2}{k_{ir} k_s}\right)^{1/2} \chi^{NL}\mathscr{E}_p\right\} \quad (1)$$

In deriving Eq. (1) it has been assumed that the wave vectors of the interacting waves exactly satisfy the momentum conservation equation or *phase-matching condition*, i.e., $\mathbf{k}_s = \mathbf{k}_p + \mathbf{k}_{ir}$. If we allow for a small momentum mismatch $\Delta \mathbf{k} = \mathbf{k}_s - \mathbf{k}_p - \mathbf{k}_{ir}$ and also assume that $\mathscr{E}_p$ is small enough for the sine function in (1) to be replaced by its argument, we find that the expression for $\mathscr{E}_s$ becomes

$$\mathscr{E}_s = \frac{2\pi}{c^2}\frac{\omega_s^2}{k_s}\chi^{NL}\mathscr{E}_{ir}\mathscr{E}_p\frac{\sin(\frac{1}{2}\Delta k L)}{\frac{1}{2}\Delta k} \quad (2)$$

The relationship between the amplitude of an electromagnetic wave in a dielectric of refractive index n (expressed in electrostatic units) and its corresponding photon flux density is given by

$$N\left(\frac{photon}{cm^2/sec}\right) = \frac{10^{-7}cn}{4\pi}\frac{\bar{E}^2}{\hbar\omega} \quad \text{where} \quad E^2 = \frac{\mathscr{E}^2}{2}$$

This enables us to rewrite (2) in the form of an expression for the photon conversion efficiency from the infrared to the sum frequency:

$$\eta_{uc} = \frac{N_s}{N_{ir}} = \frac{128\pi^3\omega_s\omega_{ir}d^2}{10^{-7}c^3 n_s n_p n_{ir}}P_p\frac{L_c^2}{A_c}\text{sinc}^2\left(\frac{1}{2}\Delta k L_c\right) \quad (3)$$

Table I

Effective Quantum Efficiencies for Several Up-Converter Situations

Material	θ_p (deg)	λ_{ir} (μm)	λ_p (μm)	λ_s (μm)	Sum detector		$\eta_{uc}{}^a$	$\eta_{uc}\eta_{det}$
					Cathode	η_{det}		
Ag$_3$AsS$_3$	26	10.6	0.6943	0.6516	S20	4.5×10^{-2}	2.0×10^{-7}	9×10^{-9}
	20	10.6	1.06	0.9636	S1	1.9×10^{-3}	1.5×10^{-7}	2.8×10^{-10}
ZnGeP$_2$	83	10.6	1.06	0.9636	S1	1.9×10^{-3}	1.1×10^{-6}	2.1×10^{-9}
	83	10.6	1.06	0.9636	InGaAs	10^{-2}	1.1×10^{-6}	1.1×10^{-8}
LiNbO$_3$	90	1.7	0.6943	0.4930	Super S11	1.8×10^{-1}	1.4×10^{-7}	2.5×10^{-8}
	45	3.5	1.06	0.8136	Extended S20	2×10^{-2}	0.5×10^{-7}	1.0×10^{-9}
Ba$_2$NaNb$_5$O$_{15}$	45	3.5	1.06	0.8136	Extended S20	2×10^{-2}	2×10^{-7}	4×10^{-9}

^{a} Values from Eq. (3) with $P_p = 1$ W, $L_c = 1$ cm, and $A_c = 1$ cm^2.

where ω is the circular frequency (rad/sec), $\mathbf{k}$ the wave propagation vector (cm^{-1}), N the photon flux (photon/sec/cm^2), η_{uc} the up-conversion photon conversion efficiency, n the refractive index, d the effective second harmonic generation coefficient (esu),* P_p the pump laser power (W), L_c the length of nonlinear mixing crystal (cm), A_c the effective aperture of mixing crystal (cm^2), and $\mathrm{sinc}(x) = \sin x/x$. The last factor in (3) has a maximum value of 1 when $\Delta k = 0$ and falls to zero when $\frac{1}{2}\Delta k L_c = \pi$. It is evident therefore that efficient sum-frequency generation will only take place if Δk is kept very small; this requirement leads to a small spectral bandwidth of infrared frequencies and a small cone of infrared directions which can be efficiently up-converted together. The properties of up-conversion which are due to this phase-matching necessity are considered in the next subsection (Section II, A, 2).

The overall quantum efficiency of an up-converter detection system is given by the product of the photon conversion efficiency, the quantum efficiency of the sum-frequency detector, and the transmission factors of any filters used to discriminate between the sum frequency and pump laser beams

$$\eta = \eta_{uc}\,\eta_{det}\,T_{filt} \tag{4}$$

Typical values of η_{uc} for a 1-W laser pumping a 1-cm cube of nonlinear material are given in Table I. When designing an up-converter system for a given infrared frequency one would normally choose the nonlinear material and pump frequency to make η as large as possible. Then η_{uc} can be increased by focusing the pump and infrared beams to make A_c smaller [see Eq. (3)]. While this is a good thing for signal up-conversion it is not so for imaging where one needs large aperture beams to maintain a good resolution figure.

2. PROPERTIES OF UP-CONVERSION DUE TO PHASE MATCHING

We have seen from Eq. (3) that efficient frequency conversion is only possible if the wave vectors of the interacting waves are such that $\mathbf{k}_p + \mathbf{k}_{ir} - \mathbf{k}_s = 0$. This *phase-matching condition* cannot be met in isotropic crystals because dispersion ensures that the refractive indices of the three waves are different. Crystals of lower symmetry (uniaxial and biaxial crystals) exhibit an anisotropy of refractive index which can be exploited to enable the phase-matching condition for up-conversion (and also for second harmonic generation and parametric oscillation) to be met. This possibility was first demonstrated by

---

† d is related to the nonlinear susceptibility written in Eqs. (1) and (2) by the formula $\chi^{NL} = 2d$ [see, for example, Appendix 3 of Boyd and Kleinman (1968)]. We use d rather than χ^{NL} in Eq. (3) because d is the nonlinear coefficient most frequently measured. The value used in Eq. (3) includes trigonometrical factors due to the point group of the particular crystal and the phase-matching angle (Midwinter and Warner, 1965).

Giordmaine (1962) and by Maker *et al.* (1962) who generated the second harmonic of ruby laser radiation in potassium dihydrogen phosphate crystals under phase-matched conditions.

In the next paragraphs the concept of phase matching in uniaxial crystals is reviewed, illustrating the differences between tangential and collinear phase matching. The way in which phase matching influences the field of view, spectral bandwidth and spectral tunability of an up-converter is also discussed.

(a) *Wave Propagation Vector Surfaces for Uniaxial Crystals.* The phase velocity of a wave in a crystal is characterized by its wave propagation vector $\mathbf{k} = (\omega n/c)\mathbf{i}$, where ω is the circular frequency, c/n the phase velocity of the wave in a crystal of refractive index n, and $\mathbf{i}$ is a unit vector perpendicular to the wavefront. In anisotropic crystals two waves with the same $\mathbf{i}$ but with orthogonal polarizations will, in general, experience different refractive indices and will therefore have different values of $\mathbf{k}$. The surfaces formed by the locus of the tips of $\mathbf{k}$ for all values of $\mathbf{i}$ are spherical for ordinary waves and ellipsoids of revolution for extraordinary waves. The two surfaces at a given frequency touch at the optic axis. As the frequency increases the $\mathbf{k}$ surfaces expand.

It may be shown (Midwinter and Warner, 1965) that phase matching for sum-frequency generation is only possible if n_s is the smaller of the two refractive indices at the sum frequency (extraordinary refractive index for negative uniaxial crystals); n_p and n_{ir} would normally be the larger of the alternatives at ω_p and ω_{ir} (i.e., ordinary refractive index for negative uniaxial crystals). If the difference between the ordinary and extraordinary indices is large enough, then it may be possible to achieve phase matching with n_{ir} (refractive index at lowest of interacting frequencies) as the smaller of its alternative values. Having established the vibration directions for the three waves in the crystal the value of the extraordinary indices are adjusted by altering the wave propagation directions until $\Delta \mathbf{k}$ is 0. Under phase-matched conditions therefore the vector triangle $\mathbf{k}_p + \mathbf{k}_{ir} - \mathbf{k}_s$ is closed as illustrated in Fig. 1 which represents sections of the $\mathbf{k}$ surfaces involved in sum-frequency up-conversion.

(b) *Tangential Phase Matching for Maximum Field of View.* Under the assumptions of plane wave theory the phase-mismatch vector $\Delta \mathbf{k}$ will lie parallel to the wave that is generated by the nonlinear interaction, i.e., the sum frequency in this case. We may therefore write the magnitude of the phase mismatch between the waves when the pump beam is inclined at θ_p to the optic axis and the infrared beam is at ψ_{ir} from the pump beam (inside the crystal) as

$$|\Delta \mathbf{k}| = k_s(\theta_p + \psi_s) - k_p(\theta_p)\cos\psi_s - k_{ir}(\theta_p + \psi_{ir})\cos(\psi_{ir} - \psi_s) \qquad (5)$$

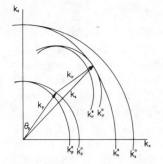

Fig. 1. k diagram illustrating tangential phase matching in a negative uniaxial crystal. Type I matching is illustrated with the infrared beam polarized as an ordinary wave.

where ψ_s is the angle between $\mathbf{k}_p$ and $\mathbf{k}_s$ and, for example, $k_s(\theta_p + \psi_s) \equiv |\mathbf{k}_s|$ when $\mathbf{k}_s$ lies at $\theta_p + \psi_s$ from the optic axis. For small values of ψ_{ir} we may write $\psi_s = k_{ir}(\theta_p)/k_s(\theta_p)$ and express $k(\theta_p + \psi)$ in terms of $k(\theta_p)$ and ψ using Taylor's theorem. This enables us to rewrite Eq. (5) in the following form where powers of ψ_{ir} greater than two have been neglected (Warner, 1969a):

$$\psi_{ir} = -\psi_0 \pm [\psi_0^2 - A(k_0 - |\Delta\mathbf{k}|)]^{1/2} \tag{6}$$

where

$$\psi_0 = \frac{[k_{ir}\, dk_s/d\theta - k_s\, dk_{ir}/d\theta]\, k_s}{(k_s k_p k_{ir} + k_{ir}^2\, d^2k_s/d\theta^2 - k_s^2\, d^2k_{ir}/d\theta^2)}$$

$$A = \frac{2k_s^2}{(k_s k_p k_{ir} + k_{ir}^2\, d^2k_s/d\theta^2 - k_s^2\, d^2k_{ir}/d\theta^2)}$$

and

$$k_0 = k_s - k_p - k_{ir}$$

Note that in the conditions on Eq. (6), k_s, k_p, k_{ir}, and their differentials with respect to θ have all been evaluated at the phase-matching angle θ_p; θ_p is the angle between the optic axis and $\mathbf{k}_p$, A and ψ_0 are slowly varying functions of both θ_p and frequency k_0, the collinear phase mismatch, determines the exact mode of phase matching, $k_0 = 0$ corresponds to exact phase matching with collinear beams, $k_0 = \psi_0^2/A$ to tangential phase matching when the sum frequency and infrared $\mathbf{k}$ surfaces are tangential, and $k_0 = (\psi_0^2/A) - 2.78$ gives a situation in which the center of the infrared field of view is a half-power saddle point.

The tangential phase-matching situation is illustrated in Fig. 1; this allows the greatest tolerance on infrared direction. Let the half-cone angle of infrared radiation for which $|\Delta\mathbf{k}| = 2\pi/L_c$ [corresponding to sinc $\frac{1}{2}(\Delta k/L_c) = 0$ in Eq. (3)] be $\Delta\psi_{ir}$. It can then be shown from Eq. (6) that

$$\Delta\psi_{ir}^2 = 2\pi A/L_c \tag{7}$$

Thus the solid field of view $\Delta\Omega$ of an up-converter varies inversely with the length of the mixing crystal ($\Delta\Omega = \pi\, \Delta\psi_{ir}^2$). It is possible to increase the field of view beyond the normal limit set by phase matching. The ability to do this would be of importance in image up-conversion where a large field of view might be a design requirement. Andrews (1969) demonstrated a method of increasing the field of view of an image up-converter looking at mono-chromatic infrared objects. His method is to alter repetitively the phase-matching angle (either mechanically as he did, or via the electrooptic effect). This produces a succession of phase-matched surafces of an annular nature, approximately centered on the tangentially phase-matched direction, the net result being a larger field of view seen on the average. The situation may be visualized with the aid of Fig. 2a. Altering θ_p causes the surface to move bodily up and down the $\bar{v}_{ir}$ axis. Therefore on the average a much wider field of view would be seen at wave number $\bar{v}_i$ than if the graphs were stationary. Andrews (1969) achieved an aperture of more than 20° by this technique when up-converting 1.06-μm radiation in KDP. Instead of scanning a single crystal one could arrange several crystals in tandem at slightly different phase-matching angles.

(c)   *Spectral Bandwidth and Tunability.*   An increase in infrared frequency causes both the sum and the infrared **k** surfaces to expand. Dispersion of the mixing crystal will usually make the surfaces expand at different rates and they will eventually separate by more than $2\pi/L_c$. The dispersion of the mixing crystal is therefore responsible for the narrow spectral bandwidth which is typical of up-conversion. Approximate expressions for the phase-matched spectral bandwidth may be obtained by considering the tangentially phase-matched infrared wave of an up-converter set to phase-match accurately the frequency ω_{ir} [i.e., $k_0 = \psi_0^2/A$ from Eq. (6)]. If the infrared frequency is

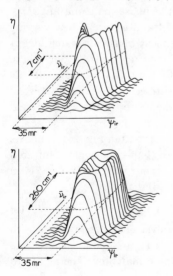

Fig. 2. η_{uc} as functions of infrared angle and frequency at two pump wavelengths for a LiNbO$_3$ up-converter operating in the 3.5-μm band. (a) With a ruby laser pump the phase-matching condition is not dispersion matched. The spectral bandwidth is approximately 7 cm^{-1} and the field of view inside the crystal is 35 mrad. (b) With a neodymium laser pump the phase-matching condition becomes dispersion matched. The spectral bandwidth is increased to about 260 cm^{-1}, while the field of view remains at about 35 mrad.

now allowed to increase to ω_{ir}', giving a new value of collinear phase mismatch $\mathbf{k}_0'$, then the phase mismatch at the center of the field of view ($\psi_{\text{ir}} = -\psi_0$) will be $\Delta\mathbf{k}'$ where, from Eq. (6),

$$\psi_0{}^2 - A(\mathbf{k}_0' - \Delta\mathbf{k}') = 0$$

where $\Delta\mathbf{k}' = \mathbf{k}_0' - \mathbf{k}_0$. If we put $\Delta\mathbf{k}' = 2\pi/L_c$ and use Taylor's series to expand $\mathbf{k}_0'$ in terms of $dk_0/d\omega$ and a frequency shift $\Delta\omega$ we find that

$$\Delta k = \frac{2\pi}{L_c} = \frac{dk_0}{d\omega}\Delta\omega + \frac{1}{2}\frac{d^2k_0}{d\omega^2}\Delta\omega^2 \tag{8}$$

Under normal conditions $dk_0/d\omega \, \Delta\omega \gg \frac{1}{2}(d^2k_0/d\omega^2)\,\Delta\omega^2$ and the phase-matched spectral bandwidth is given approximately by

$$\Delta\omega(\text{normal}) = \frac{2\pi}{L_c}\left[\frac{dk_s}{d\omega} - \frac{dk_{\text{ir}}}{d\omega}\right]^{-1} \tag{9}$$

Under special circumstances $dk_0/d\omega = 0$, in which case the spectral bandwidth would be given by the second differential as

$$\Delta\omega(\text{dispersion matched}) = \left(\frac{4\pi}{L_c}\right)^{1/2}\left\{\frac{d^2k_s}{d\omega^2} - \frac{d^2k_{\text{ir}}}{d\omega^2}\right\}^{-1/2} \tag{10}$$

Midwinter (1969) demonstrated that dispersion matching results in a much larger spectral bandwidth for the up-converter. He chose lithium niobate as the mixing crystal and achieved dispersion-matched phase matching at room temperature using a 1.064-μm laser pump and an infrared source at 3.5 μm. Other combinations of nonlinear materials and optical wavelengths that should give dispersion matched operation are given by Andrews (1970).

The differences between ordinary and dispersion-matched phase matching are illustrated in Fig. 2 where the conversion efficiency of a lithium niobate up-converter operating in the 3.5-μm band has been plotted as a function of the infrared angle to the pump beam and of the infrared wavenumber. With a ruby laser pump beam (Fig. 2a) the sum frequency is such that $dk_s/d\bar{v} \gg dk_{\text{ir}}/d\bar{v}$. Consequently as the infrared frequency is decreased, the $\mathbf{k}$ surfaces approach and then pass through each other, η increases to a single maximum which occurs when the two surfaces are tangential and then splits into two separating peaks as the surfaces interesct. The internal field of view and the spectral bandwidth at the half-power points are 35 mrad and 7 cm^{-1}, respectively. Graphs similar to Fig. 2a have been obtained experimentally by Klinger and Arams (1969b) for a proustite up-converter at 10.6 μm using a neodymium laser pump. The use of a neodymium laser in a LiNbO$_3$ up-converter gives the situation that $dk_s/d\bar{v} = dk_{\text{ir}}/d\bar{v}$ at an infrared wavelength of about 3.5 μm. For Fig. 2b the phase-matching angle is set so that at 3.5 μm there is a slight overlap of the **k** surfaces. As the infrared wavelength is increased toward 3.5 μm (frequency decreasing), the **k** surfaces approach and finally intersect; the two surfaces then move together with increasing wavelength. Further increases in infrared wavelength cause the infrared surface to contract more rapidly and the surfaces separate once again. The field of view remains at about 35 mrad but the spectral bandwidth has been increased by about 40 times to 260 cm^{-1}

B. Up-Conversion and Signal Detection

1. Introduction

The narrow, tunable, spectral bandwidth and the small field of view are attractive features of an up-converter system for detecting monochromatic infrared signals. These properties restrict the background radiation which is present in the absence of the signal beam to a very small level, making for a low noise detection system. In a conventional photoconductive or photovoltaic infrared detector such filtering would have to be done with cooled aperture stops and bandpass filters. With an up-converter the detection of the infrared beam is actually made in the visible spectrum with a photomultiplier or photodiode, neither of which need to be cooled; up-conversion therefore offers the possibility of room-temperature detection of infrared radiation with a low noise detector of rather small effective quantum efficiency. In this section we will examine the characteristics of up-conversion for the detection of weak infrared signals and identify what is to be gained over direct detection of the infrared signal using a photoconductive or photovoltaic detector.

2. Noise Performance

An important parameter of any infrared detector is its noise equivalent power (NEP); this is a measure of the amount of noise present in the detector electrical output when the infrared signal has been removed. The NEP is defined as the rms power of a 100% depth sine wave modulated infrared beam required to give an output equal to the noise level when measured in a 1-Hz bandwidth. The fundamental noise sources in an up-converter and photomultiplier detection system are:

(a) *parametric fluorescence* generated within the mixing crystal when photons at ω_p spontaneously split into photons at ω_{ir} and $\omega_p - \omega_{ir}$, a difference-frequency mixing process;

(b) the noise associated with the dark current of the photomultiplier; and

(c) sum-frequency radiation which is generated from that portion of the background radiation which is phase matched in infrared frequency and direction.

Other noise sources such as inadequate filtering of the pump-frequency radiation could be eliminated by suitable technology. It must be noted, however, that as the infrared wavelength gets longer it becomes increasingly difficult to filter the up-converted light from the intense pump beam. Takatsuji (1966) considers this to be the dominant noise source for far infrared detection by optical mixing.

Tang (1969) has shown that for sum-frequency up-conversion, where it is unusual for parametric fluorescence to be phase matched, the contribution to the noise by non-phase-matched generation of infrared photons which are up-converted is negligibly small. For difference frequency up-conversion (lower sideband mixing) the parametric fluorescence is always phase matched and large amounts of noise will be present in this case. Kleinman and Boyd (1969) discussed this difference between sum- and difference-frequency up-conversion in a comprehensive theoretical study of infrared detection by optical mixing.

For sum-frequency up-conversion therefore the photomultiplier dark current may contribute the most to the noise level. This situation has been achieved in several experiments (see Table II). Certain photomultipliers have very low dark noise currents and it is conceivable, although not yet practically realized, that background thermal radiation (within the phase-matched spectral bandwidth and field of view) would limit the noise performance. Since the spectral bandwidth and field of view of an up-converter are normally very small this limiting noise figure could be very small indeed.

Let us find an expression for the NEP of an up-converter–photomultiplier detection system in which the up-converter is tangentially phase matched but not dispersion matched. Such an up-converter might employ a ruby laser pump and a single crystal of proustite as the nonlinear material to detect radiation in the 10-μm band. The total noise current will be given by the sum of the mean square dark current fluctuations and the mean square fluctuations of the cathode current due to the up-converted background flux, i.e.,

$$\langle i_n^2 \rangle = 2e\,\Delta f(\langle i_d \rangle + \langle i_b \rangle) \tag{11}$$

Let the spectral radiance of the background be $B_\omega{}^b$ W/cm^2/sr/rad frequency. The up-converter field of view and spectral bandwidth will be given by Eqs. (7) and (9). Therefore $\langle i_b \rangle$ will be given by

$$\langle i_b \rangle = (B_\omega{}^b \Delta\omega\,\Delta\Omega/\hbar\omega_{ir})\eta e \tag{12}$$

Table II *Summary of Experiments on Up-Conversion of Infrared Radiation*

| Date | Signal beam | | Pump beam | | Crystal | Sum or difference mixing |
	λ (μm)	Power	λ (μm)	Power or energy		
1962	0.546	0.02 W	0.6943	10^3-W pulses	KDP	Sum
Oct. 1963	1.15	2.5 mW	0.6328	1.2-mW cw	ADP	Sum
Aug. 1966	0.9–2.3	—	—	—	KDP	—
Feb. 1967	1.7	$\sim 10^{-6}$ W	0.6943	2-MW/cm^2 Q-switched pulse	LiNbO$_3$	Sum
Feb. 1967	2.5–2.65	$\sim 10^{-6}$ W	0.6943	0.2-J pulses	LiNbO$_3$	Sum
Nov. 1967	3.39	2.75 mW	0.6328	4.43-mW cw	LiNbO$_3$	Sum
Mar. 1968	10.6	45 mW	0.6943	0.2-J pulses	Ag$_3$AsS$_3$	Sum
Sep. 1968	10.6	—	0.6328	1-mW cw	HgS	Difference
Jan. 1969	2–4	—	1.06	Pulsed	LiNbO$_3$	Sum
Mar. 1969	10.6	—	1.06	Pulsed	Ag$_3$AsS$_3$	Sum
Oct. 1969	10.6	—	1.06	cw	Ag$_3$AsS$_3$	Sum
Oct. 1969	10.6	0.1 W	1.06	1-W cw	Ag$_3$SbS$_3$	Sum
Sep. 1970	2.5–4.5	—	0.5145	3-W cw	LiNbO$_3$	Sum
May 1971	10.6	—	1.064	—	ZnGeP$_2$	Sum
Aug. 1971	6.5–12.5	—	1.064	1-kW peak 100-mW mean	Ag$_3$AsS$_3$	Sum

NEP (WHz$^{-1/2}$)	η_{uc}	Comments	Reference
—	—	—	Smith and Braslau (1962)
—	5.7×10^{-12}	—	Adams and Schoeffer (1963)
—	3×10^{-4}	Abstract of Paper 7c11	Johnson and Duardo (1966)
5×10^{-6}	10^{-2}	Unidentified laser-induced source responsible for large noise present	Midwinter and Warner (1967a)
10^{-7}	—	Infrared source is high-pressure xenon arc	Midwinter and Warner (1967b)
3×10^{-8}	5×10^{-7}	TEM$_{00}$ beams, noise due to inadequate filtering	Miller and Nordland (1967)
—	4.4×10^{-6}	—	Warner (1968a)
9×10^{-6}	4×10^{-10}	Dominant noise is fluctuations in pump power	Boyd et al. (1968)
—	—	Experimental check on dispersion matching [see Section II,A,2,(c)]	Midwinter (1969)
—	—	Cooled S1 photomultiplier Dominant noise was light leakage	Klinger and Arams (1969a)
—	—	Experimental check on tangential phase matching [see Section II,A,2,(6)]	Klinger and Arams (1969b)
8×10^{-7}	6×10^{-9}	PM dark current is dominant noise	Gandrud and Boyd (1969)
10^{-14}	10^{-4}	PM dark current is dominant noise. NEP smaller than obtainable with InSb detector	Smith and Mahr (1970)
—	—	This new material phase matches near to 90° from optic axis thereby offering better focusing potential and larger η_{uc} than other 10.6-μm materials. Good crystals not yet available	Boyd et al. (1971)
—	—	First reported detection of room-temperature blackbody radiation by up-conversion	Falk and Yarborough (1971)

where η is the overall quantum efficiency of the system as given by Eq. (4). The rms anode noise current leaving the photomultiplier is therefore

$$I_n = G[2e\,\Delta f(\langle i_{\mathrm{d}}\rangle + \langle i_{\mathrm{b}}\rangle)]^{1/2} \tag{13}$$

where G is the gain of the dynode chain in the phtomultiplier. The noise equivalent power NEP is defined as the rms optical power required to produce a rms current through a 1-Hz bandwidth filter which is equal to the rms noise current passing through that filter (Kruse $et\ al.$, 1962)

$$\mathrm{NEP} = I_n/\mathcal{R}\,\Delta f^{1/2}, \qquad \Delta f = 1\ \mathrm{Hz} \tag{14}$$

where $\mathcal{R}$ is the responsivity of detector (e.g., A/W). Now $\mathcal{R} = \eta Ge/\hbar\omega_{\mathrm{ir}}$, therefore we may express the NEP of the up-converter detection system as

$$\mathrm{NEP} = \left(\frac{\hbar\omega_{\mathrm{ir}}}{\eta e}\right)\left[2e\left(\langle i_{\mathrm{d}}\rangle + \frac{\eta e B_\omega{}^{\mathrm{b}}\,\Delta\omega\,\Delta\Omega}{\hbar\omega_{\mathrm{ir}}}\right)\right]^{1/2} \tag{15}$$

3. Figure of Merit for Signal Detection

A convenient figure of merit for comparing infrared detectors for a given application is the product of the signal-to-noise ratio for power and the electrical bandwidth required. Thus,

$$B = (\langle I_s{}^2\rangle/\langle I_n{}^2\rangle)\,\Delta f \tag{16}$$

A certain value of B is required to receive satisfactorily certain infrared signals. For example a TV channel with a bandwidth of 4 MHz and a signal-to-noise ratio of 53 dB needs a detector capable of operating with $B = 8 \times 10^{11}$. At the other end of the scale a detector for a spectrometer application may only need to operate with $B = 1$ ($S/N = 1$ and $\Delta f = 1$ Hz). The signal level, and therefore B, depends on the incident infrared flux. The better detector will be that which attains the required value of B with the smaller infrared power.

If we assume that the infrared signal radiation is 100% amplitude modulated, then the mean square signal current from the photomultiplier is

$$\langle I_s{}^2\rangle = \tfrac{1}{2}[G(e/\hbar\omega_{\mathrm{ir}})\eta P_{\mathrm{ir}}]^2 \tag{17}$$

where G represents the multiplier gain and η the effective up-converter quantum efficiency given by Eq. (4). The mean square noise current $\langle I^2{}_n\rangle$ is given by the sum of the mean square fluctuations of the dark current and of the currents due to the background radiation and to the signal radiation. The dark current fluctuations and background fluctuation are often expressed in terms of the noise equivalent power [(Kruse $et\ al.$, 1963; see also Eq. (15)].

It can be shown that

$$\langle I^2_n \rangle = (G^2 e^2 \eta^2 / \hbar \omega^2_{ir}) [\text{NEP}^2 + (4\hbar \omega_{ir}/\eta) P_{ir}] \tag{18}$$

and therefore

$$B = \frac{P^2_{ir}}{\text{NEP}^2 + (4\hbar \omega_{ir}/\eta) P_{ir}} \tag{19}$$

Graphs of B versus P_{ir} are plotted in Fig. 3 for up-converter detector systems and also for direct detection systems employing a photoconductor such as copper-doped germanium.

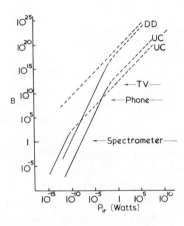

Fig. 3. The figure of merit $B = (S/N) \Delta f$ as a function of the available signal power for: DD, direct detection in a copper doped germanium photoconductor; UC, up-conversion ($\eta_{uc} = 9 \times 10^{-6}$, background noise neglected) with an S-20 photomultiplier; and UC′, up-conversion with a silicon avalanche photodiode. Also shown is the theoretical quantum limit for unit quantum efficiency. The dark noise regime is solid and the noise-in-signal regime is dashed. Also shown at appropriate levels of B are typical requirements for a spectrometer, a telephone, and a television channel.

At very high infrared powers there is sufficient signal for the shot noise fluctuations in the signal level to dominate all other noise sources. Equation (19) then reduces to $B = P_{ir}/4\hbar \omega_{ir}$ and this regime of operation is characterized by a slope of 1 in Fig. 3 (dashed portions of curves). The curves lie parallel to, and below, the quantum limit by an amount inversely proportional to the overall quantum efficiency. In this regime the better detector has the larger quantum efficiency. At very low infrared powers the signal-independent noise characterized by the NEP dominates; Eq. (19) reduces to $B = P^2_{ir}/\text{NEP}^2$ giving a steeper slope (of 2) in the graphs of Fig. 3 (solid portions of curves). The "knee" of the curves, indicating the region where the noise-in-signal is equal to the sum of the dark and background noises is situated at an infrared power proportional to the square of the NEP.

A study of Fig. 3 will reveal that it is the excellent noise characteristics of a photomultiplier and the effectiveness of the up-converter in filtering out much of the thermal background radiation which gives a region of superiority to the up-converter system in the low light level, low bandwidth quarter. The small quantum efficiency of an up-converter will prevent its efficient use as a fast infrared detector despite the small time constant of, say, an S-20 photocathode.

4. Summary of Experimental Achievements

Signal up-conversion has been experimentally studied by a number of researchers whose experiments are briefly summarized in Table II. A variety of infrared wavelengths from 1 to 10.6 μm have been up-converted into or near the visible spectrum. It has been demonstrated that, with cw pumps at least, it is possible to achieve a noise equivalent power that is set by the dark current of the photomultiplier. In other experiments the degree of optical filtering screening the pump frequency beam from the phtomultiplier was inadequate. In one experiment (Midwinter and Warner, 1967a) using a Q-switched ruby laser to pump a crystal of lithium niobate a large amount of anomalous noise was observed. This was found to be due to an unidentified source of infrared radiation. It was suggested that this spurious radiation most probably came from ruby laser-induced fluorescence, either in the lithium niobate crystal itself or in one of the filters that was exposed to the intense ruby laser beam. Optical parametric fluorescence (Tang, 1969) was not phase matched for this case. Pulsed operation of an up-converter whose noise equivalent power is limited by photomultiplier dark current has been demonstrated recently by Falk and Yarborough (1971). They used a continuously pumped, repetitively Q-switched laser, and a single crystal of proustite to up-convert thermal radiation emitted by a blackbody source at temperatures down to 27°C.

Detection of 10.6-μm radiation (the carbon dioxide laser wavelength) is of practical interest for optical communications. The curves of Kleinman and Boyd (1969) (see also Fig. 3) suggest that a large quantum efficiency is necessary. This is best achieved by optimumly focusing the pump and infrared beams into the crystal to give the largest effective pump power density, and therefore conversion efficiency, for a given laser power. If the dispersion of the refractive indices of the crystal permit phase matching without double refraction ($\theta_p = 90°$), then optimum focusing will produce much higher conversion efficiencies than if double refraction were present to limit the overall interaction volume (Boyd and Kleinman, 1968).

Most nonlinear optical crystals which have been studied for optical mixing involving a 10.6-μm beam have too large a birefringence to permit double-refraction-free phase matching. Boyd *et al.* (1971) recently reported phase-matched up-conversion of 10.6-μm radiation in zinc germanium phosphide (ZnGeP$_2$) in which the phase-matching angle is close to 90°, drastically reducing the limiting effects of double refraction. They state that the sum-mixing efficiency is 140 times superior to that of proustite (1.06-μm pump and 1-cm crystal length). This would be mostly due to the tighter focusing which is permitted in ZnGeP$_2$ (very little double refraction) since the nonlinear coefficients themselves are different only by a factor of about 5 (Kurtz, 1975). Much remains to be done to improve the crystal quality to the level enjoyed by proustite which can readily be obtained in large strain free pieces. Efficient signal up-conversion of 10.6-μm radiation therefore awaits the development of materials like ZnGeP$_2$ into good optical quality specimens. The situation is somewhat better at shorter infrared wavelengths where phase matching can be achieved in the xy plane ($\theta_p = 90°$) with materials such as LiNbO$_3$ which has received considerable attention from crystal growers. The up-converter reported by Smith and Mahr (1970) has a smaller noise equivalent power than other detectors that operate in the 2–4 μm band. Its quantum efficiency is, however, much smaller than, say, an InSb detector; so the up-converter will only be superior in the small bandwidth–small signal-to-noise ratio regime, such as infrared astronomy.

C. Up-Conversion and Image Detection

1. INTRODUCTION

Image information is transferred from an infrared beam to the sum frequency in up-conversion because of phase matching. Let us consider how an up-converter with a plane wave pump beam acts on the infrared radiation received from a distant object illuminated by a monochromatic source. Each plane wave Fourier component of the infrared wave fronts at the crystal will mix with the pump wave to produce a polarization wave along the direction of $\mathbf{k}_{ir} + \mathbf{k}_p$. An electromagnetic wave at the sum frequency will radiate in this direction with an intensity proportional to the infrared intensity multiplied by $\text{sinc}^2(\frac{1}{2}|\Delta\mathbf{k}|L_c)$. All plane wave components of the infrared wave which are incident on the crystal within its phase-matched field of view will therefore produce corresponding sum-frequency components with proportional intensities and directions. The net effect is that image information will be transferred to the sum frequency with an angular demagnification given (for a parallel-faced slab of mixing crystal) by the ratio of the infrared to sum frequencies.* The image up-converter, when acting on infrared radiation from distant objects, behaves in a very similar manner to a Galilean or lens-erecting telescope. It is an afocal system with a positive magnifying power of $M = \omega_{ir}/\omega_s$, and an aperture equal to the diameter of the pump laser beam (assuming that the latter is smaller than the clear aperture of the crystal). The field of view of the up-converter is set by the requirements of phase matching. It should be emphasized that the up-converter itself does not produce an image; that function must be carried out by some other optical component such as a convex lens or the human eye lens. The up-converter merely changes the frequency of the light from an object so that a visible image may be formed.

2. IMAGE RESOLUTION

The pump wave passing through a practical up-converter will not be a plane wave and the resulting sum-frequency wave front will not simply be an angularly demagnified version of the infrared wave; visible images will nevertheless be produced, but with degraded resolution and/or a shift in image position. The finite thickness of nonlinear crystal can also affect the attainable resolution. The variety of optical geometries for image up-conversion are considerable. The pump laser beam could be Gaussian or multimode, and either collimated through the crystal, focused into the crystal, or focused elsewhere so that spherical waves are used. The infrared beam could be passed through a telescope to alter the effective infrared field of view or it could even be imaged on to the crystal. The optimum position of the sum-frequency imaging component and the attainable resolution depends on these choices and several authors have discussed the possibilities using both geometrical optical theory and Fourier transform theory (Firester, 1969; Warner, 1969b; Andrews, 1970; Firester, 1970; Voronin et al., 1970, 1971; Chiou, 1971; Hulme and Warner, 1972). We will be content here to summarize the most important points that can be drawn from these studies.

(1) It is possible to choose a geometry in which the finite thickness of mixing crystal does not degrade the image resolution. The recipe is to locate the object and the pump beam focus in the same plane. The apparent location of

---

* It can be seen from Fig. 1 that the ratio of the angles that k_{ir} and k_s make with k_p is given by $\psi_s/\psi_{ir} = |k_{ir}|/|k_s|$ in the small angle approximation. Let us assume that the mixing crystal has plane parallel entrance and exit faces, and that the pump beam is sent into the crystal at or near normal incidence. The external angles between the infrared and pump beams and between the sum and pump beams will be given (for small angles) by $\alpha_{ir} = n_{ir}\psi_{ir}$ and $\alpha_s = n_s\psi_s$. Since $|k| = \omega n/c$ it follows that $\alpha_s/\alpha_{ir} = \omega_{ir}/\omega_s$.

the object at the sum frequency is then also in that plane irrespective of the distance between the object and the crystal. A particularly convenient geometry therefore would be to locate the object at infinity, using a plane wave pump beam and locating the sum-frequency image in the focal plane of a suitable convex lens. Consider the optical geometry depeicted in Fig. 4 which shows an infrared object of height H coplanar with the focus 0 of the pump beam. We consider the sum mixing that takes place in the incremental volume of crystal at C, distance z from 0 in the direction α from the axis. If ψ_{ir} is the angle between the pump wave normal and the infrared wave normal from the object point A, then the angle ψ_s between the pump and corresponding sum-frequency wave normal is given by $\psi_s = (\omega_{ir}/\omega_s)\psi_{ir}$. It may be shown by applying the sine rule to Fig. 4 that

$$\sin\psi_{ir}/H = \cos(\alpha-\psi_{ir})/z \qquad \text{and} \qquad \sin\psi_s/h = \cos(\alpha-\psi_s)/z \qquad (20)$$

In the small angle approximation Eq. (20) reduces to

$$h/H = (\omega_{ir}/\omega_s)[1 - \tfrac{1}{2}(\alpha-\psi_{ir})^2]/[1 - \tfrac{1}{2}(\alpha-\psi_s)^2] \qquad (21)$$

and for small values of α, h/H is independent of z and α, OA' therefore represents a true sum-frequency image which is not dependent on the crystal location or size.

(2)   The ultimate resolution will be set by diffraction of the infrared through the effective aperture of the crystal, i.e., the area that is simultaneously illuminated by both the pump and infrared beams.

(3)   Diffraction-limited resolution will only be obtained if both of the following points are observed:

(a)   the pump beam must be Gaussian (TEM$_{00}$);

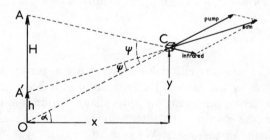

Fig. 4. Ray diagram to illustrate the insensitivity of image resolution to the location of mixing crystal when the pump source is in the object plane. The origin of coordinates coincides with the point source of pump radiation at 0. An elemental volume of mixing crystal is considered at C. Infrared radiation from an object point A is up-converted to produce a sum-frequency ray along $A'C$. It is shown in the text that in the paraxial limit all sum rays pass through A' whatever values of x and y are chosen. Therefore A' is the sum-frequency image point of A irrespective of crystal size or position.

(b)   the optical geometry must be such that the pump beam waist is located in the object plane and the crystal is in the far field of the pump beam.

For the special case of a distant object (effectively at infinity) an alternative geometry would be to center the pump beam waist on the crystal, arranging the confocal parameter to be much greater than the crystal length. These alternatives are illustrated in Fig. 5.

Experiments by Firester (1970) and by Weller and Andrews (1970) confirmed that image resolution is improved by adopting one of the geometries illustrated in Fig. 5. Both groups were using potassium dihydrogen phosphate crystals pumped with neodymium laser radiation at 1.06 μm to up-convert object radiation at 1.06 or 1.15 μm.

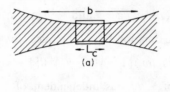

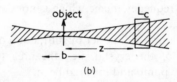

Fig. 5. Optical geometries necessary to achieve diffraction-limited resolution with a TEM_{00} pump beam. With the object at infinity [part (a)] the pump laser is focused at the mixing crystal with a confocal parameter much greater than the crystal length. Otherwise the laser is focused in the plane of the object [part (b)] which must be at a much greater distance from the mixing crystal than the confocal parameter.

3. CHROMATIC EFFECTS

We have seen that the effective transverse magnifying power of an up-converter is given by the ratio of the infrared to sum frequencies. A change in infrared frequency, producing an equal change in the sum frequency will therefore alter the up-converter's magnifying power and give rise to a chromatic blurring of the image. We may adopt the Rayleigh criterion for resolution to specify what spread in infrared frequencies can be tolerated before image blurring is excessive.

Let us assume that the infrared radiation from a point on the object (assumed to be at infinity) is up-converted into a cone of half-angle Δ. According to the Rayleigh criterion we can just discern the change in sum-frequency direction with angle if the center of the new cone of sum-frequency directions lies on the periphery of the old sum-frequency cone. We therefore write

$$\Delta = \psi_s(\omega_s + \delta\omega) - \psi_s(\omega) = \psi_{ir}\left[\frac{\omega_{ir} + \delta\omega}{\omega_s + \delta\omega} - \frac{\omega_{ir}}{\omega_s}\right] \qquad (22)$$

Provided that $\Delta\omega/\omega_s \ll 1$, we may write $(\omega_s + \Delta\omega)^{-1}$ as $(1/\omega_s)(1 - \Delta\omega)$; this reduces Eq. (22) to

$$\delta\omega = \omega_s^2 \Delta/\psi_{ir}\omega_p \qquad (23)$$

If $\delta\omega$ given by Eq. (23) is greater than the phase-matched spectral bandwidth [Eqs. (9) or (10)], then the up-converter bandwidth ensures that chromatic aberrations do not seriously degrade image resolution. This situation normally applies in image up-conversion unless dispersion matching is achieved [see Section II,A,2,(c)] in which case chromatic effects would cause serious blurring.

The chromatic blurring introduced by phase-matched up-conversion of wide-band infrared radiation can be corrected by the use of a compensating element either in the infrared beam or in the sum-frequency beam. Figure 6a shows a **k** matching diagram for collinear waves at ω_{ir} and ω_{ir}' being up-converted to produce diverging waves at ω_s and ω_s', thereby introducing chromatic aberrations. Figure 6b shows that if we employ an infrared element

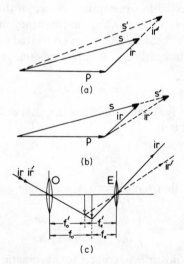

Fig. 6. Ray diagrams illustrating the cause and a possible cure with an infrared telescope of chromatic aberration in broadband up-conversion. Diagram (a) shows how an infrared ray containing two frequencies is up-converted into diverging sum-frequency rays. Correction could be achieved by spreading the infrared directions according to frequency as in diagram (b). A telescope with appropriately dispersive elements [diagram (c)] could, in principle, provide the required spread.

which alters the infrared direction according to frequency such that ω_s and ω_s' are made collinear, then the chromatic aberration will be well corrected. Figure 6c shows that an astronomical telescope of magnifying power m could be designed to give the required correction. The lens materials would have to be chosen so that the dependence of focal lengths on frequency is as indicated in the diagram. The frequency dependence of the telescope would be such that $d/d\omega(Mm) = 0$ where M is the up-converter magnifying power $(M = \omega_{ir}/\omega_s)$. We therefore conclude that chromatic correction would be achieved in an astronomical telescope if $dm/d\omega = -m\omega_p/\omega_{ir}\omega_s$. Now $m = f_o/f_e$, where f_o and f_e are the focal lengths of the objective and eyepiece lenses of the telescope (Smith, 1966). Since the lenses are separated by $f_o + f_e$ it can be shown that the required dispersions of the objective and eyepiece lenses are

$$\frac{1}{f_e}\frac{df_e}{d\omega} = +\frac{f_o}{f_e+f_o} \cdot \frac{\omega_p}{\omega_{ir}\omega_s} \tag{24}$$

$$\frac{1}{f_o}\frac{df_o}{d\omega} = -\frac{f_e}{f_e+f_o} \cdot \frac{\omega_p}{\omega_{ir}\omega_s} \tag{25}$$

Such dispersions could in principle be obtained with convex–concave doublets with one highly dispersive element and one very low dispersive element.

The chromatic aberrations of a dispersion-matched up-converter might more simply be corrected by appropriate choice of the dispersion character-istics of the lens which produces the sum-frequency image. Assuming that a lens of focal length f is used for image up-converted radiation from a distant object we write the image height h corresponding to an infrared direction ψ_{ir} as

$$h = \psi_{ir} Mf \tag{26}$$

Chromatic aberration is absent if $dh/d\omega = 0$, i.e., if $d(Mf)/d\omega = 0$. This leads us to the requirement that

$$\frac{1}{f}\frac{df}{d\omega} = -\frac{\omega_p}{\omega_{ir}\omega_s} \tag{27}$$

It may be shown that the dispersion of simple glass convex thin lens is given by

$$\frac{1}{f}\frac{df}{d\omega} = -\frac{1}{(n-1)}\frac{dn}{d\omega} \tag{28}$$

It would therefore be possible to correct for chromatic aberration if

$$dn/d\omega = (n-1)\omega_p/\omega_s\omega_{ir} \tag{29}$$

The dispersion of the lens material given by Eq. (29) has its smallest value when $\omega_{ir} = \omega_p$ (those cases in which $\omega_{ir} > \omega_p$ do not represent practical up-converter situations) but even this value of $(n_s-1)/2\omega_p$ is at least two orders of magnitude greater than the dispersion of optical glasses in the visible. It is clear therefore that the chromatic blurring caused by the up-converter cannot be corrected by any single-element lens.

A lens doublet gives us more choice in dispersive power. The power $1/f$ of a compound thin lens is given by the sum of the powers of its components, $1/f_1$ and $1/f_2$. The dispersive power of a thin lens doublet is therefore

$$\frac{1}{f}\frac{df}{d\omega} = -\left[\frac{f}{f_1(n_1-1)}\frac{dn_1}{d\omega} + \frac{f}{f_2(n_2-1)}\frac{dn_2}{d\omega}\right] \tag{30}$$

If we make the first lens convex using a highly dispersive material and the second element concave, made out of a material with small dispersion, then we can make a compound lens which is effectively convex (i.e., will produce a real, inverted image). The correct ratio of focal lengths of the lens elements may be obtained by equating Eqs. (27) and (30):

$$\frac{f_1}{f_2} = \frac{\omega_{ir}\,\omega_s}{\omega_p(n_1-1)}\frac{dn_1}{d\omega} - 1 \tag{31}$$

Let us take as a numerical example the lens required to correct chromatic aberrations that would be serious in a dispersion-matched lithium niobate up-converter as described by Midwinter (1969) [see also Section II, A, 2, (c)]. The wavelength of the sum-frequency radiation resulting from up-converting 3.5-μm radiation with a 1.06-μm laser is 0.814 μm. Equation (31) is satisfied if a lens doublet is made with a convex arsenic trisulphide lens of focal length 2.45 cm and a concave glass lens of -2.58 cm. The focal length of the combination would be $+50$ cm. In making these calculations we have assumed that the refractive index and dispersion of arsenic trisulphide at 0.814 μm is $n = 2.516$ and $dn/d\bar{v} = 3 \times 10^{-3}$ cm^{-1}, respectively, and that glass has a low dispersion and a refractive index of 1.5.

4. ACHIEVEMENTS IN MONOCHROMATIC IMAGE UP-CONVERSION

The first experimental study in image up-conversion was carried out by Midwinter (1968a) who converted image information at 1.6 μm to the visible by mixing it with highly collimated ruby laser radiation in lithium niobate. He used a multimode laser and achieved only 50 lines resolution in the up-converters field of view. Other experiments, concerned with improving the resolution, extending the infrared wavelength used and demonstrating the feasibility of holography by up-conversion, are summarized very briefly in Table III. We shall consider in a little more detail two achievements which stimulate the most interest for possible development: holographic up-conversion and imaging of objects illuminated at 10.6 μm.

Table III

Summary of Experiments on Image Up-Conversion

Wavelength (μm)			Crystal	Resolution	Comment	Reference
λ_{ir}	λ_p	λ_s				
1.7	0.694	0.493	$LiNbO_3$	50 lines in 60 mrad	First practical demonstration of image up-conversion (resolution grid-pulsed ruby laser)	Midwinter (1968a)
1.15	1.06		KH_2PO_4	70 lines/in.	cW up-conversion of grid pattern. Image reconstructed using data from a scanned photomultiplier	Gampel and Johnson (1968)
10.6	0.6943	0.6516	Ag_3AsS_3	15 mrad according to Rayleigh criterion	First demonstration of image up-conversion from 10.6 μm. Two point sources of variable angular separation formed the object	Warner (1968b)
1.06	0.6943		KH_2PO_4	Not measured	Demonstration of increasing field of view of up-converter by mechanically scanning the crystal orientation	Andrews (1969)
1.06	1.06	0.53	KH_2PO_4 (30 mm)	Not quoted	Demonstration of holographic image up-conversion. Object and reference beams at 1.06 μm up-converted to green. Real and virtual images reconstructed from green hologram plate. The polarizations of λ_p and λ_{ir} were orthogonal so that phase matching for second harmonic generation was not achieved	Voronin et al. (1969)

1.06	0.53	KH_2PO_4	500 line pairs/in. of object, increased to 750 line pairs/in. under optimum geometry	Demonstrated that the sum-frequency image position is a function of object position for a well collimated pump. Best resolution obtained with object at infinity	Firester (1970)
1.06	0.53	KH_2PO_4	150 × 150 elements within field of view	TEM$_{00}$ Q-switched laser. Object located close to crystal thereby reducing resolution well below diffraction limit	Voronin et al. (1970)
10.6	0.6516	Ag_3AsS_3	25 × 25 elements within field of view	50 pps ruby laser giving "real time" viewing of 10.6 μm illuminated objects (see Fig. 7 and text)	Warner (1970)
1.15, 0.633	1.06	KH_2PO_4	Within factor of two of diffraction limit	Experiments to check location of sum frequency image [the thin lens equation (Firester, 1969)] and resolution. The latter was found to be the best when the object and pump were at infinity	Weller and Andrews (1970)
CO$_2$ laser	0.6328	Ag_3AsS_3	Not quoted	Brief report of image up-conversion of interferometer rings of various laser transitions—presumably phase-matched for difference frequency up-conversion	Riccius et al. (1971)
10.6	1.06	Ag_3AsS_3	3-mrad resolution in 12° field of view	Near diffraction-limited resolution measured using cw TEM$_{00}$ Nd:YAG laser as pump	Chiou and Pace (1972)
10.6	0.6943	$AgAsS_3$ (10 mm)	100% modulation depth bar chart reduced to 25% at 20 mrad/cycle	Multimode pulsed laser-limited resolution	Lucy (1972)

Holograms of infrared illuminated objects may be produced by up-converting both the signal and a reference infrared beam to the visible to form a hologram in the usual (photographic) way with the up-converted signal and reference beams.

The feasibility of this idea has been demonstrated by Voronin *et al.* (1969) who up-converted signal and reference beams at 1.06 μm in a KDP crystal to record a hologram at 0.53 μm. Real and virtual sum-frequency images were subsequently reconstructed by reilluminating the hologram plate with 0.53-μm radiation.

The successful photographic recording of holograms with sum-frequency beams originating from carbon dioxide laser illumination would be of special interest in optical metrology where the wavelength of the CO_2 laser is more convenient than visible wavelengths. As well as the convenience of photographic recording, the up-converter, by frequency changing before detection, solves the aberration problems that are introduced in holography by recording at one wavelength and reconstructing at a much shorter wavelength. (The sum wavelength resulting from mixing 10.6 with 0.6943 μm (ruby laser) is 0.652 μm—conveniently close to the helium–neon laser at 0.6328 μm.)

The author knows of no serious study of the attainable resolution in holographic up-conversion. It would appear that the fundamental limit would be set by diffraction of infrared waves through the pumped aperture of mixing crystal. Since this has to be of moderate size due to crystal growing difficulties it seems that only moderate resolution will be possible.

The carbon dioxide laser is an attractive source of infrared illumination. It is a very efficient laser and can be made to operate either cw or pulsed, delivering high infrared powers at a wavelength (10.6 μm) for which the atmosphere is exceptionally transparent. Image up-conversion of objects illuminated at 10.6 μm is therefore of special interest.

There are few nonlinear optical crystals which have a sufficiently wide transmission band and sufficient birefringence to permit phase-matched up-conversion from 10.6 μm to the visible. Proustite (Ag_3AsS_3), pyragyrite (Ag_3SbS_3), cinnabar (HgS), silver gallium sulfide ($AgGaS_2$), and zinc germanium phosphide $AgGeP_2$ are examples. Attempts to grow synthetic crystals have been most successful in the case of proustite (Bardsley and Jones, 1968) where crystals of good optical quality 3 cm in diameter by 10 cm in length have been grown. Large aperture crystals are required for good image resolution and, so far, proustite is the only material that has been used for image up-conversion of 10.6-μm illuminated objects.

Figure 7 shows the sort of image resolution that has been obtained by the author (Warner, 1969c) in a laboratory demonstration of up-conversion of 10.6-μm radiation from a distant object using a ruby laser pumped crystal of proustite. The ruby laser, operating at 50 pulses per second was multitransverse mode with a divergence of 1.2 mrad when collimated to an 8-mm diameter

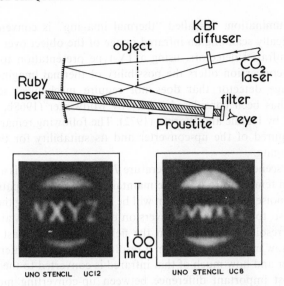

Fig. 7. Photographs of up-converted images of plastic lettering stencils illuminated at 10.6 μm. Infrared radiation scattered from the object was concentrated into a crystal of proustite where it was phase matched for sum-frequency mixing with a ruby laser beam. The up-converted wavelength was 0.652 μm.

beam. A cw carbon dioxide laser beam was scattered by a roughened disk of potassium bromide. This provided an extended source of 10.6-μm radiation against which objects, including the plastic lettering stencils shown in Fig. 7, could be viewed in silhouette. A concave mirror served the double function of condensing the infrared through the proustite crystal and, if the object were placed in the focal plane of the mirror, putting the object at optical infinity. Sum-frequency radiation was filtered from the pump light and observed with a relaxed eye or recorded with a camera focused at infinity.

Notice that the bright disk in the photographs of Fig. 7 represents the field of view set by phase matching, other apertures in the system were larger than this. The resolution was limited by the multimode divergence of the laser beam and would be improved if a TEM_{∞} mode laser beam were used. Chiou and Pace (1972) recently reported near diffraction-limited operation of a neodymium laser pumped proustite up-converter operating at 10.6 μm. A TEM_{00} mode laser was used and a resolution of 3 mrad and field of view of 12° were obtained.

5. POSSIBILITIES FOR THERMAL IMAGING BY UP-CONVERSION

There is considerable interest in detecting and imaging objects by virtue of their own thermal emission rather than by reflection of either sunlight or artificial illumination. So-called "thermal imaging" is conventionally done by mechanically scanning an infrared image of the object over a linear array of cooled infrared detectors, using a TV-type presentation to build up the image. Up-conversion offers the possibility of thermal imaging with an uncooled image detector that does not require mechanical scanning. This possibility has been considered briefly by Midwinter (1968b, 1969) and in more detail by Hulme and Warner (1972). The following remarks summarize what is required of the up-converter and its suitability for the purpose of thermal imaging.

Given a scene at ambient temperature T the up-converter is asked to detect difference in temperature ΔT while maintaining a specified spatial resolution. Clearly the noise level of the system will be a factor governing the temperature resolution, so too will be the conversion efficiency, the crystal size, and the number of resolvable positions in the field of view. The last three factors determine how much infrared from a resolution element is intercepted by the up-converter and how much of that infrared is converted to the visible.

The most important difference between up-converting monochromatic images and up-converting thermal images which are essentially broad band (e.g., 8–13 μm atmospheric window) is that phase-matching requirements, which impose strict limitations on the field of view in the monochromatic case, do not in general restrict the field of view of an up-converter when looking at thermal radiation. This is because the up-converter will phase match different infrared frequencies in different infrared directions. This is illustrated in Fig. 2a. The total infrared frequency spread will depend on the field of view chosen. This must not be allowed to exceed the bounds of the atmospheric window used. Notice from Fig. 2b, however, that phase matching will restrict the field of view if dispersion-matched phase matching is achieved. It may be shown (Hulme and Warner, 1972) that the phase-matched spectral bandwidth is almost independent of infrared direction although the center of the band might phase match at 10 μm at the center and at 12.5 μm at the periphery of a 1-rad field of view. The maximum amount of infrared flux is therefore collected from the thermal scene by using an infrared telescope to match the required field of view in the object plane to the maximum possible field at the crystal.

The ultimate temperature resolution of a thermal image up-converter would be achieved when the system were "noise-in-signal" limited (see Section II, B, 2). Let us assume that an image intensifier is employed to detect the sum-frequency image. The temperature resolution is given by equating the change in photoelectron flux from one resolution element due to a change in object temperature ΔT to the shot noise fluctuation of the photoelectron flux arising from the infrared radiation recieved from one resolution element at ambient temperature T. This leads to the following equation for the ultimate temperature resolution (Hulme and Warner, 1972);

$$\Delta T^2 = \frac{h_c \, v_{ir} \, N_v}{\eta_{int} \, \eta'_{uc} \, P_p \, \tau \, \Delta v_{ir} \, \mu^2 L_c (dN_v/dT)^2 \pi \, \Delta \alpha^2} \tag{32}$$

where v_{ir} is the infrared wave number (reciprocal wavelength cm^{-1}), Δv_{ir} the phase-matched spectral bandwidth for a 1-cm-long crystal [see Eqs. (9) or (10) which give $\Delta \omega_{ir} = 2\pi c \, \Delta v$], N_v the spectral radiance of scene [W sr^{-1} cm^{-2} (cm^{-1})$^{-1}$], η_{int} the intensifier photocathode quantum efficiency, η'_{uc} the conversion efficiency of 1-cm cube of crystal with 1 W of pump power [see Eq. (3) and Table I], τ the integration time (time allowed between "frames"), μ the magnifying power of infrared telescope preceding up-converter, and $\Delta \alpha$ the half-cone angle of one resolution element in the infrared field of view. If the field of view of the system has to contain p resolution elements across its diameter, then the half-cone angle of radiation entering the mixing crystal $\phi = p\mu \, \Delta \alpha$. Equation (32) may therefore be rewritten as

$$\Delta T = pA/(P_p \tau)^{1/2} L_c^{1/2} \phi \tag{33}$$

where for convenience we write

$$A^2 = h_c \, v_{IR} \, N_v / \eta_{int} \, \eta'_{uc} \, \Delta v_{IR} (dN_v/dT)^2 \tag{34}$$

The ratio $N_v/(dN_v/dT)^2$ has a minimum value, corresponding to maximum

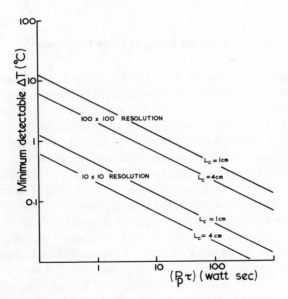

temperature sensitivity, at $\nu_{IR} = 1000$ cm^{-1} ($\lambda_{ir} = 10$ μm) for a blackbody source at 300°K. Figure 8 shows graphs of ΔT plotted against $P_p \tau$ according to Eq. (33) where $\phi = 1$ rad and A has been evaluated for a ruby laser pumped proustite up-converter followed by an S-20 image intensifier (e.g., EMI type 9694).

We therefore conclude that temperature resolutions of the order of a degree or so should be attainable while maintaining a spatial resolution of about 100×100 resolution elements. A powerful ruby laser system would be required and filtering all but the sum-frequency radiation from the intensifier would be a problem. It is important to note that the threshold contrast ratio of the eye of an observer would be greater than the contrast ratio corresponding to the threshold temperature increment of a degree or so. Some sort of contrast enhancement must therefore be applied to the image before it is presented to the observer. Such enhancement is inherent in the pyroelectric vidicon tube (Putley et al., 1972) since the pyroelectric effect is only produced by temperature variations. This tube, operating at room temperature, detects thermal images by the use of a pyroelectric material as the cathode of a vidicon tube. As in image up-conversion, mechanical scanning of the image is not required. The rapid development of pyroelectric image tubes weakens the case for developing up-converters for thermal imaging since the latter are inherently more complicated and apparently offer only marginal performance at great effort (in terms of laser energy required).

D. Up-Conversion to Extend the Range of Laser Frequencies

One might have deduced from the preceding paragraphs that optical up-conversion is only suited to changing infrared frequencies prior to detection in a more sensitive wave band. In fact up-conversion has been exploited experimentally to produce tunable laser like beams at the visible end of the spectrum just as difference-frequency generation has provided a method of generating frequency tunable sources in the infrared spectrum.

Experiments by Carmen et al. (1967) and by Martin and Thomas (1967) showed that a continuously tunable pulsed visible source of narrow linewidth (~ 8 cm^{-1}) with power levels in the kilowatt range could be derived from a neodymium : glass laser. The technique used sum-frequency generation in either KDP (Carmen et al., 1967) or lithium niobate (Martin and Thomas, 1967). One of the frequencies was that of the neodymium : glass laser while the other was selected by phase matching from an infrared continuum produced by self-focusing and small scale trapping in carbon disulfide or other similar Raman active liquids. The laser beam intensity in the trapped filaments was so great that efficient iterative stimulated Brillouin, Raman, and Rayleigh wing scattering took place causing a redistribution of the laser energy mainly toward lower frequencies.

The tunability of the sum frequency stems from the phase matching. Angle tuning the KDP crystal resulted in a tuning range from 4820 to 5790 Å. Carmen *et al.* found considerable variation in sum-frequency power over the tuning range, from 10 kW at 5300 Å to 0.1 kW at 5790 Å and from 0.1 kW just below 5300 Å to about 1 W at 4820 Å. Martin and Thomas had a wider tuning range in lithium niobate, from 0.5 to 0.8 μm, achieved by altering the refractive indices by temperature tuning the crystal (LiNbO$_3$, being a ferro-electric material, has temperature-sensitive refractive indices). A considerable variation of sum-frequency power over the tuning range was found in this case also.

A different method of using up-conversion to extend the range of tunable laserlike wavelengths in the visible spectrum has been demonstrated by Campillo and Tang (1971). They used the output of a doubly resonant lithium iodate parametric oscillator to provide the infrared beam for a lithium iodate up-converter which was pumped with the same ruby laser beam as the para-metric oscillator. In this case the infrared beam was not a continuum and both oscillator and up-converter crystals had to be correctly set for phase matching the following frequency changes:

(1) oscillator: $\omega_{\text{Laser}} = \omega_{\text{ir}} + \omega_{\text{idler}}$,

(2) up-converter: $\omega_{\text{Laser}} + \omega_{\text{ir}} = \omega_{\text{sum}}$.

Continuously tunable sum-frequency outputs in the range 0.42–0.51 μm were obtained at substantial power levels. In their first experiments the sum-frequency powers varied from 0.5 to 10 kW. This was thought to be due to variations in oscillator mirror reflectivity (and consequently in oscillator output power) with wavelength. A better design of oscillator mirrors should yield a more uniform sum-frequency output. Campillo and Tang predicted that the coupled parametric oscillator up-converter system could be capable of producing 100-kW output powers in the 0.4 to 0.5 μm range using a 1-MW ruby laser pump and a pair of lithium iodate crystals (oscillator and up-converter) within the oscillator resonator. The mirrors forming this resonator would then be designed to have a high reflectivity at the oscillator wavelengths 1–2 μm, but sufficiently low reflectivity in the 0.4 to 0.5 μm range to couple out an optimum amount of power.

III. DIFFERENCE FREQUENCY GENERATION

Difference frequency generation is a three-wave nonlinear optical mixing process with a different emphasis from up-conversion, namely one of gener-ating radiation rather than detecting it. It would appear to be complementary to the parametric oscillator which has sought to provide tunable infrared radiation at near infrared wavelengths (to 2.56 μm). It is expected (Hanna *et al.*, 1971) that difference frequency generation between a ruby laser and a tunable dye laser could give narrow-band infrared radiation that would be tunable over the wide band from 2.5 to 12.5 μm. This is a wave band for which there are no convenient laser frequencies to pump a parametric oscillator (Chapter 9).

In this section we will very briefly outline the essentials of the theory of optical difference frequency generation and go on to review experimental achievements in the field. We will outline the theory in terms of a plane wave analysis in which the effects of double refraction and diffraction are ignored. Boyd and Kleinman (1968) showed that, under certain conditions, difference-frequency mixing of focused Gaussian laser beams can be described by their theory for second harmonic generation using Gaussian beams; the plane wave theory is, however, adequate to illustrate the characteristic features of the mixing.

A. Theory

As in the case of sum-frequency up-conversion we are concerned with finding a solution of the coupled differential equations which describe the interplay of field amplitudes of the three interacting waves subject to the boundary conditions appropriate to difference frequency generation. These are (1) that the intensity of the generated frequency is much less than either of the other frequencies, (2) the intensities of the generating beams do not vary appreciably throughout the crystal (i.e., small conversion efficiency), and (3) that the energy ($\equiv$ frequency) and momentum ($\equiv$ phase matching) conservation equations apply (i.e., $\omega_1 = \omega_3 - \omega_2$ and $\mathbf{k}_1 = \mathbf{k}_3 - \mathbf{k}_2$). The difference frequency field amplitude then grows linearly throughout the crystal length and emerges as

$$\mathcal{E}_1 = (4\pi\omega_1{}^2/k_1 c^2)\, d\mathcal{E}_3 \mathcal{E}_2 L_c \tag{35}$$

The intensity of a wave in a dielectric of refractive index n is related to the field strength $\mathcal{E}$ (in electrostatic units) by the equation

$$I = 10^{-7} c\mathcal{E}^2 n/8\pi \quad \text{W/cm}^2 \tag{36}$$

We may use Eq. (36) to obtain the following expression for the intensity of the difference frequency wave inside the mixing crystal:

$$I_1 = \frac{128\pi^3 \omega_1{}^2 d^2 L_c{}^2}{10^{-7} c^3 n_1 n_2 n_3} I_2 I_3 \tag{37}$$

Equation (37) is only valid if phase matching for difference frequency generation has been satisfied. An extra factor of $\operatorname{sinc}^2(\tfrac{1}{2}|\Delta\mathbf{k}| L_c)$ where $|\Delta\mathbf{k}| = k_3 - k_2 - k_1$, should be included if there is any phase-mismatch $\Delta\mathbf{k}$.

The most probable experimental situation is one in which visible or near infrared laser frequencies, preferably in a Gaussian (TEM$_{00}$) mode, are mixed to produce an infrared difference frequency. Following the procedure of Boyd and Ashkin (1966) one can modify Eq. (37) to be applicable to the mixing of Gaussian beams where effects due to double refraction and diffraction have been ignored. If we also include the transmission coefficients T of the crystal faces to relate the powers available outside the crystal, we find (Hanna et al., 1971) that

$$P_1 = \frac{2^4 w_1{}^2 d^2 \times 9 \times 10^9 P_3 P_2 L_c{}^2 T_1 T_2 T_3}{n_1 n_2 n_3 c^3 (w_2{}^2 + w_3{}^2)} e^{-\alpha l/2} \quad \text{(W)} \qquad (38)$$

The last factor in this expression allows for any absorption of the crystal at the difference frequency, and w_2 and w_3 are the spot radii of the incident Gaussian beams.

B. Experimental Achievements

Early work on the generation of medium infrared radiation by difference frequency mixing was restricted by the lack of phase-matchable crystals to producing fairly weak (300 μW) beams at a few fixed wavelengths around 10 μm. Martin and Thomas (1966) used a neodymium : glass laser to excite Raman emission from several liquids. Difference-frequency radiation was generated from the laser beam and the Raman-shifted emissions in single crystals of CdS and CdSe. Neither of these crystals had sufficient birefringence to phase match and consequently the generation efficiency was rather poor (10 MW of neodymium radiation to 300 μW of 10-μm radiation).

More recent experiments have used tunable dye laser radiation and a ruby laser beam as the two beams to beat in crystals that phase match. The dye laser was pumped by a part of the ruby laser beam and so the whole scheme for providing tunable infrared consists of the ruby laser, the dye laser, and a phase-matchable nonlinear crystal. Dewey and Hocker (1971) used a concentrated DTTC Iodide/DMSO dye solution in a transversely pumped dye laser which was tunable from 0.84 to 0.89 μm with the aid of a diffraction grating. This output was mixed with the ruby laser beam at 0.694 μm in a crystal of LiNbO$_3$. Simultaneous tuning of the dye laser diffraction grating and the orientation of the crystal for phase matching resulted in 6-kW pulses of infrared radiation in the 3 to 4 μm band from about 4MW of ruby laser power.

LiNbO$_3$ is opaque beyond 4 μm and is therefore unsuitable for generating infrared radiation in the 10-μm band. Hanna et al. (1971) have used proustite as the mixing crystal to generate radiation around 5 and 11 μm. They too used a ruby laser and a ruby laser pumped dye laser. An M cryptocyanine dye was used to provide radiation tunable from 0.734 to 0.743 μm which were the limits of the free spectral range of the Fabrey–Perot resonator used to provide the tuning. TEM_{00} mode beams were used and the measured powers agreed well with the predictions of Eq. (38); 100 mW of radiation at 10.6 μm was generated from 45 kW at 0.6943 μm and 500 W at 0.743 μm.

Good agreement between experiment and theory led Hanna et al. (1971) to consider what infrared powers might be generated using larger crystals and more powerful lasers. One possibility was to stay with ruby lasers and proustite as the mixing crystal. A TEM_{00} mode ruby laser of 0.5 MW and a ruby laser-pumped dye laser of 0.5 MW power in TEM_{00} mode bandwidth 0.02 cm^{-1} could produce, in a 10-mm cube of proustite, tunable infrared power ranging from 5 kW to 2.5 μm to 150 W at 12.5 μm with a bandwidth of the order of 0.02 cm^{-1}. Several changes of dye solution would be necessary to cover the whole tuning range.

A more attractive solution, but one which awaits the development of suitable dye lasers and mixing crystals of adequate quality would be to use two rhodamine 6G flash lamp pumped dye lasers and a crystal of silver thio gallate ($AgGaS_2$). TEM_{00} powers of 50 kW, one fixed at 0.590 μm and the other tunable from 0.619 to 0.633 μm would produce infrared powers ranging from 300 W at 8.7 μm to 100 W at 12.6 μm using a 3-mm cube of $AgGaS_2$. The calculated laser intensities in both examples would be below the known laser damage thresholds for these materials.

These achievements in difference frequency generation of tunable dye laser radiation, coupled with the expected development of the dye lasers themselves, suggests that the technique is worthy of serious consideration as a comparatively simple method of generating tunable infrared radiation in a spectral region of considerable importance (for example, atmospheric pollution measurement).

REFERENCES

Adams, N. I., and Shoeffer, P. B., (1963). *Proc IEEE* **51**, 1366.
Andrews, R. A. (1969). *IEEE J. Quant. Electron.* **QE5**, 548
Andrews, R. A. (1970). *IEEE J. Quant. Electron.* **QE6**, 68.
Armstrong, J. A., Bloembergen, N., Ducuing, J., and Pershan, P. S. (1962). *Phys. Rev.* **127**, 1918.
Bardsley, W., and Jones, O. (1968). *J. Crystal Growth* **3**, 268.
Boyd, G. D., and Askin, A. (1966). *Phys. Rev.* **146**, 187.
Boyd, G. D., and Kleinman, D. A. (1968). *J. Appl. Phys.* **39**, 3597.
Boyd, G. D., Bridges, T. J., and Burkhardt, E. G. (1968). *IEEE J. Quant. Electron.* **QE4**, 515.
Boyd, G. D., Grandrud, W. B., and Buehler, E. (1971). *Appl. Phys. Lett.* **18**, 446.
Campillo, A. J., and Tang, C. L. (1971). *Appl. Phys. Lett.* **19**, 36.
Carmen, R. L., Hanus, J., and Weinberg, D. L., (1967). *Appl. Phys. Lett.* **11**, 250.
Chiou, W. C. (1971). *J. Appl. Phys.* **42**, 1985.
Chiou, W. C. and Pace, F. P. (1972), *Appl. Phys. Lett.* **20**, 44.

Dewey, C. F., and Hocker, L. O. (1971). *Appl. Phys. Lett.* **18**, 58.

Falk, J., and Yarborough, J. M. (1971). *Appl. Phys. Lett.* **19**, 68.

Firester, A. H. (1969). *Optoelectronics* **2**, 128.

Firester, A. H. (1970). *J. Appl. Phys.* **41**, 703.

Gampel, L., and Johnson, F. M. (1968). *IEEE J. Quant. Electron.* **QE4**, 354.

Gandrud, W. B., and Boyd, G. D. (1969). *Opt. Commun.* **1**, 187.

Giordmaine, J. A. (1962). *Phys. Rev. Lett.* **8**, (No. 1), 19.

Hanna, D. C., Smith, R. C., and Stanley, C. R. (1971). *Opt. Commun.* **4**, 300.

Hulme, K. F., and Warner, J. (1972). *Appl. Opt.* **11**, (12) 2956.

Johnson, F. M., and Duardo, J. A. (1966) *IEEE J. Quant. Electron.* **QE2**, 296.

Kleinman, D. A., and Boyd, G. D. (1969). *J. Appl. Phys.* **40**, 546.

Klinger, Y., and Arams, F. (1969a). *IEEE Spectrum* **6**, 5.

Klinger, Y., and Arams, F. (1969b). *Proc. IEEE* **57**, 1797.

Kruse, P. W., McGlauchlin, L. D., and McQuistan, R. B. (1962). "Elements of Infrared Technology." Wiley, New York.

Kurtz, S. K. (1975). *In* "Quantum Electronics" (H. Rabin and C. L. Tang, eds.). Vol. IA. Academic Press, New York.

Lucy, R. F. (1972). *Appl. Opt.* **11**, 1329.

Maker, P. .D, Terhune, R. W., Nisenoff, M., and Savage, C. M. (1962). *Phys. Rev. Lett.* **8**, (No. 1), 21.

Martin, M. D., and Thomas, E. L. (1966). *IEEE J. Quant. Electron.* **QE2**, 196.

Martin, M. D., and Thomas E. L. (1967). *Phys. Lett.* **28A**, 637.

Midwinter, J. E., and Warner, J. (1965). *Brit. J. Appl. Phys.* **16**, 1135.

Midwinter, J. E. (1968)a. *Appl. Phys. Lett.* **12**, 68.

Midwinter, J. E. (1968b). *IEEE J. Quant. Electron.* **QE4**, 716.

Midwinter, J. E. (1969). *Appl. Phys. Lett.* **14**, 29.

Midwinter, J. E., and Warner, J. (1967a). *J. Appl. Phys.* **38**, 519.

Midwinter, J. E., and Warner, J. (1967b). *Bull. Amer. Phys. Soc* **12**, 61

Miller, R C , and Nordland, W A. (1967). *IEEE J. Quant. Electron.* **QE3**, 642.

Putley, E. H., Watton, R., and Ludlow, J. H. (1972). *Ferroelectrics (GB)* **3**, 263.

Riccius, H. D., Siemsen, K. J., and Pfitzer, E. K. (1971). *Phys. Canada* **27**, 57.

Smith, A. W., and Braslau, N. (1962). *IBM J. Res. Develop.* **6**, 361.

Smith, H. A., and Mahr, H. (1970). *Int. Quant. Electron. Conf. Kyoto, Japan* paper 5.10.

Smith, W. J. (1966). "Modern Optical Engineering." McGraw-Hill, New York.

Takatsuji, M. (1966). *Jap. J. Appl. Pyhs.* **5**, 389.

Tang, C. L. (1969). *Phys. Rev.* **182**, 367.

Voronin, E. S., Divlekeyev, M. I., and Il'inski, Yu. A. (1969). *JETP Lett.* **10**, 108.

Voronin, E. S., Divlekeyev, M. I., Il'inski, Yu. A., and Solomatin, V. S. (1970). *Sov. Phys.-JETP* **31**, 29.

Voronin, E. S., Divlekeyev, M. I., Il'inski, Yu. A., and Solomatin, V. S. (1971). *Optoelectronics* **3**, 153.

Warner, J. (1968a). *Appl. Phys. Lett.* **12**, 222.

Warner, J. (1968b). *Appl. Phys. Lett.* **13**, 360.

Warner, J. (1969a). *Optoelectronics* **1**, 25.

Warner, J. (1969b). *Proc. Joint Conf. Lasers Optoelectron. IERE* Conf. Proc. No. 14, p.25.

Warner, J. (1969c). *IEEE J. Quant. Electron.* **5**, 354.

Warner, J. (1970). *Optoelectronics* **3**, 37.

Weller, J. F., and Andrews, R. A. (1970). *Optoelectronics* **2**, 171.

Index

(Combined Index for Parts A and B)

A

Absorption spectroscopy
  design considerations in, 330–331
  experimental apparatus for, 328–335
  principle of, 328–330
  two-photon, *see* Two-photon absorption
    spectroscopy
  typical setup in, 331–333
ADP ($NH_4H_2PO_4$) crystals, 482, 547, 589
  properties of, 592
ADP intercavity doubled dye laser, 663
ADP optical parametric oscillator, 660–669
  Nd:YAG laser and, 665
  as second harmonic generator, 661
Amorphous solids, nonlinear susceptibilities
  of, 191
Angular aperture effect, limiting due to, 535
Angular spectra, 526–527
Anisotropic media, harmonic generation in,
  525–532
Anisotropic solids
  phase and energy propagation directions
    in, 223–225
  wave propagation in, 222–230
Anisotropy parameter, 529
Anomalous dispersion
  in birefringent crystals, 487
  phase matching in homogeneous isotropic
    media with, 498
Anthracine, two-photon absorption in, 340–
  341
Aperture effects, in optical multipliers ex-
  cited with laser pulses, 554–557
Arbitrary incidence, in nonlinear effects in
  nonlinear media, 504
Argon-ion laser, cw radiation and, 575
Atomic Coulomb field, 4
Atomic orbitals, for polyatomic molecules,
  122
Atoms
  third-order effects in, 131–145

two-photon absorption in, 302 315, 338–
  345
variational method in, 134

B

Backward-stimulated Raman scattering,
  462–465 *see also* Raman scattering
Bandwidth ratio, for parametric amplifier,
  597
Benzene
  Raman-scattering cross section for, 260
  two-photon cross sections for, 310
Birefringent crystals
  anomalous dispersion in, 487–492
  efficiency of optical multipliers with, 532–
    535
  phase-matching condition in, 489–490
Bloch description, for electronic susceptibilities, 147
Bloch Hamiltonian, 147
Bond additivity, for polyatomic molecules,
  121–122
Born–Oppenheimer approximation, 90–94
  for condensed media, 146
  in infrared dispersion of nonlinear sus-
    ceptibilities, 155–159
  for molecules, 145
  third-order dipole polarizability and, 101
Born–Oppenheimer states, for free atoms,
  145
Born–Oppenheimer wave functions, 90, 110
Brillouin scattering, 4, 420–421
Broad-band radiation, in two-photon absorption spectroscopy, 300–301

C

Cadmium-germanium arsenide crystals, parametric figure of merit for, 685
Calcium carbonate crystal, phase-matched
  interactions in, 490

A 5
B 6
C 7
D 8
E 9
F 0
G 1
H 2
I 3
J 4